Disclaimer

The publisher of this book is by no way associated with the National Institute of Standards and Technology (NIST). The NIST did not publish this book. It was published by 50 page publications under the public domain license.

50 Page Publications.

Book Title: Developing Emergency Communication Strategies for Buildings

Book Author: Erica D. Kuligowski; Steve M. Gwynne; Kathryn M. Butler; Bryan L. Hoskins; Carolyn Sandler;

Book Abstract: The purpose of this document is to provide the foundation for the development of a guidance document on emergency communication message content and dissemination strategies. The document answers three major questions regarding emergency communication systems: 1) What technology exists or is proposed for use in emergency notification? 2) What approaches are currently being used to disseminate messages? 3) How does the public respond to different types of information and information sources? The document begins with a discussion of the technology that exists or is proposed for use in emergency notification, along with the positive and negative aspects of each system. The ways in which social media tools can be used to provide warnings in emergencies are included. Second, the document discusses the various types of emergencies for which warnings are needed, the range of protective actions that are taken by building occupants in emergencies, and the nature of the information required based upon the emergency type. The emergency communication systems installed in two different college campuses are described as examples of approaches used to disseminate warnings during emergencies. Finally, a comprehensive literature review is presented on how the public responds to various types of information and information sources both in emergency and non-emergency conditions. A summary list of the relevant findings from each literature source is assembled in Appendix A to identify the most effective ways to create or disseminate messages to achieve optimal occupant response. Detailed annotations for each source are presented in Appendix B.

Citation: NIST TN - 1733

Keywords: emergency communication; mass notification; decision-making; disasters; emergencies; warnings

Technical Note 1733

Developing Emergency Communication Strategies for Buildings

Erica D. Kuligowski, Steven M.V. Gwynne, Kathryn M. Butler, Bryan L. Hoskins, and Carolyn R. Sandler

NIST National Institute of Standards and Technology • U.S. Department of Commerce

2

Technical Note 1733

Developing Emergency Communication Strategies for Buildings

Erica D. Kuligowski, Steven M.V. Gwynne*, Kathryn M. Butler, Bryan L. Hoskins, and Carolyn R. Sandler

Fire Research Division
Engineering Laboratory

*Hughes Associates UK
London, UK

March 2012

U.S. Department Of Commerce
John E. Bryson, Secretary

National Institute Of Standards and Technology
Patrick D. Gallagher, Under Secretary of Commerce for Standards and Technology and Director

3

Certain commercial entities, equipment, or materials may be identified in this document in order to describe an experimental procedure or concept adequately. Such identification is not intended to imply recommendation or endorsement by the National Institute of Standards and Technology, nor is it intended to imply that the entities, materials, or equipment are necessarily the best available for the purpose.

National Institute of Standards and Technology Technical Note 1733
Natl. Inst. Stand. Technol. Tech. Note 1733, 361 pages (March 2012)
CODEN: NTNOEF

Acknowledgements

The U.S. Department of Homeland Security, Science and Technology Directorate sponsored the production of this material under Interagency Agreement HSHQDC-07-X-00723 with NIST. The authors would also like to thank the Fire Protection Research Foundation (FPRF) for sponsorship of the Project Technical Panel. The members of the Technical Panel are listed here:

- Kathleen Almand, Fire Protection Research Foundation
- Oded Aron, Port Authority NY and NJ
- June Ballew, Cooper Notification
- Don Bliss, NI^2 Center for Infrastructure Expertise (NH)
- Bob Boyer, UTC Edwards
- Robert Chandler, University of Central Florida
- Joe Collins, Dallas Fort Worth airport
- Rita Fahy, National Fire Protection Association's Fire Analysis Division
- Daniel Finnegan, Siemens Fire
- Dave Frable, U.S. General Services Administration
- Denis Gusty, U.S. Department of Homeland Security
- Edwina Juillet, NFT/LSPwD
- Matthew Kelleher, Montgomery County Fire & Rescue
- David Killian, Walt Disney Parks and Resorts
- Amanda Kimball, Fire Protection Research Foundation
- Scott Lacey, Lacey Fire Protection Engineering
- Chris Maier, National Weather Service
- Derek Mathews, Underwriters Laboratories
- Philip Mattson, U.S. Department of Homeland Security
- Dennis Mileti, University of Colorado
- Wayne Moore, Hughes Associates, Inc.
- Rodger Reiswig, SimplexGrinnell
- Lee Richardson, National Fire Protection Association's staff liaison
- Robert Schifiliti, RP Schifiliti and Associates

Guidance from the Technical Panel, as well as NFPA 72's Emergency Communications Systems Technical Committee, was integral in the development of this document. Also, thank you to Andrew Flynn for his significant contributions to this document, and to Shonali Nazare, Richard Peacock, Jason Averill, and Anthony Hamins, with special thanks to Aoife Hunt, for a comprehensive review of the document before publication.

Thank you to the Industry Project Sponsors: Siemens Industry, Inc., SimplexGrinnell, Underwriters Laboratories, Inc., and xMatters, as well as to the Contributing Industry Project Sponsors: Bosch, Cooper Notification, and United Technologies.

Executive Summary

When an emergency occurs, it is not always sufficient to simply initiate alarm bells. Individuals may not know what the alarm bell means and as a result may respond inappropriately to its sound. Many buildings have installed mass notification or emergency communication systems, based upon requirements provided in model codes and standards, that can be used to disseminate information in the event of an emergency. However, there is a lack of guidance on how to use current emergency communication systems in the most effective manner. As the use of newer technologies such as mobile devices or social networking tools grows, guidance on message content and dissemination becomes ever more critical for ensuring effective and safe response of building occupants during an emergency.

The purpose of this report is to provide the foundation for the development of a document that provides recommendations on emergency communication message content and dissemination strategies. This report answers three major questions regarding emergency communication systems:

1) What technology exists or is proposed for use in emergency notification?
2) What approaches are currently being used to disseminate messages?
3) How does the public respond to different types of information and information sources?

The document begins with a discussion of the technology that exists or is proposed for use in emergency notification, along with the advantages and disadvantages of each system. The ways in which social media tools can be used to provide warnings in emergencies are included. Second, the document discusses the various types of emergencies for which warnings are needed, the range of protective actions that may be taken by building occupants in emergencies, and the nature of the information required based upon the emergency type. The emergency communication systems installed in two different college campuses are described as examples of approaches used to disseminate warnings during emergencies. Finally, a comprehensive review of the scientific literature from different disciplines is presented on how the public responds to various types of information and information sources both in emergency and non-emergency conditions. A summary list of the relevant findings from each literature source is assembled in Appendix A to identify the most effective ways to create or disseminate messages to achieve optimal occupant response. Detailed annotations for each source are presented in Appendix B.

Table of Contents

1 Introduction

At present, many buildings and building campuses in the United States are installing mass notification or emergency communication systems to improve communication given by the building or emergency officials to the public. The National Fire Alarm and Signaling Code (NFPA 72), 2010 edition, provides requirements for the application, performance and installation of emergency communication (or mass notification) technology (NFPA 2010a). However, NFPA 72 provides little guidance on how to use these systems for effective emergency communication. Additionally, many countries use British Standard BS 5839 -- Fire Detection and Fire Alarm Systems for Buildings – Part 8 (BS 5839-8:2008), in which chapters 20 and 21 discuss emergency messages and dissemination techniques.

In the United States, there is little guidance outside of the building codes regarding the content and dissemination strategies for emergency messages. The people providing the messages (i.e., message providers) during an emergency may not have the necessary tools, techniques, guidance, and training to effectively provide information to the public when an emergency event is imminent or unfolding. This problem exists across all modes of notification, whether visual, audible, tactile, or social systems are employed. Irrespective of the mode used, it is necessary for emergency communication to be effective in order to facilitate the procedural measures employed and the desired response.

The purpose of this research project, which was funded by the U.S. Department of Homeland Security, Science and Technology Directorate, relates specifically to the effectiveness of emergency notification or communication. This report discusses the range of communication technologies and approaches that are used to alert or warn the public of an emergency. This report also includes a literature review of representative material examining the effectiveness of visual, audible and tactile notification systems and emergency message design. The review draws from several different subject areas, including ergonomics/human factors, communications, fire, disasters, sociology, acoustics, lighting, transportation, and language study, among others. Given the range of subject areas examined, it was not practical to provide an exhaustive analysis of all of the material publicly available. The approach has therefore focused on examining a representative range of material from a broad range of sources, and the criteria used to identify these sources will be described later in the report. The intention of this review is to gather together information to support the development of regulatory guidance.

The review includes 160 sources. A summary of the contents of each source is presented in Appendix B. The key findings relevant to this report were extracted from the examined material. The findings were compiled and organized within a general framework, which appears as Appendix A. The compiled findings are intended to inform the later stages of this research project during the development of specific guidance for the design and implementation of notification during emergencies.

Many of the original sources do not relate specifically to notification systems. Rather, they relate to presentational, perceptual, and design issues, which may be translated to the design of mass notification. It is assumed that lessons can be learned from these issues and practices when implementing the systems as part of an overall procedural response.

2 Historical Perspective: Misunderstanding the Need for Notification

Until very recently, it was widely assumed that telling people about an emergency incident would lead to panic (Tierney 2003; Quarantelli and Dynes 1972). This assumption was commonplace, both with the lay population and with many emergency responders. The view that people would panic in response to an incident (and specifically to information describing the incident) influenced both the notification procedures employed and the language used to report the exhibited behavior. This assumption influenced a difficult and harmful cycle consisting of the following steps: people report that they had panicked, emergency officials continue to believe that panic is a normal response, emergency information is withheld in the next disaster, human response is delayed and inefficient, the situation becomes more dire, and human response becomes more desperate.

Over the last 25 years, this point of view has been slowly replaced with the recognition that people need detailed information as early as possible in order to initiate and inform their response. The availability of this information encourages people to accept the emergency procedures and to improve their familiarity with the required response, and later informs the decision-making process that determines their response. People need information in order to act. Detailed information by no means guarantees the desired response; however, without this information an uninformed approach (ignorant of the conditions and the options available) is much more likely.

It is now broadly accepted that depriving evacuees of information is more likely to lead to an inefficient and inappropriate response; e.g., misinterpreting the incident and the threat it poses, delaying their response, and ignoring safe egress routes. During an incident, people will seek information regarding the nature of the incident and what they should do in response to it. Unfortunately, this information may not always be easy to find, reliable, consistent or accurate. It is critical that an information vacuum is avoided and that accurate, credible information is provided.

The provision of information does not guarantee that a population will respond immediately or appropriately. Even when information is provided, there will inevitably be delays and possibly misinterpretations. Once people decide that they must act, there is usually a delay as they prepare to act (e.g., shutting down a PC or collecting a coat). To minimize these delays, clear and concise information must be provided to convince the population of the need to act and what to do. Ideally, this information should be accompanied by active, assertive and informed members of staff; i.e., technical resources supported by human resources, with both sets of resources providing consistent information. The expected performance of numerous, assertive, well-trained, and well-informed staff members is a benchmark for a notification system. The adaptability and credibility of such staff members is more likely to encourage the desired response from a population than a solely technological response. However, the use of a sophisticated notification system can complement staff activities and provide a safety net for performance shortfalls or situations in which staff is absent. In such situations, these technical systems can at least provide sufficient information to allow people to make well-informed decisions.

The effectiveness of human and technical warning systems also depends on the level of information already present in the target population. A well-prepared population that is familiar with the procedures and has been subjected to well-planned drills will be better placed to respond to the instructions provided by staff members or a technological notification system. However, in some occupancies (e.g., airports), this preparation may not be possible. This places a heavier burden upon the notification system and on the staff for direction in case of an emergency.

In contrast to the panic model previously assumed, the response of an individual during an incident can be better characterized as a decision-making process in which people receive information from their environment, interpret that information, and respond based upon their interpretations (Lindell and Perry 2004). Understanding this process and the factors that influence certain interpretations (as shown in Figure 1) can aid in the development of better guidance on emergency messages and dissemination systems. This decision-making process will be described in Section 6 of this report and serves as the basis for this literature review on human response to emergency communication systems and messages.

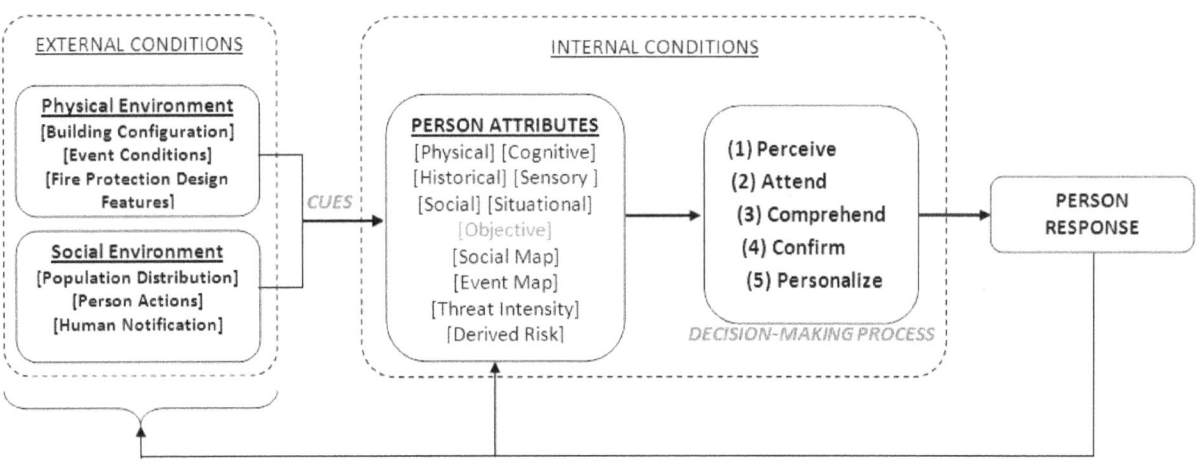

Figure 1: Simplified Decision-Making Process based on Lindell and Perry (2004).

3 Purpose

The purpose of this literature review is to summarize the current state-of-the-art in emergency communication technology, approaches, and literature on human response to public warnings. This is the first step in a larger project to standardize strategies for communicating emergency information to building occupants through the development of a guidance document for message providers. The guidance document will focus on emergency communication strategies for relevant hazard and threat scenarios in buildings and building campuses in the United States.

The literature review contains answers to three distinct questions:
  1) What technology exists or is proposed for use in emergency notification?
  2) What approaches are currently being used to disseminate messages?
  3) How does the public respond to different types of information and information sources?

What technology exists or is proposed for use in emergency notification?
To answer question one, a review has been carried out of existing and proposed notification technologies, including those technologies that are not currently mentioned in NFPA 72. Traditionally, NFPA 72 covers fire alarm systems and other mass communication devices, such as television, radio broadcast, and stationary public announcement systems. However, more and more message providers, including emergency managers and building safety officials, are acknowledging benefits of newer technologies such as computers, smart phones, and other mobile, internet-based devices. Section 4 of this document addresses the technology already being used for emergency communication, emerging technologies for future use in emergencies, and devices used for other purposes that could disseminate messages during an emergency.

What approaches are currently being used to disseminate messages and when are they used?
To answer question two, Section 5 presents a discussion on the ways in which emergency notification technology can be used in disseminating emergency communication. The various types of building emergency scenarios for which emergency communication is necessary are discussed, along with the kinds of emergency information that are appropriate for each scenario type. The characteristics of communication systems that are most beneficial during an emergency are also introduced. Finally, two examples of current approaches used by university campuses are presented, highlighting the use of new emergency dissemination approaches (e.g., social media) and specialized approaches necessary to reach vulnerable populations such as the hearing impaired.

How does the public respond to different types of information and information sources?
To answer question three, a review has been completed covering the various types of research on human behavior in response to public warnings before and during emergencies. Engineering and social science research, including human factors and ergonomics, communications, psychology, sociology, and fire protection engineering, has been surveyed to identify the ways in which people respond to various types of emergency

messages and dissemination techniques, including both audible and visual technologies. Sections 6 through 8 and Appendices A and B report on 160 studies published over the last 60 years on public response to emergency and non-emergency messages disseminated via various types of technology.

4 Warning Systems: Modes of Notification

Warning and notification systems represent a means to influence the information available to the population and the manner in which this information is perceived. The technology used determines the type of information that can be provided and how this information is provided. NFPA 72's Technical Committee on Emergency Communication Systems has accepted a proposal to categorize emergency communications or mass notification systems by four layers: In-Building systems, Wide-Area systems, Notification through Individual Measures, and Notification through Public Measures. These layers describe the method by which the individual is likely to receive the message. Several different technologies categorized using this classification scheme are presented in Table 1.

Table 1: Categorization of emergency notification systems.

Mass Notification Layer	Notification Technology/ Approach	Affected Sense (main)	Pros [A]	Cons [A]
In-Building systems	Public address system (voice notification)[B]	Aural	Already installed in most cases; Allows for voice message dissemination	Quality of sound may be low
	Alarm, Horn, Bell	Aural	Widespread; Already installed in most cases	Lack of message content
	Visible notification appliances (e.g., strobes)[B]	Visual	Already installed in most cases	Lack of message content
	Textual / Digital Signage, Displays[B]	Visual	Widespread use; Can include audible signals/lights	Quality of lighting/ graphics may be low
Wide-Area systems	Sirens	Aural	Widespread; Quick to sound	Lack of message content; Loud, may interfere with other communications
Notification through Individual Measures	Short message service (SMS) text via cell phone	Visual/Tactile	Widespread use; Quick dissemination	Can be turned off; Limits to reception; Overloading of system; Limited message
	Vibrating Pagers	Tactile	Back-up device	Some cell service/phone required
	Computer pop-ups	Visual	Reach networked monitors	May be unattended
	Email broadcast (internet)	Visual	Widespread use; Quick dissemination	Overloading of system; Spam filters may influence receipt; Can be ignored; May be limited to a select audience
	Automated voice dialing and text messaging	Aural / Visual/ Tactile	In place in most homes	Can be ignored; Overloading of system

Mass Notification Layer	Notification Technology/ Approach	Affected Sense (main)	Pros [A]	Cons [A]
Notification through Public Measures	Satellite/AM/FM radio broadcasts	Aural	In place in most cities/towns	May be unattended; Limited to a select audience
	Satellite/off-air television broadcasts	Aural/Visual	In place in most cities/towns; Large amount of info	May be unattended; Limited to a select audience
	Websites/news/blogs (internet)	Visual	Consistent message; Can be updated and time-stamped; Large amount of info	Can be ignored or not used
	Tone alert radio	Aural	Provides specific weather-related or emergency info	May not be able to provide specifics about the building

[A] National Clearinghouse for Educational Facilities. "Mass Notification for Higher Education," Accessed December 2010. http://www.ncef.org/pubs/notification.pdf
[B] These technologies can also serve as Wide-Area systems.

Certain characteristics link the technologies associated with each mass notification layer. In-Building systems are those that are likely to be installed inside a building and are controlled only by authorized users. Examples of In-Building systems are public address systems, alert tones (i.e., alarms, horns and bells), visible notification (i.e., strobes), and electronic or digital signage. Wide-Area systems provide notification to individuals on the exterior of a building and are controlled only by authorized users. Examples of these systems are similar to In-Building systems, with the exception of outdoor siren systems used to alert individuals who are located outside of buildings to seek additional information about a possible impending disaster (e.g., tornadoes or hurricanes). Layer three consists of technologies that provide notification through individual means. In other words, an individual can receive alerts or warnings on a personal device based upon pre-emergency registration in the alert system or geographical location in reference to the emergency event[1]. Examples of these technologies include short message service (SMS) texts provided via cell phones, computer monitor pop-up services, email broadcasts via the internet, and automated voice dialing and text message services. The fourth layer, notification through public measures, includes those technologies that can broadcast emergency information to a large population. Examples such as radio and television broadcasts, tone alert radios, and internet-based news and blogs differ from other technology categories because they require individuals to actively turn to these technology systems for information.

Table 1 also shows how the emergency notification technologies differ by affected sense. Although many of these systems can be extended to provide information that can be received by more than one sense, the table categorizes each notification

---

[1] Please see the following website for more information about CMAS, or the Commercial Mobile Alert System, a system by which commercial mobile service providers can transmit emergency alerts to subscribers based upon the subscribers' geographic location: http://transition.fcc.gov/pshs/services/cmas.html.

technology/approach based on the main sense to which they are targeted: aural, visual or tactile.

Finally, there are obvious pros and cons associated with each notification system, related to their use in emergency communication. Some examples of the pros and cons for each system, provided by the National Clearinghouse for Educational Facilities, are also included in Table 1.

Relevant to this project, technology can be distinguished based upon whether it is capable of providing an emergency message with significant content or is instead limited to providing an alert that can attract attention but not disseminate a large amount of information. In the cons column, there are a few technologies that are labeled as providing "a lack of message content" and others labeled as providing "a limited message." Technologies listed as lacking message content can only disseminate a sound or visual notice. Individuals must understand the meaning of these signals before they can respond. Technologies that are listed as containing a limited message are able to provide a limited number of characters per emergency message. The best use of these technologies may be as an alerting system that leads individuals to resources or links where additional information can be provided.

Advances in Social Media

A new form of personal alerting that is not covered in the current edition of NFPA 72 is social media. According to Garton, Haythornthwaite, and Wellman (1997:2), a social network is "a set of people (or organizations or other social entities) connected by a set of social relationships, such as friendship, co-working or information exchange." Online social media sites have been developed that allow people to maintain virtual connections via the internet. These web-based services enable people to construct a profile, develop a list of other individuals with whom they share a connection, and view/follow their connections as well as the connections with others in the system (White et al. 2009). Examples of online social media channels include Facebook, YouTube, MySpace, Twitter, Yahoo Groups, LinkedIn, Flickr, Yelp, Foursquare, Google+, and many other forums and blogs (American Red Cross 2010a). These sites and others like them are increasingly being used for providing information (i.e., photos, videos, graphics, and text) before, during and after emergencies. This can include maps, evacuation instructions, evacuation site/shelter locations, and directions and evacuation routes (American Red Cross 2010b).

There are several reasons why social media sites are being used more frequently for disseminating information before and during an emergency. First, a majority of the current U.S. population is participating in social media regularly (American Red Cross 2010a), and participation is constantly increasing (Heverin and Zach 2010; Lenhart and Fox 2009). According to a survey conducted by the American Red Cross (2010a) of over 1,000 individuals in the U.S. aged 18 and over, as of the time of the survey (2010), data show that almost 75 % of individuals participate in at least one online community or social network, with Facebook being the most well-populated channel. Of these individuals, 82 %

participate in social media at least once a week, and almost half participate once a day or more. The survey also looked at anticipated social media use by age group. Individuals aged 18 to 34 were found to be more likely to make use of social media during both routine and emergency situations; whereas older respondents were more likely to consult the television news.

Social media sites also provide certain advantages over traditional approaches. Most current emergency communication modes are one-way, in that warning providers send information to a population with little opportunity to receive feedback from or to monitor the actions of the affected individuals. Social media provides the potential for interaction and two-way communication between emergency officials and the affected public (Latonero and Shklovski 2010). Social media sites allow the general public to post information or photos about what is going on during a disaster, giving emergency response officials on-the-ground information on the hazards as well as on the actions of the general public. As such, messages can be formulated (and even posted back onto social media sites) based upon real-time, current, and accurate information and can reach the public from a much wider set of sources.

Social media sites have been used in actual emergencies as a form of communication and community interaction, including wildfires (Sutton, Palen and Shklovski 2008), police emergencies (Heverin and Zach 2010), campus shootings (Kavanaugh et al. 2010;Vieweg et al. 2008), and hurricanes (Hughes and Palen 2009). In the majority of these events, members of the general public have used social media following the crisis to share crisis-related information, update others on their own personal safety status, and search for the status of loved ones and friends. These types of participation show the potential for using social media as an emergency information dissemination tool *before* a disaster occurs. Additionally, Brian Humphery (emergency manager for the city of Los Angeles, California), a strong advocate for the use of social media in disasters, stresses that social media should be used as a tool not only for disseminating information but also for having a conversation with the general public. In his opinion, it is important to understand the needs of the public, by monitoring the messages posted by the general public, for example, in order to provide them with the right information on which they can respond before and during an emergency (Humphrey 2011).

As in the case of the emergency communication technology and services included in Table 1, there are also pros and cons associated with the use of social media, as shown in Table 2. First, there are several benefits to the use of social media in emergency situations (Sutton and Mileti 2009). Using social media sites and tools for emergency communication can provide quick dissemination of information, especially for the younger population who participate in these sites on a regular and frequent basis. Another benefit of social media is the ability for emergency managers and other message providers to monitor public response, allowing them to craft messages that are appropriate for the current emergency as well as to provide follow-up messages that are relevant to the affected population and can correct or redirect actions if necessary. Although there has been a concern about the perpetuation of false information via social media sites, research has found that users often correct false information, such that social media in emergencies becomes self-correcting

(Sutton and Mileti 2009). Another benefit of social media is its facilitation of the milling process. In any emergency, individuals are likely to spend time discussing the emergency with others to decipher what is going on and what should be done about it. By posting information on social media sites, the milling process takes place in a virtual forum, which can reduce the time spent in this process and allow individuals to respond in a quicker manner (Sutton and Mileti 2009). Finally, social media provide a very fast, cheap, and relatively easy way for mass distribution of communications (White et al. 2009).

Table 2: Categorization of social networking tools.

Mass Notification Layer	Notification Technology/ Approach	Affected Sense (main)	Pros	Cons
Notification through Public Measures	*Social networking tools (internet)*	*Visual*	• *Quick dissemination;* • *Ability to monitor public response;* • *Self-correcting;* • *Facilitates the milling process;* • *Free and relatively easy to use*	• *Time and personnel resources to monitor sites;* • *Some limit message;* • *May be limited to a select audience;* • *Technology or software failure*

However, there are limitations for using social media to disseminate emergency information. Agencies may be required to spend a significant amount of time filtering and verifying incoming information both before and during a disaster (Humphrey 2011; Latonero and Shklovski 2010; White et al. 2009). Depending upon the length of the disaster and the size of the affected community, this may not be a feasible avenue for two-way emergency communication. Additionally, depending upon the specific social networking tool, there may be limits to the character size of the message that can be provided (Humphrey 2011b), resulting in limitations similar to those for some technologies identified in Table 1. For example, Twitter messages are restricted to 140 characters. In some situations, therefore, Twitter may be better suited to provide a link to or a resource for additional information about the emergency. Another limitation to social media is that it is relatively confined to certain subpopulations in the United States, for example, individuals younger than 35 years old or populations with access to the internet (American Red Cross 2010c). Finally, as with any other technology that relies on the internet, electricity, and computer software, natural and human-caused disasters represent a risk of technology or software failure (or security breach) (White et al. 2009).

5 Warning Dissemination Approaches: Incident Scenarios and Procedural Measures

Now that the emergency technologies and systems have been identified, it is important to discuss the ways in which these technologies and systems can be used to disseminate information in emergencies. The most appropriate and efficient ways to use these technologies depend upon the nature of the emergency, and in particular the timeframe of events.

The timeframe of the emergency influences the information that is needed by the affected public and the ways in which this information should be provided. For the sake of this project, emergencies have been categorized into two different timeframes: rapid-onset and slow-onset emergency events. Rapid-onset events are those that occur with no or almost no (in the case of minutes) notice. Examples of rapid-onset events, taken directly from NFPA 1600, the Standard on Disaster/Emergency Management and Business Continuity Programs (NFPA 2010b), are included in Table 3. Slow-onset events, on the other hand, are emergencies in which we are aware of their occurrence hours or even days in advance. In this project, we focus specifically on the ways in which to provide messages for rapid-onset events; with only a brief description of the types of information appropriate for slow-onset emergencies or educational campaigns.

Table 3: Potential Hazards categorized here as rapid-onset incidents.

Naturally occurring hazards	Human-caused events
<ul><li>Geological hazards<ul><li>Earthquake</li><li>Tsunami</li><li>Volcano</li><li>Landslide/mudslide</li><li>Glacier, iceberg</li></ul></li><li>Meteorological hazards<ul><li>Flood, flash flood</li><li>Fire (forest, range, urban, wildland, UI)</li><li>Avalanche</li><li>Windstorm, cyclone, tornado, dust storm or sandstorm, water spout</li><li>Lightning strike</li><li>Geomagnetic storm</li></ul></li></ul>	<ul><li>Accidental hazards<ul><li>Hazardous material spill/release</li><li>Explosion/fire</li><li>Transportation accident</li><li>Building/structure collapse</li><li>Water control structure/dam/levee failure</li></ul></li><li>Intentional hazards<ul><li>Terrorism (chemical, biological, radiological, nuclear, cyber)</li><li>Sabotage</li><li>Civil disturbances</li><li>Enemy attack, war</li><li>Insurrection</li><li>Physical or information security breach</li><li>Workplace/school/ university violence</li></ul></li></ul>

Research shows that a successful warning message, regardless of emergency type, must provide specific information to elicit appropriate and effective emergency response (Mileti and Sorensen 1990). The message must provide the who, what, when, where, and why of the disaster (listed here in appropriate order for display within the message):
- Who: the name or title of the source that is providing the information,
- Why: information on the danger/hazard/disaster,
- Where: a description of the location of the risk or hazard,
- What: guidance on what people should do about it,
- When: an idea of when people need to act,

However, the information provided under each heading differs depending upon the type of hazard, especially the "what" or the guidance on what people should do about the disaster and the "when" or the guidance on when people need to act. In the case of rapid-onset events, there are three main actions associated with these types of emergencies that the public would be asked to perform, shown in Table 4. These actions are to remain in place, perform a total evacuation from the building or the community to an area of safety, or relocate from their original location to a designated area within the same building (or area within the community). Additionally, with rapid-onset events, the "when" or an idea of when people need to act is often immediately or "now."

Table 4: Potential Procedural Responses.

Potential Procedural Measures	System Requirements
Total evacuation	The affected population is simultaneously notified and then evacuated from the building (or the community).
Relocation (or partial evacuation)	The affected population is simultaneously notified that they are required to evacuate their current position and travel to another, safer (pre-determined) location within the building or community.
Shelter-in-place	A (sub-) population is required to remain in their original location.

In the case of messages for slow-onset events, the "what" or the guidance on what people should do about the disaster can consist of a variety of different actions. For example, in the case of a hurricane or snow storm, where information can be given days before the storm hits, guidance can be provided to individuals to "stay tuned" or keep watching or listening for further information on what to do next. Additionally, guidance can be provided to individuals to begin preparation activities in case they are told to evacuate or relocate. Often, in slow-onset events, the warning guidance on "when" individuals need to act is not as urgent as would be the case for rapid-onset events. Guidance may be provided to individuals to "stay tuned," however, they should do so often over the course of hours or days.

Additionally, emergency communication is used to prepare the public for what *could* happen in the future. These messages are often referred to as public education campaigns. Educational messages can be provided to individuals telling them how to prepare for various types of emergencies, even before one is imminent. In these types of messages, guidance should still be provided on the source of the message, information on the dangers possible, and what people should do about it; although guidance on the location of the hazard and the timing of the hazard (i.e., when people should act) is less relevant when an emergency is not taking place.

When a disaster is imminent, there are certain characteristics of technology that are most beneficial when eliciting public response (Chandler 2010). The first characteristic is

11

immediacy, especially in the case of rapid-onset events. The faster the message is delivered, and assuming that the message is heard and processed, the more time is provided for individuals to respond and take actions to achieve safety. Examples of technology that can provide warning information in a faster manner are those devices that specialize in alarm tones or non-textual visual notification, including sirens, strobes, horns, and bells. Automated notification communication can also be provided quickly, for example, pre-recorded building-wide voice notification messages or pre-recorded messages sent via personal devices, such as email, SMS, or cell and landline-based phone messages. Especially with messages sent to personal devices, systems can track whether the messages were successfully delivered, they can be sent via multiple devices for the same person, and they can be initiated from different places to increase reliability in the face of emergencies (Chandler 2010). However, it should be noted that a great deal of preparation is needed upfront to develop pre-recorded messages that are relevant in future emergencies.

The second characteristic of technology to be used in emergencies is accessibility. Messages disseminated via emergency technology should reach a large, diverse audience, including individuals who have hearing or visual disabilities. One way to reach a wider audience is to combine communication technology in the development of a system in which both visual and audible notification is used. An example of this is providing both strobe lights and voice notification in buildings. Another example is providing warnings via the television that are both spoken and presented using closed captioning. Additionally, disseminating messages via personal alerting technology, that is widely used by a community or building population, can increase the chances that a member of the population receives the warnings. The building can institute an emergency opt-out program, in which all members of the population will receive an automated alert when an emergency occurs via a means previously provided by the building, unless they specifically choose to be excluded.

Third, it is important for the technology to be able to disseminate the message in a persistent manner. The more often the message is provided, the more likely the building or population will perceive (i.e., see or hear) it. If messages are only provided once and the individuals were not tuned into the technology originally, then they are likely to miss the information altogether. Examples of technology that provide persistent messages are social networking tools and the internet, where messages can be displayed for an infinite amount of time, and visual signage where messages can be repeated in a scrolling fashion.

Technology should also be reliable in the case of emergencies. In emergencies, technology that has the capability of providing power even when the main power source is not available can be categorized as reliable. Examples of these technologies are indoor and outdoor public address systems, digital displays, and alarms/sirens/bell systems. Also, technology that can typically operate even with saturation of communication lines can also be categorized as reliable.

Especially in rapid-onset events, message providers should employ technology that is obvious. The use of push communications or a combination of push and pull communications provide information to individuals without requiring them to take extra effort to seek it out (Chandler 2010). Push communications can take the form of a phone call or text message delivered to the individual, a computer pop-up at his/her desk, a horn or sirens that sounds in his/her office, or a radio or television broadcast that interrupts normal programming (given that the individual already has the television or radio turned on). Not included in push communications are websites or telephone hotlines or call centers, in which individuals must make a concerted effort to seek out additional information without specific prompting (referred to as pull communications).

Finally, it is also important for emergency technology to be interactive. Two-way communication between the message provider(s) and the affected population allows the message provider(s) to monitor what the population thinks, knows, believes, as well as their responses, including any misunderstandings or rumors about the situation or what should be done in response to the situation (Chandler 2010). The technology types that allow for two-way communication are social networking sites, instant messenger, email (if it is monitored by someone), call centers, and two-way radios or in-building intercom systems.

Two examples of emergency communication systems are provided here to show the ways in which various types of technology can be used to account for immediacy, accessibility, persistence, reliability, obviousness, and interactivity. College campuses typically provide good examples of emergency communication systems for a number of reasons: the variety of population types they are required to reach, the number of buildings and outdoor areas that need to be covered, and the various types of emergencies to which they are susceptible. Additionally, alerting systems have been instituted in colleges and universities nationwide, spurred both by the 2007 shooting incident that resulted in 33 deaths at Virginia Polytechnic Institute and State University and the Higher Education Opportunity Act of 2008, which requires the immediate notification to campus communities upon confirmation of significant emergencies or dangerous situations. The two college campuses featured in this report are Gallaudet University, a college for the deaf and hard of hearing, and Virginia Polytechnic Institute and State University, a university that has established a new and robust emergency communication system since the 2007 shooting incident.

Gallaudet University, located in Washington, DC, is an institution of higher education that "ensures the intellectual and professional advancement of deaf and hard of hearing individuals though American Sign Language and English" (Gallaudet University 2011). The university, under the Department of Public Safety, has designed and implemented an extensive emergency communication system for the campus, allowing students to receive emergency information via a variety of methods. First, the university manages an alert system to students and faculty who elect to subscribe to the system, which is called Gallaudet Alert. Gallaudet Alert is described as an email-based system in that the University President or other designated individual notifies the Department of Public Safety of a situation, the Department of Public Safety then sends a message to the listserv, and any individuals who have subscribed to this listserv receive the announcement immediately. Individuals can receive alerts via email, pager, cell phone (with text capability), laptop, personal computer, and other personal digital devices. In addition to Gallaudet Alert, the university also disseminates information about an emergency condition via fire bells and strobe lights throughout the campus, a campus-wide email (aside from Gallaudet Alert), emergency notification beacons that display textual messages, an emergency broadcast announcement that is displayed as a pop-up message on computers, flashing blue lights at Blue Emergency Button Stations, cable television interrupt using an emergency broadcast message, person-to-person communication, emergency signage on exit doors, and orange flags on Department of Public Safety bicycles and vehicles across campus. Finally, students and staff are also encouraged to subscribe to Alert DC, which provides notification about emergencies that occur within Washington, DC.

In the wake of the crisis that occurred on the campus of Virginia Polytechnic Institute and State University (VT) on April 16, 2007 in Blacksburg, Virginia, where a student opened fire and killed 33 individuals, the university has established a new and robust emergency communication system (National Research Council 2011; Virginia Polytechnic Institute and State University 2010). The VA Tech Emergency Notification System (ENS) consists of the following messaging channels: VT phone alerts, VT desktop alerts, digital signage in key academic classrooms and laboratories, broadcast emails to VT email addresses, posts to the VT homepage, outdoor sirens and public address systems, and voicemail to VT campus phones. Whereas all other channels provide information to all registered students and employed faculty, VT phone alerts and VT desktop alerts are opt-in systems. VT Phone alerts require users to sign up and provide non-campus phone numbers and email addresses, and in the event of an emergency, delivers emergency messages via text (i.e., SMS and email) and voice (i.e., cell and landline phones) to subscribers. Users can select up to three channels by which they will be notified by this service. In the event of its use, users are asked to confirm the receipt of the emergency messages, and if messages are not confirmed, the system attempts to reach the individual via other registered numbers provided by the user. The VT desktop alerts opt-in system posts an outbound message on the screen of all computers that are logged on to the internet and have downloaded the VT desktop alert module. Additionally, VT provides access to a hotline where a recorded message plays for in-bound callers (most commonly used for weather information). As far as the message provided, Virginia Tech focuses on providing three main pieces of information in an emergency message: the nature of the incident, the location where the incident has occurred, and the action that needs to be taken by the message receiver. Any subsequent messages are given when additional information is required or when other sources can be accessed to obtain additional or more detailed information.

6 Human Response to Emergency Warning

6.1 Processing Information

Over the last 50 years, numerous empirical studies have sought to systematically chart the social processes involved in human responses to emergency incidents (Tierney, Lindell and Perry 2001; Mileti and Sorensen 1990; Drabek 1986). Of these, the Protective Action Decision Model (PADM) is selected here to structure the analysis of the material gathered

(Lindell and Perry 2004). This is selected given the appropriateness of the model and to remain consistent with other publications of the project (Kuligowski 2011a).

The PADM provides a framework that describes the information flow and decision-making that influences protective actions taken in response to natural and technological disasters (Lindell and Perry 2004), and has been recently used to analyze behavioral response to the 2001 World Trade Center disaster (Kuligowski 2011b). This model will be used as a basis to examine and categorize the material available regarding human response to warning systems.

Specific to public warnings and emergency information, the PADM asserts that the process of decision-making begins when people are first presented with warning messages. The introduction of these messages initiates a series of processes that must occur in order for the individual to perform protective actions, split into pre-decisional processes (PRE-DEC, which determine whether a decision-making process commences), and decisional processes (DEC – the key components of the decision-making process itself). A simplified version of this process is presented below

- PRE-DEC_1: the individual must perceive or receive the cue(s); e.g., a visual signal must be seen.
- PRE-DEC_2: the individual must pay attention to the cue(s); i.e., given that it is possible for the signal to be seen, the occupant actually takes note of the signal.
- PRE-DEC_3: the individual must comprehend the cue(s) and the information that is being conveyed; i.e., given that the signal is noted, that the information is understood.
- DEC_1: the individual must feel that the incident suggested by the cues and/or information is a credible threat.
- DEC_2: the individual must personalize the threat (i.e., feel that the incident is a threat to them) and feel that protective action is required; i.e., something needs to be done.
- DEC_3: the individual searches for what this action might be and establishes options.
- DEC_4: the options identified are assessed (given the information available) and a final action selected.
- DEC_5: the individual determines whether the protective action needs to be performed immediately.

Initially, the individual needs to receive a cue, pay attention to it, and comprehend the meaning associated with the cue (i.e., that it indicates an event). These represent the three pre-decisional stages of the PADM (PRE-DEC1-3) – the stages that determine whether external information is processed such that it can inform the decision-making process (Lindell and Perry 2004). Given that this information is processed, it then needs to be assessed to determine whether the information provided is credible (DEC_1). At this stage, the individual decides if there is actually something occurring that may require action, sometimes referred to as warning belief (Mileti 1974). If the individual's answer is yes, then he or she is said to believe the threat, and subsequently moves on to consider the next question in the process.

The individual next tries to determine whether the threat is relevant to him/her (DEC_2), known as personalizing the threat (or risk). Research has shown that a person's perception of personal risk, or "the individual's expectation of personal exposure to death, injury, or property damage" is highly correlated with disaster response (Lindell and Perry 2004:51). In this stage, also known as personalizing risk (Mileti and Sorensen 1990), the individual determines the likelihood of personal consequences that could result from the threat and asks the following: "Do I need to take protective action?" Essentially, at this point, which is also discussed in human factors research as "situation awareness" (Groner 2009), the individual tries to gain insight on the potential outcomes of the disaster and what those potential outcomes mean to his or her safety. The more certain, severe, and immediate the risk is perceived to be, the more likely the individual is to perform protective actions (Perry, Lindell and Greene 1981). If the cues are deemed to relate to them, the individual then determines whether it is relevant and pressing. This then requires the individual to determine the nature of the response required at that point in time.

At this stage, the individual engages in a decision-making process to identify 1) what can be done to achieve protection, and 2) the best available method for achieving protection. This consists of a search for protective actions, and the outcome of this stage is a set of possible protective actions from which to choose. After establishing at least one protective action option, an individual engages in protective action assessment. This involves assessment of the potential option(s), evaluating the option(s) in comparison with taking no action and continuing with normal activities, and then selecting the best method of protective action. In this simplified model, these last actions are combined in DEC_3.

The PADM describes the perception and decision-making from the individual's perspective. As part of this work, an attempt is made to better understand the factors that can influence this decision-making process given the provision of emergency information. Table 5 shows the six main stages of the PADM as they relate to receiving and subsequently acting upon information provided via an emergency communication system.

6.2 Inhibiting Factors

Passage through the stages of the decision-making process can be problematic. There are several factors which could prevent an individual from perceiving the warning signal/ message, paying attention to it, comprehending the message, believing the message, personalizing the risk, and developing an appropriate line of action(s) to perform.

The first factor type that can inhibit human response to warnings are source-related (technological) factors, or factors that originate from the source disseminating the message. As discussed earlier, there are a variety of communication modes that can be used to disseminate messages and signals. Many of these communication modes contain inherent limitations that may inhibit occupant response to a warning message. For example, some of the communication modes, such as strobes, do not provide any message material at all (see Table 1). While they may alert individuals to the fact that an emergency is taking place, such as the use of strobes for the emergency-trained hearing impaired

Table 5: Categorization of the impact of visual notification systems.

PRE-DEC_1	⟶	**Pc**	Perception: Whether it is possible for the information to be perceived
PRE-DEC_2	⟶	**At**	Attentiveness: Whether the information available is noticed
PRE-DEC_3	⟶	**Co**	Comprehension: Whether the information noticed is understood.
DEC_1	⟶	**Cr**	Credibility: Whether the information that is understood is deemed to be credible.
DEC_2	⟶	**Ps**	Personalization: Whether the credible information is deemed to be pertinent.
DEC_3 – DEC_5	⟶	**Ac**	Action: Whether the pertinent information indicates an appropriate action.

population, they provide no additional information on what is going on and what should be done about it.

The second factor that may inhibit a population's safe and effective response to a warning message are external factors - the physical structure around them and the environmental conditions that they face before and during the incident. In some cases, buildings have systems in place that provide non-emergency information to building occupants on a regular basis. For instance, passengers waiting at airline gates at an airport are often bombarded by visual (and audible) messages about departure information, seat changes, and delays. These regular messages can interfere with the ability to provide emergency messages as well as the message's ability to grab occupants' attention since the occupants may not differentiate the emergency message from the regular messages. Another environmental concern is the dynamic nature of the hazard itself. Changing environmental conditions inside or outside the warning area may inhibit the passage of information to individuals and require them to take different actions for safety than were previously suggested to them.

Finally, there are factors related to the warning receivers, i.e., the population, that can inhibit an effective response to an emergency message. There are distinct sections of any population that are likely to have difficulties following the steps of the decision-making process in order to respond to a warning. Gwynne (2007), in his research on ways in which to optimize fire alarm notification for high risk groups, identified various occupant types that could be vulnerable in relationship to emergency warnings. These occupant types include the following: individuals with sensory disabilities, such as those with hearing impairments, hearing loss, visual impairments, vision loss, and cognitive, thinking or learning disabilities; the aging population; children; individuals in large groups; people who are alone; untrained or unprimed people; people who are asleep; individuals who are intoxicated or are experiencing the same symptoms, such as those who are sleep deprived; non-native speakers; and people who are committed to a particular activity when the alarm or warning begins. In addition, there are situational factors that can inhibit the (previously unhindered) individuals' likelihood of receiving cues and the manner in which they interpret them. For instance, the actions in which individuals are engaged, their alertness, attention, location, etc.

The disaster condition itself can also induce vulnerabilities for the general population. For example, stress and anxiety during an emergency has been shown to reduce our capacity for paying attention and processing information (Chandler 2010; Keselman, Slaughter and Patel 2005). In effect, the emergency scenario itself can interfere with human response. Additionally, when people spend a great deal of time in the same situation, for example, their workplace, and are used to receiving the same information, sounds, smells, etc., they can sometimes neglect to pay attention to new information. In essence, people screen out messages based on previous habits and conditioning (Chandler 2010).

Even when people do pay attention and receive information, they are unlikely to understand messages containing highly technical terms, ambiguous language, an overloading of information, and categories or codes without proper training or explanation (Mileti and Sorensen 1990). Finally, once people understand the message, they are unlikely to believe it or personalize the risk because their first assumptions, regardless of the intensity of the information received, are often that 1) nothing unusual is occurring, and 2) if it is, I am not at risk. The first assumption is referred to in the literature as normalcy bias (Drabek 1986; Okabe and Mikami 1982) and the second is known as optimistic bias (Kunreuther 1991). These assumptions are further complicated if the at-risk population is frequently exposed to false alarms in the building because it makes it even easier to disbelieve the warning.

If, at any stage in the decision-making process, the individual is uncertain about the threat and/or the risk, he or she engages in additional information-seeking actions. The greater the ambiguity involved in the situation, the more likely that individuals will search for additional information that can guide their actions (Fahy and Proulx 1997; Mileti and O'Brien 1992). Information seeking is especially likely to occur when individuals think that time is available to gain additional insight on the question at hand. If information seeking is successful, in that the person at risk judges he or she has received enough information to assess the risk and/or respond, he or she will do so. However, if the information-seeking action is unsuccessful, then there will be additional searching for information as long as he or she is optimistic that other sources or channels can help clarify the situation (Lindell and Perry 2004). This could result in a significant amount of time spent seeking information, thus risking potential exposure to unsafe conditions and causing the individual to deviate from the desired procedural response.

7 System Effectiveness: What questions need to be addressed to improve system design?

As discussed earlier, it is possible for notification technologies to address some of the key behavioral components. In order for this potential to be realized, however, a number of key questions need to be addressed. Table 6 presents these key questions, along with the related component(s) of the decision-making process (from Table 5).

Table 6: Key design questions and their potential impact on the decision-making process.

Pc	Does the notification system have sufficient coverage? Can information be received from the notification system throughout the target areas?
At	Does the alert or message gain and retain the population's attention? Can the signal/message draw the target population's focus from the current task in order to focus them on the information being provided?
Pc At Co	Is the signal/information intelligible? Can the information be distinguished from background information that may potentially pollute the emergency information?
Co	Can the alert/information/message provided be understood? Does the target population comprehend the message content?
Cr	Does the population believe the information provided? Is the source known and trustworthy?
Cr	Does the information conflict with other information provided? Does the notification/messaging system complement other information systems active during the incident by providing supporting information via a different channel (staff, etc.)?
Cr	Does the system reliably identify incidents that require a response? Is the system prone to false alarms or misinformation?
Ps	Does the population believe that the information is meant for them? Does the target population personalize the message and the risk to themselves and/or others around them?
Cr Ps Ac	Is the information provided accurate and current? Is the information provided updated and informed by current 'intelligence'?
Ac	Is the information sufficient for the population to understand the nature of the incident and determine what response is required of them? Are instructions on the desired response provided?
Cr Ps Ac	Does the information provided relate to the specific incident at hand, or does it reflect a general response?

	Is the system able to collect and distribute information quickly enough for it to be of value? Is the response time quick enough for the procedure being employed?
Cr Ps Ac	

8   Review Method

A literature review has been conducted in order to inform the development of guidance related to the design and implementation of emergency communications systems. The review has been based on sample material drawn from the following subject areas: acoustics/audiology, buildings/engineering, crisis management/disasters, disability, ergonomics/human factors, human behavior in fire, illumination/lighting, language, media/communications, psychology/cognition, standards, and transportation (see Table 7). The selected material is intended to present a representative – as opposed to exhaustive – view into research and best practices. All material was drawn from publicly available resources, published in English.

Table 7: Subject Areas.

Subject Area	Icon
Acoustics / Audiology	A
Buildings / Engineering	B
Crisis Management / Disasters	C
Disability	D
Ergonomics / Human factors	E
Fire	F
Illumination / Lighting	I
Language	L
Media / Communication	M
Psychology / Cognition	P
Standards	S

Transportation	T

The review process followed the schematic outlined in Figure 2. Initially, source material was collected and scanned to determine its relevance. A significant amount of material was rejected on the basis that it could not be related to the notification modes addressed. The accepted material was then examined to develop an annotation for each source.

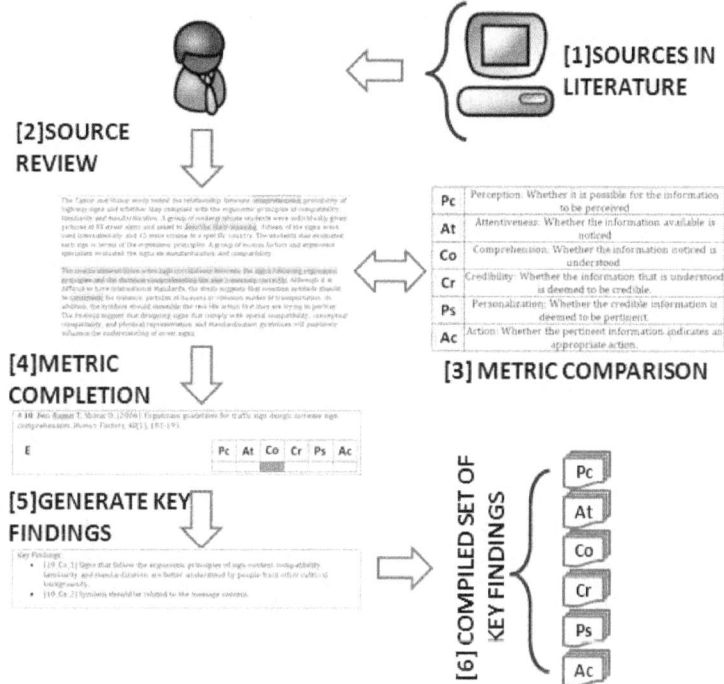

Figure 2: Review and presentation process.

The Protective Action Decision Model (PADM) (Lindell and Perry 2004) was used as a framework for categorizing the source material that was reviewed. As each source was reviewed, excerpts from the source were highlighted to identify which of the components of the decision-making model, shown in Table 5, were addressed by the literature source. The key stages identified earlier in Table 5 were used as a metric for categorizing each source, shown here in Table 8.

Table 8: Review Metric derived from Lindell and Perry (2004).

Pc	Perception: Whether it is possible for the information to be perceived
At	Attentiveness: Whether the information available is noticed
Co	Comprehension: Whether the information noticed is understood.

Cr	Credibility: Whether the information that is understood is deemed to be credible.
Ps	Personalization: Whether the credible information is deemed to be pertinent.
Ac	Action: Whether the pertinent information indicates an appropriate action.

A metric was then completed for each source identifying the components addressed. For example, the article by Ben-Bassat and Shinar (2006) was deemed to address issues relating to comprehension (see Figure 3(a)). In contrast, the article by Tang, Lin and Hsu (2008) addressed comprehension and action (see Figure 3(b)). This enabled the relevant subject matter of each source to be more easily identified by the reader and the key findings to be compiled by decision-making component.

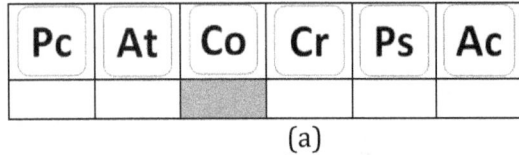 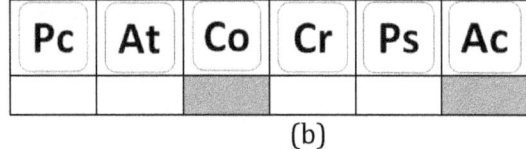

(a)   (b)

Figure 3: Metric used to identify the decision-making components addressed in each source.

A general format was produced and completed to develop an annotation for each literature source (see Table 9). Each annotation lists the source reference, identifies the subject area of the literature in which the source was found and the components of the decision-making process that were covered in the material, summarizes the contents, and lists the key findings.

Table 9: Review Format

SOURCE REFERENCE	
SUBJECT AREA	ASPECTS OF DECISION-MAKING PROCESS ADDRESSED BY SOURCE
SOURCE SUMMARY	
KEY FINDINGS	

The key findings from each of the reviews are collected and compiled in Appendix A. The annotation for each literature source is presented in Appendix B. An example of a formatted source review is shown in Figure 4. It should be noted that the detailed text is removed from this report to simplify the material presented.

9 Conclusions

This report provides a comprehensive review of emergency communication strategies for building emergencies. The focus of this report has been on the technology, approaches and human response to warnings provided for rapid-onset emergency events. This report has answered three main questions regarding emergency communication. First, what technology exists or is proposed for use in emergency notification? Second, what approaches are currently being used to disseminate messages? And third, how does the public respond to different types of information and information sources? This report of the ways in which people respond to emergency and non- emergency warnings and alerts

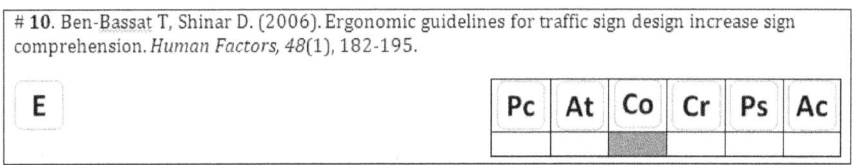

 # 10. Ben-Bassat T, Shinar D. (2006). Ergonomic guidelines for traffic sign design increase sign comprehension. *Human Factors, 48*(1), 182-195.

| E | | Pc | At | Co | Cr | Ps | Ac |

The Tamar and Shinar study tested the relationship between comprehension probability of highway signs and whether they complied with the ergonomic principles of compatibility, familiarity and standardization. A group of undergraduate students were individually given pictures of 30 street signs and asked to describe their meaning. Fifteen of the signs were used internationally and 15 were unique to a specific country. The students also evaluated each sign in terms of the ergonomic principles. A group of human factors and ergonomic specialists evaluated the signs on standardization and compatibility.

The results showed there were high correlations between the signs following ergonomic principles and the students comprehending the sign's meaning correctly. Although it is difficult to have international standards, the study suggests that common symbols should be consistent; for instance, pictures of humans or common modes of transportation. In addition, the symbols should resemble the real-life action that they are trying to portray. The findings suggest that designing signs that comply with spatial compatibility, conceptual compatibility, and physical representation and standardization guidelines will positively influence the understanding of street signs.

Key Findings:
- [10_Co_1] Signs that follow the ergonomic principles of sign-content compatibility, familiarity and standardization are better understood by people from other cultural backgrounds..
- [10_Co_2] Symbols should be related to the message content.

Figure 4: Use of the metric.

has resulted in a literature review of 160 literature sources, including annotations of each source (Appendix B). Findings from each literature source have been extracted and listed in Appendix A, which will serve as the foundation for the development of a future guidance document on emergency communication message content and dissemination strategies.

10 References

American Red Cross. (2010). *Social Media in Disasters and Emergencies.* Accessed January 2011 at http://www.redcross.org/www-files/Documents/pdf/SocialMediainDisasters.pdf

The American Red Cross. (2010b). "The Case for Integrating Crisis Response with Social Media." *Discussing the Future of Social Disaster Response.* Proceedings of the Emergency Social Data Summit, Washington, DC. Comp. The American Red Cross. Accessed January 2011 at http://www.slideshare.net/americanredcross/the-case-for-integrating-crisis-response-with-social-media.

American Red Cross. (2010c). "The Path Forward: A Follow up to 'The Case for Integrating Crisis Response with Social Media.'" *Proceedings from the Emergency Social Data Summit*, August 12, 2010, Washington, D.C. Accessed June 2011 at http://www.scribd.com/doc/40080608/The-Path-Forward-ARC-Crisis-Data-Summit-Wrap-Up.

Ben-Bassat T, Shinar D. (2006). Ergonomic guidelines for traffic sign design increase sign comprehension. *Human Factors, 48*(1), 182-195.

British Standard BS 5839-8:2008. (2008). Fire Detection and Fire Alarm Systems for Buildings – Part 8: Code of practice for the design, installation, commissioning and maintenance of voice alarm systems.

Chandler, R. (2010). *Emergency Notification.* Santa Barbara: Praeger.

Drabek, T. E. (1986). *Human System Responses to Disaster: An Inventory of Sociological Findings.* New York, NY: Springer-Verlag.

Fahy, R. F. and G. Proulx. (1997). Human Behavior in the World Trade Center Evacuation. In Y. Hasemi (Ed.). *Fire Safety Science -- Proceedings of the Fifth International Symposium* (pp. 713-724). London: Interscience Communications Ltd.

Gallaudet University. (2011). Accessed September 2011 at http://www.gallaudet.edu/dps/emergency_preparedness_guide.html.

Garton, L., C. Haythornthwaite, and B. Wellman. (1997). Studying Online Social Networks. *Journal of Computer-Mediated Communication* 3 (1) June 1997.

Groner, N. (2009). A Situation Awareness Analysis for the Use of Elevators During Fire Emergencies. *In 4th International Symposium on Human Behavior in Fire: Conference Proceedings* (pp. 61-72). London: Interscience Communications Ltd.

Gwynne, S. M. V. (2007). Optimizing Fire Alarm Notification for High Risk Groups. Quincy, MA: The Fire Protection Research Foundation.

Heverin, T. and L. Zach. (2010). "Microblogging for Crisis Communication: Examination of Twitter Use in Response to a 2009 Violent Crisis in the Seattle-Tacoma, Washington Area." *Proc. of the 7th International ISCRAM Conf.* Seattle, USA. 5 pages (pdf).

Hughes, A. L. and L. Palen. (2009). "Twitter Adoption and Use in Mass Convergence and Emergency Events." *Proc. of the 6th International ISCRAM Conf.* Gothenburg, Sweden. 10 pages (pdf).

Humphery, Brian. (2011). "Brian Humphrey on Best Practices for Social Media." *YouTube.* 4 Mar. 2011. Accessed: 27 June 2011. <http://www.youtube.com/watch?v=Zt-_n54wPE&feature=related>.

Humphrey, Brian. (2011b). "Los Angeles Fire Department on Twitter." *Los Angeles Fire Department on Twitter.* Los Angeles Fire Department, Accessed: 18 July 2011 at http://twitter.com/#!/LAFD.

Kavanaugh, A., S. D. Sheetz, F. Quek, B. J. Kim. (2010). "Cell Phone Use with Social Ties During Crisis: The Case of the Virginia Tech Tragedy." *Proc. of the 7th Internatiional ISCRAM Conf.* Seattle, WA. 12 pgs (pdf).

Keselman, A., L. Slaughter and V. L. Patel. (2005). Toward a framework for understanding lay public's comprehension of disaster and bioterrorism information. *Journal of Biomedical Informatics*, 38, 331-344.

Kuligowski, E. D. (2011a). Communicating the Emergency: Preliminary findings on the elements of an effective public warning message. National Institute of Standards and Technology, Gaithersburg, MD.

Kuligowski, E. D. (2011b). *Terror Defeated: Occupant sensemaking, decision-making and protective action in the 2001 World Trade Center disaster.* Ph.D. Dissertation. Boulder, CO: University of Colorado at Boulder.

Kunreuther, H. (1991). A Conceptual Framework for Managing Low Probability Events. Philadelphia, PA: Center for Risk and Decision Processes, University of Pennsylvania.

Latonero, M. and I. Shklovski. (2010). "Respectfully Yours in Safety and Service: Emergency Management & Social Media Evangelism." *Proceedings of the 7th International ISCRAM Conf.*, Seattle, WA. 10 pgs (pdf).

Lenhart, A. and S. Fox. (2009). *Twitter and status updating.* Washington DC: PEW Internet & American Life Project. Retrieved January 14, 2011 from http://www.pewinternet.org/Reports/2009/Twitter-and-status-updating.aspx.

Lindell, M. K. and R. W. Perry. (2004). *Communicating Environmental Risk in Multiethnic Communities*. Sage Publications.

Mileti, D. S. (1974). A Normative Causal Model Analysis of Disaster Warning Response. Boulder, CO: University of Colorado Department of Sociology.

Mileti, D. S. and J. Sutton. (2009). "Social Media and Public Disaster Warnings." Accessed June 2011 at http://www.jeannettesutton.com/uploads/WARNINGS_Social_MediaMileti_Sutton.pdf.

Mileti, D. S. and P. O'Brien. (1992). Warnings during Disasters. *Social Problems*, 39, 40-57.

Mileti, D. S. and J. H. Sorensen. (1990). Communication of Emergency Public Warnings. ORNL-6609, Oak Ridge: National Laboratory.

National Clearinghouse for Educational Facilities. "Mass Notification for Higher Education," Accessed December 2010. http://www.ncef.org/pubs/notification.pdf

National Research Council. (2011). "Public Response to Alerts and Warnings on Mobile Devices: Summary of a Workshop on Current Knowledge and Research Gaps," Committee on Public Response to Alerts and Warnings on Mobile Devices, Washington, DC.

National Fire Protection Association. (2010a). NFPA 72 US National Fire Alarm and Signaling Code. Quincy: National Fire Protection Association.

National Fire Protection Association. (2010b). NFPA 1600, Standard on Disaster/Emergency Management and Business Continuity Programs. Quincy: National Fire Protection Association.

Okabe, K. and S. Mikami. (1982). A Study on the Socio-Psychological Effect of a False Warning of the Tokai Earthquake in Japan. A Paper presented at the Tenth World Congress of Sociology, Mexico City, Mexico.

Perry, R. W., M. K. Lindell, and M. R. Greene. (1981). *Evacuation Planning in Emergency Management*. Lexington, MA: Lexington Books.

Quarantelli, E. L. and R. R. Dynes. (1972). "When Disaster Strikes (It Isn't Much Like What You've Heard and Read About)" *Psychology Today* 5(February): 67-70.

Sutton, J., L. Palen, and I. Shklovski. (2008). "Backchannels on the Front Lines: Emergent Uses of Social Media in the 2007 Southern California Wildfires." *Proc. of the 5th International ISCRAM Conf.*, May 2008, Washington, DC. 9 pages (pdf).

Tang C. H., C. Y. Lin, Y. M. Hsu. (2008). Exploratory research on reading cognition and escape-route planning using building evacuation plan diagrams. *Applied Ergonomics 39*(2), 209-217.

Tierney, K. J. (2003). "Disaster Beliefs and Institutional Interests: Recycling disaster myths in the aftermath of 9-11." Pp. 33-51 in *Terrorism and Disaster: New Threats, New Ideas (Research in Social Problems and Public Policy, Volume 11)*, edited by Ted I. K. Youn. Bingley, UK: Emerald Group Publishing Limited.

Tierney, K. J., M. K. Lindell, and R. W. Perry. (2001). *Facing the Unexpected: Disaster Preparedness and Response in the United States.* Joseph Henry Press.

Vieweg, S., L. Palen, S. B. Liu, A. L. Hughes, and J. Sutton. (2008). "Collective Intelligence in Disaster: Examination of the Phenomenon in the Aftermath of the 2007 Virginia Tech Shooting." *Proceedings of the 5th International ISCRAM Conf.,* Washington, DC. 11 pgs (pdf).

Virginia Polytechnic Institute and State University. (2010). "Emergency Notification System Protocols." Accessed September 2011 at http://www.police.vt.edu/static/VT_ENS_Protocols_2010FEB.pdf.

White, C., L. Plotnick, J. Kushman, S. R. Hiltz, and M. Turroff. (2009). "An Online Social Network for Emergency Management." *Proceedings of the 6th International ISCRAM Conf.,* Gothenburg, Sweden. 9 pages (pdf file)

APPENDIX A   Key Findings

The purpose of this section is to capture the key findings of each of the annotated reviews included in Appendix B. No effort has been taken here to weight or compare findings; rather, each finding may be seen as a piece of evidence that might inform future analysis. Each time a recommendation or finding was listed in one of the literature sources in Appendix B, it was labeled and relisted in this section, where it has been organized according to the component in the decision-making process that it addresses. These components, Perception (Pc), Attentiveness (At), Comprehension (Co), Credibility (Cr), Personalization (Ps), and Action (Ac), are derived from the Protective Action Decision Model (PADM), as described in Section 6 and listed in Table 8. Many findings address more than one component – these findings were assigned multiple labels and included under each component to which they pertain.

After being compiled within each component in the decision-making process, the key findings were categorized as pertaining to visual warnings only, to auditory warnings only, or to warnings in general. To bring further order to this large list, findings were additionally aggregated into subcategories. The subcategories were suggested by the findings themselves; as the findings were collected, certain themes emerged and were given titles. Table 10 presents the entire list of categories and subcategories. Although there is plenty of overlap among these groupings, each finding appears only once within each component. Any application of the material compiled under one of the components should therefore keep in mind that information throughout the list of findings may be of use.

The label for each finding is coded as [SourceNumber_ComponentCode_FindingNumber]. For example, the key finding labeled [97_Pc_2] can also be found in Appendix B within Source Review #97 under the same label, as the second finding in this review pertaining to Perception. In this manner, the reader may trace each finding back to its original source.

As indicated in Table 10, shades of gray are used throughout this Appendix to delineate findings that pertain to the categories of General, Visual, and Auditory warnings, respectively. Subcategories that did not contain any findings under one of the components were removed.

Table 10: Subcategories for organization of key findings

General warnings	Visual warnings	Auditory warnings
Design	Design	Design
Positive messages	Content	Alarm tones
Message ordering	Word choice	Synthesized speech
Validation	Symbols	Content
Content	Diagrams	Word choice
Word choice	Message length	Message length
Complexity	Complexity	Complexity
Timeliness	Urgency	Urgency
Communication channels	Color	Pitch
Alternative communication modes	Contrast	Sound level
Multiple alarms	Font	Timing and delivery rate
Alarm reliability	Visibility distance	Clarity
Messenger	Placement	Cocktail party effect
Public perception	Angle	Ambient noise
Stress	Message display rate	Individual and group characteristics
Cognitive load	Lighting – Static	Bilingual and non-native English speakers
Disruption of activity	Lighting – Dynamic	Sleep arousal
Repetition	Lighting – Wayguidance systems	
Familiarity and training	Smoke	
Individual and group characteristics	Surroundings and context	
Non-native English speakers	Individual and group characteristics	
Effects on behavior	Bilingual messages	
	Sleep arousal	

29

A.1 Perception (Pc)

Pc	Summary Findings - Perception
General Warnings	*Design* • [91_Pc_1] An extensive literature is available on the communication of hazard information to the public and the public response during disasters. • [139_Pc_3] Effective risk communication requires the engagement of communication representatives in designing and evaluating educational materials, emergency messages, and dissemination practices, including language resources and accessibility. *Timeliness* • [127_Pc_2] Captioning of news broadcasts should begin immediately as soon as audio information begins. • [127_Pc_3] Important new emergency information should be identified in captions as well as in the news broadcast. • [128_Pc_7] Paid staff should be on call at broadcast stations to provide emergency broadcasts at any time. *Communication channels* • [97_Pc_1] All television broadcasters, cable operators, and satellite television services that provide local emergency news are required by the FCC to relay all essential emergency information via captioning or other visual means in order to reach the deaf and hard of hearing population. • [97_Pc_2] Federal, state, and local governments are legally required to make their emergency notification systems accessible to deaf and hard of hearing people. • [97_Pc_3] Textual information from NOAA National Weather Service radio broadcasts is available through modified radio receivers, which provide an alert through a strobe light, an auditory signal, and, if requested, a pillow vibrator or bed shaker to awaken the person. • [97_Pc_4] The full text of a National Weather Service warning is available through receivers connected to a satellite feed or through a pager system. • [97_Pc_5] A powerful strobe light added to a civil defense siren tower can aid in alerting individuals but is limited to those who are awake and in direct line of sight from the tower. • [97_Pc_7] Reverse 911 systems, which enable an emergency notification agency to initiate TTY calls, can be used to alert deaf or hard of hearing populations. • [98_Pc_2] Innovative technologies and programs, such as adapted weather radios, text messaging, and weather pagers, promote wider dissemination of warning information. • [98_Pc_3] Social networks, including work units, neighborhood associations, and community organizations, can be arranged in advance to help warn disabled populations. Preparedness plans should include alternative backup networks. • [99_Pc_2] Camera operators and editors need to include in the picture any sign language interpreter next to the official spokesperson.

Pc	Summary Findings - Perception
General Warnings (cont.)	• [99_Pc_3] Emergency hotlines must include TTY/TDD (text telephone or telecommunication device for the deaf) numbers when available, or the instruction "TTY callers use relay." • [99_Pc_4] Make sure that the emergency website is accessible to those with disabilities. • [99_Pc_5] Provide information in alternate formats based on population needs, including Braille, audio recording, large font, text messages, and emails. • [100_Pc_1] The wide usage of cell phones by the U.S. population and the ability to target messages to a cell phone's actual location makes the development of the Commercial Mobile Alert Service (CMAS) an attractive opportunity to more precisely target those individuals who would be most at risk in a crisis situation. • [100_Pc_2] Some technologies are better suited for delivering alerts (e.g., sirens or CMAS), and others are better suited for warnings (e.g., broadcast radio or television). • [100_Pc_3] For time-sensitive disasters, sirens and other immediate alerting systems are required. • [113_Pc_1] Take advantage of the public's willingness to participate in the warning process, including encouraging the installation of in-home warning devices. • [128_Pc_4] It would be useful for Spanish-language media websites to contain local news, information about developing emergency situations, and up-to-date lists of emergency contact persons. • [128_Pc_12] Strategies for social media include sending multiple language audio and/or text messages to cell phones with brief guidance on actions to be taken. • [139_Pc_2] Different communities obtain information from different sources, including social networks, ethnic media, and places of worship. *Alternative communication modes* • [130_Pc_2] Pillow and bed shakers at high intensity are much more effective than strobe lights at waking individuals who are hard of hearing. • [130_Pc_3] Although they are effective at waking sober adults, pillow and bed shakers are not effective at waking those impaired by alcohol. • [130_Pc_4] A 520 square wave auditory signal at 75 dBA or above is more effective than either strobe lights or bed shakers at waking both hard of hearing and alcohol impaired individuals. *Multiple alarms* • [2_Pc_1] The temporal pattern for the ANSI standard audible emergency evacuation signal can also be applied to visual and tactile signals to aid occupants with impaired hearing and occupants in zones with intense background noise. • [14_Pc_1] Even with light intensities on the face exceeding those from commercially available visual alarms, strobe lights cannot be relied upon to wake a sleeping person. Tactile devices should be used with the flashing strobes to wake up hard of hearing people during a fire. • [89_Pc_8] Multiple modes of warning message transmission are advised, both to reinforce the message and to make sure that at least one message is perceived and understood.

Pc	Summary Findings - Perception

General Warnings (cont.)

- [98_Pc_5] Messages need to be provided using a wide variety of media sources through both audible and visual means.
- [106_Pc_2] The route decision-making of a visual wayfinding system was improved through the addition of a tactile component.
- [106_Pc_3] The slow evacuation time for a tactile wayfinding system was improved by adding a visual component.
- [127_Pc_1] Television broadcasters and internet providers should provide accurate, real-time, and verbatim captioning of all news broadcasts, with contrasting background to ensure readability and nothing blocking the message.
- [127_Pc_4] Sign language interpreters should be positioned onscreen during emergency broadcasts and not blocked by captions or other visual material.
- [127_Pc_5] Televisions in public places should have captions turned on at all times.
- [127_Pc_6] Text messaging systems should be interoperable in order to provide multiple communications channels in an emergency.
- [127_Pc_7] All audio announcements should be broadcast with a simultaneous display of text, which should provide appropriate information, including the nature of the emergency and instructions on what to do.
- [128_Pc_10] Because Spanish-speaking communities have limited computer and social media access, emergency materials should also be disseminated in print. The Spanish language print media would make a good partner.
- [128_Pc_11] Neighborhood radio, siren, and text messaging systems were recommended by community leaders.
- [139_Pc_4] Multiple modes of communication are required to reach diverse communities.
- [140_Pc_1] The ADA regulations require the use of both audible and visual signals for fire alarm systems.
- [140_Pc_2] The ADA regulations require elevator control buttons to be identified both tactilely and visually, with raised characters and Braille positioned directly to the left of the control button and using standard tactile symbols for emergency stop, alarm, door open/close, main entry floor, and phone buttons.
- [140_Pc_11] Multiple means of communicating important messages should be used when feasible. The ADA regulations require the use of both audible and visual signals for fire alarm and two-way communication systems. The use of tactile information with visual information is required for signs in elevators, detectable warning surfaces on boarding platforms, and other critical applications.
- [148_Pc_4] Communication of warning messages may be visual, auditory, olfactory, or tactile. A systems approach using multiple modalities can be used to reach people in different ways, at different times, and in different subgroups.

Alarm reliability

- [104_Pc_10] Increased maintenance and the selection of long-life illumination sources improve the reliability of exit signs.

Messenger

- [98_Pc_4] Members of the vulnerable communities should be involved in

Pc	Summary Findings - Perception
General Warnings (cont.)	planning and disseminating messages in ways that are amenable to how the vulnerable populations use the technology.

Stress

- [132_Pc_1] Emotional stress can cause a physiological hearing deficit, preventing some messages from being heard.

Cognitive load

- [132_Pc_2] The narrowing of attention in challenging environments can result in inattentional deafness, in which unexpected features are simply not noticed.
- [132_Pc_3] Message redundancy and ensuring that multiple people are listening can help overcome sensory overload issues.

Familiarity and training

- [157_Pc_2] People that were familiar with a building were less likely to notice exit signs than those who were unfamiliar with it.

Individual and group characteristics

- [86_Pc_1] Population segments differ in their access to warning messages and in their perception of the credibility of information sources.
- [98_Pc_1] Messages and practices must consider the needs, means, and abilities of vulnerable populations. There is often significant overlap among different vulnerable groups, and accommodating people with disabilities often improves response for everyone.
- [100_Pc_4] Special consideration is needed for at-risk populations, including those who are blind or have low vision, those with hearing impairments, those who are elderly, and racial and ethnic minorities and non-English speakers
- [108_Pc_1] People in vulnerable populations are often marginalized, and are less likely to receive warnings and to respond appropriately in emergency situations. Vulnerable populations include those based on age, race and ethnicity, socioeconomic status, gender, physical disability (e.g., mobility, hearing, and vision), language, work, and isolation.
- [108_Pc_2] Socioeconomically disadvantaged families may lack the ability to receive forecasts and warnings, a safe home environment and the transportation to evacuate.
- [108_Pc_3] Deaf populations may not be well served by the media despite regulations to provide messages live during emergencies through closed captioning and interpreters. Lip reading is impossible when television experts turn their backs to the camera or are not visible on screen.
- [108_Pc_4] Elderly people may have less access to social networks and technology, and may find it more difficult to evacuate due to mobility problems.
- [108_Pc_5] Children may be on their own or in settings that are unprepared to handle the emergency. Conversely, schools may provide an avenue to provide information to other family members.
- [108_Pc_6] Tourists are unfamiliar with the environment, available resources, and possibly the language.
- [108_Pc_7] Interactions among disadvantages, such as age, poverty, disability, and isolations, multiply the risk.
- [108_Pc_8] Messages should be targeted for the social networks of vulnerable

Pc	Summary Findings - Perception
General Warnings (cont.)	populations and developed by people that understand those communities. • [108_Pc_9] Research is needed on how vulnerable populations receive and disseminate information. • [128_Pc_5] Attention is needed to planning for crisis situations affecting local communities with little or no access to broadcast/cable media news and emergency information. [139_Pc_1] Emergency management professionals need to identify at-risk populations. *Non-native English speakers* • [128_Pc_1] Barriers to emergency broadcasts on Spanish-language television stations include a lack of authorization to preempt syndicated network programming, in addition to limited local news staff and technical equipment for live coverage. • [128_Pc_2] Government agencies are in general not properly staffed to produce and disseminate messages in multiple languages. Not all information on websites is translated, and content in languages other than English may be difficult to find. • [128_Pc_3] Signal strength of Spanish-language radio stations may be reduced at night, and staff members are generally not available to cover nighttime emergency broadcasts. Spanish speakers outside of metropolitan areas are particularly affected. • [128_Pc_6] Regulations are recommended that require broadcasters to transmit emergency alerts in the operating language of the licensed station, at any time of day or night. • [128_Pc_8] Arrangements should be made for English-language broadcast stations to air multilingual messages if ethnic stations are knocked off the air. • [128_Pc_9] A multilanguage emergency alert system is being developed by a national nonprofit organization for ethnic news organizations.
Visual Warnings	*Design* • [29_Pc_7] Visibility is affected by a number of sign design factors, including contrast, color, adaptation, lighted/opaque backgrounds, and luminance levels. • [79_Pc_2] Borders increase sign conspicuity. • [152_Pc_1] To be noticed and used, a sign must be as salient as possible. Salience can be enhanced by using large bold print, high contrast, colors, borders, pictorial symbols and special effects, such as flashing lights. • [153_Pc_1] The word EXIT exhibits higher visibility (requires smaller sign content height from the same viewing distance) than pictograms. • [153_Pc_3] Visibility of an exit sign with combined symbols is slightly lower than that of a sign containing only one of these symbols. *Word choice* • [148_Pc_1] The literature on visual warnings generally recommends a signal word panel (DANGER, WARNING, CAUTION, etc.) and a message panel, which can be text or symbols. • [148_Pc_2] Good warning design includes using standard colors and meanings for signal words (e.g., DANGER, WARNING, CAUTION) and symbols.

Pc	**Summary Findings - Perception**
Visual Warnings (cont.)	Symbols • [104_Pc_8] The chevron style is the most visible for the arrow symbol. The chevron should be at least 60 mm high to be visible from a distance of 30 meters. • [115_Pc_7] Clear and stable figure to ground articulation, which differentiates a figure from the background using texture, is essential. • [135_Pc_1] Well-designed symbols can be advantageous because they are quicker to read and comprehend than words, can be recognized from farther away, and are more recognizable under poor environmental conditions such as darkness and weather. • [153_Pc_2] Arrows with tails are more visible than arrows without tails. Color • [29_Pc_1] Visibility was generally best for red/white exit signs with luminances of 170, 391, 1272 cd/m² relative to other colors and designs. • [29_Pc_17] Given that many people have difficulty seeing the color red, translucent green materials are likely to appear brighter to more people than translucent red materials for the same light source. • [29_Pc_15] Under lit and dark conditions, illuminated yellow exit signs were the most readable, followed by green and red signs. • [41_Pc_2] If a sign is meant to be visible at night or with low lighting, colors in the middle of the spectrum should be used (light greens and yellows). • [61_Pc_4] The best visibility for outdoor signs during the day has been found for negative polarity with white characters on a dark green background. At night, positive polarity is preferred. The detection distance of green signs is shorter if the sign is highly reflective. • [79_Pc_1] Black on orange and white on green signs were detected at greater distances than black on white signs. • [89_Pc_5] For elderly populations, do not use similar colors, especially blue and green shades, to distinguish between two different things. • [104_Pc_5] Humans tend to be more sensitive to the yellow-green region of the spectrum, so less energy is required to operate a green light with the same perceived brightness as a red light. • [104_Pc_6] In clear air experiments, the color of exit signs (when controlled for brightness) seemed to have a marginal impact on visibility. • [112_Pc_6] Typically, translucent green materials are brighter to more people than translucent red materials for the same light source. Green signs are therefore likely to be more visible through smoke than red signs. • [112_Pc_7] Use color to aid occupants in discriminating between exit signs and other luminous sources. • [114_Pc_1] Older adults have trouble with colors in the violet, blue and green range. • [153_Pc_4] Green or black sign content is slightly more visible than red. Contrast • [41_Pc_1] For a light-reflecting sign, the contrast ratio between the text and the

Pc	**Summary Findings - Perception**

<table>
<tr><td rowspan="2">Visual Warnings (cont.)</td><td>background should be at least 30%, although recommended values are 40% in daylight and 50% at night. The required contrast also depends on the size of the letters and the viewing distance.</td></tr>
<tr><td>

- [61_Pc_1] Glare results in a measurable change in visibility due to a reduction in luminance contrasts in the human retinal image.
- [63_Pc_2] Color contrast between an icon and its background aids in the icon being noticed. In the absence of control over the background, edge definition is necessary to decrease the effect of color changes.
- [79_Pc_3] The internal contrast ratio between legend and background should be 12:1, where luminances for both are measured in cd/m^2.
- [93_Pc_1]-Sign plates must provide good color contrast against the background and surroundings.
- [93_Pc_2] Sign graphics must provide sufficient contrast against the sign plate.
- [104_Pc_7] Sign color should contrast with the background surface.
- [114_Pc_2] Text on signs should have a high contrast from the background and glare should be minimized.
- [140_Pc_6] The ADA regulations require visual messages to have a non-glare finish and high contrast between the image and background in order to improve their legibility for persons with low vision. Lighting shadows and surface glare should be minimized.
- [140_Pc_12] Visual messages require a non-glare finish and high contrast between the image and background in order to improve their legibility for persons with low vision. Lighting shadows and surface glare should be minimized.
- [141_Pc_4] The contrast between sign content and the background should be at least 60 %.
- [143_Pc_1] Contrast was identified as the most important legibility factor for color selection.

Font

- [79_Pc_4] The most legible stroke width to height ratio for both positive and negative contrast letters is 1:5.
- [89_Pc_4] Use large sans serif fonts in warning texts, especially for elderly populations.
- [114_Pc_3] Font should be at a minimum 12 point to 14 point and be sans-serif.
- [114_Pc_4] The minimum vertical type size of 6.7 point and no more than 39 characters per inch should be allowed if older people need to comprehend the sign.
- [114_Pc_5] A stroke width ratio of 1:6 to 1:8 for black letters on a white background and 1:8 to 1:10 ratios for white type on a black background is recommended.
- [115_Pc_2] For printed material, the ratio of the thickness of the stroke to the height of the letter should be 1:6 to 1:8 for black on white and 1:8 to 1:10 for white on black.
- [115_Pc_4] Lowercase letters are easier to read than all uppercase letters.
- [115_Pc_5] Close set type can be read faster than regular type.
- [115_Pc_6] On computer screens, letter sizes should have a minimum visual

</td></tr>
</table>

Pc	Summary Findings - Perception
Visual Warnings (cont.)	angle of 11 or 12 arc minutes, which converts to 0.06 inches to 0.07 inches for lower case letters. • [140_Pc_5] The ADA regulations require raised characters to be uppercase, in a plain and conventional font, and raised at least 0.8 mm above their background. • [140_Pc_13] For legibility, character fonts in warning messages should be plain and conventional, and character size, character spacing, line spacing, and message height need to take into account viewing distance for visual messages and finger legibility for tactile messages. • [140_Pc_14] The appropriate use of uppercase and lowercase letters for legible signs depends on the application. • [143_Pc_2] A bold sans-serif font was found to be slightly more legible than the serifed font in use at airports. • [143_Pc_3] The font with the widest typeface was found to be the most legible. • [153_Pc_5] Larger signs are needed for older adults than for younger adults. *Visibility distance* • [29_Pc_13] For signage legibility, calculations of visual angle suggest a minimum object size of 1.25 inches. • [34_Pc_1] Signs should be designed to ensure that the expected observation/recognition distance does not exceed $D=172\ s$, where s is the symbol size, or $D=132\ h$, where h is the overall sign height. • [41_Pc_3] The rule of thumb for letter size indicates that for each one inch of letter height the sign is legible another 50 feet farther away. Spacing between lines should be ¾ the average height of upper-case letters, and the borders above and below the text should equal the average letter height. The width of the side borders should equal the height of the largest letter. • [61_Pc_5] Characters on signs should be 15 cm to 25 cm high in order to be visible at a distance of 40 meters to 60 meters. • [71_Pc_2] The visibility of an exit sign is independent of size when the luminance is over 300 cd/m². • [93_Pc_3] People with normal sight can read text at a distance 400-500 times the height of its lower case letters. • [114_Pc_6] A ratio of 1:100 of letter size to reading distance on signs with a minimum letter size of 15 mm is recommended. • [115_Pc_3] The stroke width (Ws) for letters to be read at a specific distance d by a person with a given Snellen acuity score S is given by $Ws = 1.45 \times 10^{-5}\ d\ /\ S$. • [116_Pc_1] Under daylight viewing conditions, the legibility distance for older adults to read highway signs is 80 % of that for young adults. • [116_Pc_2] Nighttime reduces the distance at which people are able to read highway signs. Relative to the value for young adults under daylight viewing conditions, the legibility distance is reduced to 70 % for young adults and to 46 % for older adults. • [116_Pc_3] The presence of glare reduces the distance at which older people are able to read highway signs at night by about 26 %. The legibility distance is 34 % that for young adults under daylight viewing conditions • [137_Pc_1] The legibility distance of drivers from a DMS using 18 inch characters is about 800 feet during the day and 600 feet at night, based on the

Pc	Summary Findings - Perception
Visual Warnings (cont.)	highway industry standard of accommodating 85 % of drivers. • [137_Pc_2] At night, drivers can read no more than one unit for nine inch characters, even at 30 mph, and no more than two units for 10.6 inch characters. Since most messages communicate three units or more, using nine inch and 10.6 inch characters are not recommended for arterial roads. *Placement* • [34_Pc_4] For safety signs that indicate the escape route, multiple reinforcing directional signs are preferable to a single larger sign at a longer viewing distance. • [104_Pc_9] Placement on the floor or low on the wall may be beneficial given the lower smoke densities and lower ambient light levels at this height. However, smoke densities may not always be lower at the floor, and low-level signs may also be more prone to damage. Low level exit signs may therefore be of most benefit as supplemental rather than replacement signage. • [136_Pc_1] Positioning considerations for PCMS roadway signs include the distances needed for sign visibility, for legibility, and for separation of multiple PCMS. • [140_Pc_7] The ADA regulations require tactile characters to be located between 1.22 meters and 1.525 meters above the floor and visual characters to be located at a height of at least 1.015 meters. • [140_Pc_10] Visible alarms must be located within the space they serve in order to ensure that the signal is visible to the target population. • [141_Pc_1] Wayfinding signs should be consistent in height, type of location, and style. • [148_Pc_3] Task analysis is a method for determining good potential locations for the warning. *Angle* • [34_Pc_2] Safety signs should not be oriented at a displacement angle greater than 15 degrees from the direct line of sight. • [156_Pc_1] The legibility of signage is dependent upon the angle of observation. • [157_Pc_1] The probability of detecting an exit sign was less when an individual approached at an angle as compared to head on. *Lighting - Static* • [15_Pc_1] LED exit signs are more visible and legible than illuminated or non-illuminated signs, especially for visually impaired people. • [29_Pc_3] Detectability and readability increase with luminance. • [29_Pc_4] Aiming the ambient light source away from an illuminated sign reduces the negative impact on detectability, particularly when smoke is present. • [29_Pc_6] Internally lit green/white signs outperformed externally lit and edge lit signs. • [29_Pc_9] Visibility for a tritium exit sign deteriorated much faster in smoke than that for an electrically illuminated sign. • [29_Pc_11] Stencil-faced signs (with illuminated letters and opaque backgrounds) are more visible than panel-faced signs.

Pc	Summary Findings - Perception
Visual Warnings (cont.)	• [29_Pc_18] Smart fixtures that increase sign brightness and decrease ambient light in smoky conditions would improve visibility. • [29_Pc_19] Increasing the brightness, reducing ambient lighting, using stencil-faced rather than panel-faced signs, and choosing translucent green over translucent red are recommended for exit signs. • [29_Pc_16] Under clear conditions, an electrically illuminated sign was visible and legible from 300 feet, a tritium sign was legible from 75 feet, and a phosphorescent sign was legible from 25 feet. • [50_Pc_1]The luminance of LED lights decreased over time from 36.49 cd/m^2 after 30 minutes to 29.27 cd/m^2 after 90 minutes. • [50_Pc_2] The luminance levels of LED signs were typically two orders of magnitude greater than those of photoluminescent signs or tritium signs. • [50_Pc_3] The minimum luminance of exit signs should be between 8.6 cd/m^2 and 15 cd/m^2 to be effective. • [61_Pc_2] Illuminance on signs and information boards should be at least 100 lx. • [61_Pc_3] Mean luminance on signs where work is performed should be at least 10 cd/m^2. • [112_Pc_5] The readability of exit signs can be improved by limiting unnecessary sources of scatter from the sign itself. • [104_Pc_2] Bifunctional signs are available that increase sign brightness during an emergency. These may be able to address issues of both performance and comfort. • [104_Pc_4] Signs with opaque backgrounds appear to perform better in smoke than signs with trans-illuminated backgrounds. In smoke, the large luminous areas on trans-illuminated backgrounds act in a manner similar to ambient lighting – veiling or confusing the sign content. • [141_Pc_2] Lighting should be sufficiently bright that wayfinding signs can be read from at least 25 feet away. • [141_Pc_3] A white or light background should be used for signs in low lighting. • [105_Pc_1] The luminance at the position of the viewer is a more reliable indicator of visibility than the illuminance of the light source • [153_Pc_6] To be visible under emergency lighting conditions, exit sign content needs to be larger than under normal lighting conditions. For example, for people with normal eyesight, exit sign characters need to be 75 mm to 80 mm high to be confidently identified from 18 meters away under emergency lighting, compared to 50 mm to 55 mm high under normal lighting. • [155_Pc_2] A fluorescent sign with a green and white stencil and mean luminance over 1000 cd/m^2 performed the same as LED signs with 100 LEDs each having luminance of 140 mcd or 260 mcd. *Lighting - Dynamic* • [70_Pc_2] Flashing lights are more conspicuous than static lighting systems. • [71_Pc_1] Flashing lights have a more significant affect on the noticeability of a sign for smaller signs than for larger ones. • [101_Pc_1] Static flashing lights were able to attract the attention of people unfamiliar with an escape route but were unsuccessful in directing movement in a particular direction. An arrow may have helped.

Pc	Summary Findings - Perception
Visual Warnings (cont.)	• [103_Pc_1] Green flashing lights placed next to some exits in an office building were not mentioned by evacuating occupants as a factor influencing exit choice. Although more people evacuated through those exits, the difference was not significant. • [103_Pc_2] In a movie theater, all those evacuating used the exits with green flashing lights in preference to those with normal exit signs. Flashing lights may be more effective in influencing the choice of exits in less familiar places. • [117_Pc_1] Strobes can be seen both directly and indirectly – from reflective surfaces, for example. • [117_Pc_2] Blind spots were found in each of three large warehouse stores where strobe alarms could not be seen either directly or indirectly. • [117_Pc_3] The existence of blind spots in facilities with visual signaling alarm devices depends on a number of factors, including structural, procedural and population, not necessarily related to strobe performance. • [130_Pc_1] Strobe lights alone are not able to attract sufficient attention to overcome sleep. • [149_Pc_1] Safety signs using dynamic LED displays were not found to increase compliance over static LED displays. *Lighting –Wayguidance systems* • [12_Pc_1] Although it was not difficult to find the handrail in a stairwell illuminated by either PLM lighting or reduced emergency lighting, almost half the people evacuating under either condition complained that it was difficult to see. • [12_Pc_2] L-shaped markers along the sides of the stairwell alone made it hard for occupants to see the steps. • [12_Pc_3] During an emergency, PLM lighting with a one inch strip along the front edge of each step in a stairwell is preferred to emergency lighting at an average of 37 lx. • [12_Pc_4] For PLM lighting in a stairwell, the front edge of each step should be delineated so that occupants can see the steps clearly. • [101_Pc_2] Flashing lights were much more noticeable than floor markings in an escape route. • [106_Pc_1] Two continuous visual wayfinding systems (incandescent and PLM) and a combination of visual semi-continuous and tactile systems outperformed other wayfinding systems in terms of route adoption and evacuation times. • [106_Pc_4] A wayfinding system with continuous markings is superior to traditional exit signs. • [110_Pc_1] PLM should not be considered equivalent to wayguidance systems • [110_Pc_2] PLM may be considered as functionally equivalent to emergency lighting. • [110_Pc_3] PLM can be used to reduce the levels of emergency lighting. • [110_Pc_5] PLM produces favorable results when compared to emergency lighting. • [147_Pc_2] Powered (LED and miniature incandescent) wayfinding systems are visible in smoke at a significantly greater distance than photoluminescent systems.

Pc	**Summary Findings - Perception**

Visual Warnings (cont.)	• [147_Pc_3] Groups of small lights placed close together make an effective wayfinding system, enabling the accentuation of complex structural features and the use of dynamic configurations. • [154_Pc_1] Wayguidance systems with lighting tracks mounted on the wall resulted in significantly faster walking speeds in smoke than overhead lighting. This was true for stairwells, stair landings, and corridors. • [154_Pc_2] Walking speeds for a wayguidance system that illuminated at lower than recommended levels were found to exceed speeds under normal overhead lighting by more than a factor of three in a long corridor. • [154_Pc_3] Wayguidance systems can highlight architectural features such as steps and doorframes, giving people more confidence when entering a stairwell and approaching an exit door. *Smoke* • [29_Pc_2] Red exit signs were slightly less affected by the scattering effect of smoke than other colors. • [29_Pc_5] Similar smoke densities were required to obscure signs with background luminance at and above 89 cd/m^2 when viewed from 77 inches away, with less smoke needed to obscure a sign with luminance of 31 cd/m^2. • [29_Pc_10] Signs with luminance over 70 cd/m^2 are visible under much heavier smoke conditions than signs with luminance under 20 cd/m^2. • [29_Pc_14] Adding smoke to a lit corridor reduced legibility of illuminated exit signs from 125 feet to between 40 feet and 50 feet. • [70_Pc_1] Red signs have a visibility 20 % to 40 % greater than blue signs in smoke. • [104_Pc_1] In the presence of smoke, signs should be as bright as possible to overcome smoke obscuration and enhance visibility. No upper limit was found; the brighter the signs, the more subjects preferred them. • [105_Pc_3] An increase of back illumination of an exit sign by a factor of 50 compensated for a factor of 10 increase in the smoke density. • [106_Pc_5] Continuous tactile wayfinding systems are recommended for optical density exceeding 1.5, visual continuous systems are recommended for optical density between 0.1 and 1.5, and traditional exit signs are recommended only for optical density under 0.1. • [110_Pc_4] The use of PLM in stairways may reduce the veiling effect of ambient lighting when smoke is present. • [112_Pc_1] Smoke densities at detectability and readability thresholds have been measured for standard exit signs. • [112_Pc_2] Exit signs are more visible through smoke with the room lights off than with room lights on. • [112_Pc_3] Exit signs should be sufficiently bright that they are visible through smoke. • [155_Pc_1] In moderate smoke of smoke density around 0.8 m^{-1} to 1.4 m^{-1}, the viewing distance of signs is reduced by about one to two m. • [155_Pc_3] Fluorescent signs and LED signs perform better in smoke (longer visibility distances) than electroluminescent signs.

Pc	**Summary Findings - Perception**

	Surroundings and context
Visual Warnings (cont.)	• [29_Pc_8] Visibility is affected by a number of environmental factors, including smoke, dust, glare, veiling reflections, and ambient illumination. • [29_Pc_12] Scattered light from ambient illumination reduces the readability of low contrast signs more than high contrast signs. • [34_Pc_3] Safety signs should not be placed near other signs or objects of similar size and color, or where color contrast is poor. • [61_Pc_6] Glare on or near outdoor signs should be eliminated. • [61_Pc_7] Outdoor signs should be well-lit, and lighting for adjacent signs should not affect the visibility or legibility of the opposite sign. • [63_Pc_1] The effectiveness of the design of a sign is influenced by the context of the surroundings and its contrast with the sign. • [97_Pc_6] Care must be taken by television broadcasters not to block the visual emergency information, including with other captioning. • [99_Pc_1] Television stations must not run text message crawls across any area reserved for closed captioning. • [104_Pc_3] Light sources at an exit sign or between the viewer and the sign obscure the sign by scattering the light. Ambient illumination should be reduced. • [105_Pc_2] An increase of back illumination of an exit sign by a factor of 1.6 resulted in the same visibility in ambient illumination of 0.55 lx as had been achieved with the lights off. • [105_Pc_4] Ambient lighting can affect the readability of signage. • [112_Pc_4] The visibility of exit signs would be improved if scattered light from light fixtures and any other luminous sources was reduced. This can be accomplished by appropriately locating exit signs with respect to other light fixtures. • [112_Pc_8] Strategies other than color may be needed to aid discrimination of exit signs in environments where many colored light sources are used. • [147_Pc_1] Ambient lighting compromises the effectiveness of emergency signage and wayfinding systems in smoke. • [149_Pc_2] The personalized sign that was located the farthest distance away but in the least cluttered area resulted in the greatest compliance. *Individual and group characteristics* • [89_Pc_1] Decreasing visual abilities in older adults can lead to difficulties in seeing smaller size fonts, differences between objects of similar colors (especially shades of green and blue), and any image in the presence of glare. • [115_Pc_1] Visual acuity and contrast sensitivity decline with age. At age 75, the average visual acuity has declined to about 0.6, from normal vision of 20/20 = 1.0 in young adults.
Auditory Warnings	*Urgency* • [7_Pc_2] An urgent message delivered by an emotional female voice is more effecting in arousing young adults than either sounds of a house fire with breaking glass (2.5 dBA higher) or a shift between these two signals (3.75 dBA higher).

Pc	Summary Findings - Perception
Auditory Warnings (cont.)	• [68_Pc_1] An audible warning must be loud enough to attract attention and be heard over ambient noise, but not so loud that it is painful and distracting. This may leave a narrow range for manipulation to affect perceived urgency.

Pitch

- [18_Pc_1] Low frequency (400 Hz and 520 Hz) square wave alarms in T-3 patterns are significantly more effective at waking young adults than square waves at higher frequencies, white noise, whoops, a rising sequence of three pure tones or a rising sequence of three square waves.
- [18_Pc_2] Evidence suggests that the low frequency square wave is both the most detectable signal while awake and the most alerting while asleep. This is consistent with the concept of critical bandwidths.
- [19_Pc _1] A mixed frequency T-3 signal, including the fundamental tone at about 520 Hz and several harmonics, is significantly more effective than high or low frequency T-3 signals or a male voice for waking older adults.
- [89_Pc_2] For auditory messages, higher frequencies (above about four kHz) become harder to hear for older adults due to both noise-induced and age-related hearing losses, and background noise is harder to filter.
- [89_Pc_6] For elderly populations, transmit auditory warnings at frequencies below four kHz that are not affected by hearing loss.
- [140_Pc_4] The frequency range of speech messages in elevators is required by ADA regulations to be 300 Hz to 3000 Hz.

Sound level

- [2_Pc_3] The sound pressure during the 'on' phase of the ANSI standard audible emergency evacuation signal must exceed the highest level of the background noise averaged over 60 seconds and be at least 65 dBA;
- [2_Pc_4] For arousing people from sleep, the sound pressure of the ANSI standard audible emergency evacuation signal must be at least 75 dBA at the normal head position of the sleeping person, with all doors closed.
- [19_Pc _2] Significantly higher volumes (about 10 dBA) are required to wake young adults than older adults for a mixed frequency T-3 signal.
- [47_Pc_1] Alarms are easily heard when they sound at 15 dBA to 25 dBA above the masked threshold level. In normal office settings, this is achieved by setting the level of alarm between 70 dBA and 75 dBA.
- [55_Pc_1] Warnings must be designed to be audible within the environment in which they are used, without being so loud and distracting that people endanger themselves by shutting them off.
- [109_Pc_1] In buildings where over 80 % of residents reported that the fire alarm in their apartment was loud enough, the mean delay time to start evacuation was about three minutes, compared to about nine minutes when this was not the case.
- [109_Pc_2] In a nighttime fire where the alarm was inaudible, evacuation did not start for 10 minutes to 30 minutes.
- [140_Pc_3] The ADA regulations require verbal messages in elevators to be provided at sound levels 10 dB minimum above ambient, without exceeding 80 dB measured at the source.
- [140_Pc_8] Public telephones with volume control are required by ADA

Pc	Summary Findings - Perception
Auditory Warnings (cont.)	regulations to provide an adjustable gain up to 20 dB or more. Incremental volume control requires at least one intermediate step of at least 12 dB of gain. • [140_Pc_9] Assistive listening systems are required by ADA regulations to have sound pressure level capability between 110 dB and 118 dB, with a dynamic range of 50 dB on the volume control. The signal-to-noise ratio for internally generated noise must be at least 18 dB, and peak clipping is not allowed to exceed 18 dB relative to the peaks of speech. *Timing and delivery rate* • [18_Pc_3] The insertion of a 10 seconds period of silence into a 520 Hz square wave T-3 signal tends to reduce the time to awaken young adults. *Cocktail party effect* • [89_Pc_3] Information from multiple speakers in a noisy environment (e.g., from the radio, telephone, PA system, or face-to-face) can be missed by older adults. *Ambient noise* • [2_Pc_2] The ANSI standard audible emergency evacuation signal assumes the existence of an audible alarm signal that can penetrate a particular background noise pattern in a given building. • [89_Pc_7] Eliminate background noise, especially for elderly populations. *Bilingual and non-native English speakers* • [19_Pc _3] Voice alarms with English text may be unsuitable for populations that include non-English speakers. *Sleep arousal* • [6_Pc_1] Female voice and T-3 alarms are equally effective at arousing young adults from sleep while sober and after alcohol consumption. The standard Australian home alarm is significantly less effective, requiring a sound level more than 20 dBA over the background noise to awaken sober individuals. • [6_Pc_2] Alcohol consumption significantly worsens the ability to awaken to auditory alarms. The most effective alarms require about 80 dBA to arouse young adults with 0.05 BAC, compared to 60 dBA while sober. • [6_Pc_3] Young adults are more likely to sleep through an auditory alarm as alcohol level increases, with 36 % of trials at 0.05 BAC resulting in no response below 95 dBA. • [6_Pc_4] Alcohol effects when arousing young adults from sleep with auditory alarms do not depend on gender. • [7_Pc_1] Children of ages six to 17 years and sleep-deprived young adults do not reliably awaken to alarm signals.

A.2 Attentiveness (At)

At	Summary Findings - Attentiveness
General Warnings	*Design* • [4_At_1] Official evacuation notices are most effective when they are specific about the area affected, are delivered in a personalized way, describe the specific actions to be taken, and are worded aggressively and delivered in an urgent manner. • [27_At_1] A message may be ignored if there is not a sufficient means of attracting attention to it. • [91_At_1] An extensive literature is available on the communication of hazard information to the public and the public response during disasters. *Positive messages* • [32_At_4] People focus more on negative information than positive *Message ordering* • [25_At_3] Communicators should assume that the public will process (i.e., pay attention to) only the first 30 seconds (about three sentences or 30 words) of a message. Each message should be front-loaded with the most relevant and critical information at the time. (3 & 30 principle) • [30_At_1] Recall was significantly better for words at the start or end of sentences for non-native speakers. • [30_At_4] Critical components of auditory messages should be at the beginning or end of messages for non-native speakers. • [32_At_7] Present the most important messages first and last *Content* • [32_At_5] Visual aids and narratives provide redundancy and can be heard above the mental noise in a high stress situation. • [38_At_5] Provide specific information about the emergency situation, including what is happening, what is being done, and what actions need to be taken (for digital communicators). • [65_At_3] Recall of a statement about the traffic consequences of an incident was worse than recall of other information, including the nature of the incident and the action to be taken by drivers. • [75_At_1] Recall of news items is best for names and places, lower for events, and very low for specific numbers. Recall is improved when the information is relevant, repeated, local, sensational or tragic, or matches previous knowledge. • [89_At_2] An announcement that the listener should remove themselves from distractions can be helpful to older adults. • [100_At_1] A warning usually follows an alert, providing detailed information indicating who is at risk, where the risk resides, who is sending the warning message, and what protective actions should be taken. • [134_At_1] Letters were easier to remember than numbers in license plate numbers displayed on the DMS. • [134_At_2] A radio station to tune to was easier to remember than a license

At	Summary Findings - Attentiveness

General Warnings (cont.)

plate.

- [148_At_4] Priority in warning messages for multiple hazards, as demonstrated by ordering and conspicuousness, should go to hazards that are more severe, more likely to occur, unfamiliar, and of higher importance.
- [152_At_2] Effective warnings consist of four message components: a signal word to attract attention, identification of the hazard, explanation of the consequences if exposed to the hazard, and directives for avoiding the hazard.

Word choice

- [10_At_2] The choice of signal word matters. Out of five words, the word DEADLY results in the highest reported intent to be careful, followed by DANGER, WARNING/CAUTION, and NOTICE.
- [90_At_4] Words in lists consisting of all nouns are recalled better than words in mixed lists.
- [152_At_3] The signal word to attract attention should indicate the level of hazard present, such as DANGER or WARNING.

Complexity

- [73_At_2] Messages must be simple in order to represent a predominant cue that can be processed under stress.

Timeliness

- [25_At_1] The first notification message is critical for effective emergency management.
- [99_At_1] Begin transmission of information in all formats as early as possible.

Communication channels

- [86_At_1] Emergency managers should use multiple pathways for disseminating information, including hotlines to provide additional information.
- [92_At_1] Messages that are disseminated frequently and through the correct channels are more likely to achieve an appropriate public response.
- [92_At_2] To maximize warning response, provide messages in multiple languages and via multiple sources (especially those that represent vulnerable groups) and encourage dissemination of warnings through informal channels.
- [97_At_1] Textual information from NOAA National Weather Service radio broadcasts is available through modified radio receivers, which provide an alert through a strobe light, an auditory signal, and, if requested, a pillow vibrator or bed shaker to awaken the person.
- [97_At_2] A powerful strobe light added to a civil defense siren tower can aid in alerting individuals but is limited to those who are awake and in direct line of sight from the tower.
- [99_At_2] New and improved technology should be used to help those with hearing impairments.
- [99_At_3] Public officials should keep aware of new emergency alert systems.
- [100_At_2] Some technologies are better suited for delivering alerts (e.g., sirens or CMAS), and others are better suited for warnings (e.g., broadcast radio or television).
- [100_At_3] For time-sensitive disasters, sirens and other immediate alerting systems are required.

At	Summary Findings - Attentiveness
General Warnings (cont.)	• [113_At_1] Encouraging people to contact their neighbors with information will reduce the overloading of telephone systems, warn individuals that might not have received the warning, and help to confirm the warning for those who have received the message. People are less likely to seek confirmation if with family. • [136_At_1] Portable Changeable Message Signs (PCMS) are versatile, since they attract attention and can be programmed to display a wide range of messages. • [139_At_1] Multiple modes of communication are required to reach diverse communities. *Alternative communication modes* • [80_At_2] For long in-car road information messages (containing 14 to 18 units of information), visual presentations result in better recall than single auditory messages. • [80_At_4] Repeated auditory messages were preferred by 63 % of drivers for longer auditory in-car road information messages (containing 10 to 18 units of information), and visual presentations were preferred by 33 %. • [80_At_5] No clear preference was found among visual, auditory, and repeated auditory modalities for short in-car road information messages (four to nine units). • [90_At_3] Items are recalled better when presented visually rather than verbally. • [130_At_2] Pillow and bed shakers at high intensity are much more effective than strobe lights at waking individuals who are hard of hearing. • [130_At_3] Although they are effective at waking sober adults, pillow and bed shakers are not effective at waking those impaired by alcohol. • [130_At_4] A 520 square wave auditory signal at 75 dBA or above is more effective than either strobe lights or bed shakers at waking both hard of hearing and alcohol impaired individuals. *Multiple alarms* • [2_At_1] The temporal pattern for the ANSI standard audible emergency evacuation signal can also be applied to visual and tactile signals to aid occupants with impaired hearing and occupants in zones with intense background noise. • [128_At_2] Neighborhood radio, siren, and text messaging systems were recommended by community leaders. • [140_At_1] The ADA regulations require the use of both audible and visual signals for fire alarm systems. • [148_At_5] Communication of warning messages may be visual, auditory, olfactory, or tactile. A systems approach using multiple modalities can be used to reach people in different ways, at different times, and in different subgroups. *Messenger* • [4_At_2] Door-to-door communication by authority figures resulted in 97 % evacuation rates from beaches in two hurricanes. • [38_At_4] Provide a live warning announcement by a familiar authoritative source (for visual and auditory communicators). • [122_At_2] Credible sources have a wider range of language strategies to choose

At	Summary Findings - Attentiveness
General Warnings (cont.)	from, including more intense language that may attract more attention. *Stress* <ul><li>[32_At_3] Listeners under high stress have less recall for items in the middle of a message.</li><li>[66_At_1] Arousal may narrow attention, so that peripheral stimuli are ignored and peripheral tasks are performed poorly. Attention may be divided between the external demands of the task and internal monitoring (worry, self-blame).</li><li>[73_At_1] Under stress, people may become intensely focused and be able to process only major elements in the environment that are perceived as relevant to the situation. They also tend to fall back on familiar responses that may or may not serve them well in an emergency.</li><li>[75_At_2] When arousal is high, the focus of attention is very narrow, and some relevant information may be missed. Alertness is heightened and information is more readily encoded into long-term memory, although short-term memory access is diminished.</li><li>[78_At_3] Information indicating a high level of threat may cause people to be more engaged or, conversely, may result in attentional avoidance.</li><li>[87_At_2] In order to capture the attention of both high and low trait-anxious people from a normal low-anxiety state, an alarm needs to communicate a high level of danger.</li><li>[87_At_4] As the number of threatening cues in the environment increases, a person may start to focus on a small number of cues, to the detriment of processing new cues.</li><li>[96_At_1] Negative stimuli have better access than positive stimuli to processing resources, with the brain either processing the information faster or requiring less information to judge the contents.</li><li>[132_At_1] Emotional stress can cause a physiological hearing deficit, preventing some messages from being heard.</li></ul> *Cognitive load* <ul><li>[25_At_2] People are continuously faced with multiple cues during an emergency and must decide which most deserve their attention.</li><li>[77_At_2] Expertise (musical training) helped but was not fully successful in drawing the attention of those occupied in a listening task to a striking change (e-guitar improvisation) in a well-known musical piece.</li><li>[77_At_3] A striking musical modification drew the attention of 50 % of non-musicians occupied in a listening task, compared to over 80 % for those not otherwise occupied and to one out of 29 for a less obvious modification.</li><li>[132_At_2] The narrowing of attention in challenging environments can result in inattentional deafness, in which unexpected features are simply not noticed.</li><li>[132_At_3] Message senders can transpose numbers or words without realizing it when their attention is focused elsewhere and not on the message itself.</li></ul> *Disruption of activity* <ul><li>[38_At_3] Disrupt activities by ending visual distractions, turning up lights, cutting TV and internet signals, and ending group activities (for all - visual, kinesthetic, auditory, and digital communicators).</li></ul>

At	**Summary Findings - Attentiveness**
General Warnings (cont.)	• [77_At_1] A distracting task can cause listeners to overlook an unexpected sound that is obvious to others. This inattentional deafness has been demonstrated using a well-known musical piece. • [77_At_4] In order to attract everyone's attention for an auditory message, other activities that are occupying them should be halted. • [77_At_5] When people are engaged in a task, an auditory message must be an obvious change from ambient noise to draw attention. • [78_At_1] A model of attending to a new stimulus suggests three steps: 1) initial shift of attention, 2) engagement with the stimulus, and 3) disengagement from the stimulus. Attentional bias, in which the stimulus is judged to be more threatening than it is in reality, may relate to any of these steps. • [89_At_3] Intrusive alarms should be used to disrupt other tasks (including sleeping) and gain attention. • [109_At_2] Attention-getting changes should be made to the environment during an alarm, such as shutting off the movie, turning off the background music, and/or turning up the lights. *Repetition* • [32_At_8] State the three key messages three times. • [33_At_1] The phenomenon of wearout suggests the existence of an optimal level of repetition for messages, such as those used in an educational campaign. • [36_At_1] The improved effectiveness of messages that are repeated over internals (rather than massed at one time) is confirmed by a study of airborne leaflets dropped over communities. • [36_At_2] Airborne leaflets delivered over three consecutive days resulted in a better informed community than 33 % more leaflets delivered in a single day. • [36_At_3] Increasing the number of airborne leaflets delivered to a community by a factor of four resulted in almost twice as many people (64 % compared to 34 %) being aware of the message. • [36_At_4] A large number of messages repeated at intervals rather than all at once improves community awareness of the message. • [39_At_1] Repetition of key parts of a message (overlearning) helps with retention. • [44_At_1] Better driver response (faster, correct action) was seen when each of multiple message phases was displayed for a relatively short duration and repeated. • [53_At_1] The amount of information recalled, both immediately and after one or two days, increases with the number of repetitions. • [65_At_2] Recall of an auditory message generally improved from one to two presentations. The effect on recall of increasing the number of presentations from two to three is inconclusive. • [65_At_4] Internally redundant audio messages, with repetition only within the message, resulted in significantly less success in completing a set of instructions (31% of participants) than successively redundant messages, with repetition only of the entire message (57 % of participants). • [65_At_8] Repeat an auditory message at least once for improved recall. • [80_At_3] Recall of longer auditory in-car road information messages

At	Summary Findings - Attentiveness

General Warnings (cont.)

(containing 10 to 18 units of information) increases when the message is repeated.

- [90_At_1] Recall is higher when items are repeated at intervals (distributed) than when the items are repeated consecutively (massed).
- [90_At_2] The advantage of distributed practice over massed practice is especially true when the number of repetitions is large.
- [90_At_5] The recall of words that appear twice improves with increasing lag (number of words between presentations), although the effect is smaller for auditory lists of mixed word types.
- [90_At_7] Messages should be repeated at intervals rather than consecutively for maximum recall.
- [98_At_1] Multiple repetitions of warning messages may be needed.

Familiarity and training

- [33_At_2] Message recall over time is enhanced by attention and by motivation to retrieve the information.
- [39_At_2] Refresher training at about three week intervals maintains optimal performance for cognitive tasks.
- [51_At_1] Although a warning label on a product was noticed by 88 % of the subjects, only 50 % of those who noticed actually read the entire warning.
- [152_At_5] Familiarity is an important factor in warning habituation. The more often a warning is encountered, the less likely the person will notice it during following encounters. One way to reduce the negative effects of habituation is to change the content or appearance of a warning sign.
- [157_At_1] People that were familiar with a building were less likely to notice exit signs than those who were unfamiliar with it.

Individual and group characteristics

- [38_At_1] People can be categorized into four groups by communication preference: visual, kinesthetic, auditory, and digital. Tailored warning methods can be used to move each group toward evacuation.
- [78_At_2] Individuals differ in their response to threatening information, with some people alerting to lower levels of stimulus and remaining in an anxious state longer than others.
- [87_At_1] The threshold value for response to threat cues is different for each individual, and increasing fear or anxiety temporarily sets the threshold lower.
- [87_At_3] Low trait-anxious people show a tendency to avoid paying attention to threat cues so that they can complete the target task. Warnings need to break through this tendency.
- [89_At_1] Older adults have a harder time using selective attention to filter out irrelevant auditory information, and may fail to attend to the important information in a warning.

Non-native English speakers

- [128_At_1] A multilanguage emergency alert system is being developed by a national nonprofit organization for ethnic news organizations.

At	Summary Findings - Attentiveness
Visual Warnings	*Design* • [43_At_1] Some motorists (28 %) considered the dynamic provision of information by CMS a distraction. [63_At_1] An asymmetric arrangement of icons (symbols) attracts attention and forces the user to focus on the design. • [148_At_3] A warning symbol should be bold and avoid irrelevant graphical details. It should be legible from a distance and after moderate degradation. • [152_At_1] To be noticed and used, a sign must be as salient as possible. Salience can be enhanced by using large bold print, high contrast, colors, borders, pictorial symbols and special effects, such as flashing lights. • [152_At_4] Warning text should be in bullet form as opposed to continuous flowing text. *Word choice* • [148_At_1] The literature on visual warnings generally recommends a signal word panel (DANGER, WARNING, CAUTION, etc.) and a message panel, which can be text or symbols. • [148_At_2] Good warning design includes using standard colors and meanings for signal words (e.g., DANGER, WARNING, CAUTION) and symbols. *Symbols* • [69_At_1] Compliance with safety instructions was higher for a warning message that combined a pictograph with text than for either pictograph or text alone. • [150_At_2] Adding a pictorial to a written warning in instructions for a chemistry experiment was not found to increase compliance. • [159_At_1] Pictorials, icons, color, and the signal word WARNING increase the likelihood of a warning being noticed. *Message length* • [80_At_1] An inversely proportional relationship was found between the length of in-car road information messages and recall. The recall of a message with four units of information was 100 %, whereas the score for a message with 14 to 18 units dropped to 48.4 %. *Color* • [104_At_2] Exit sign color may be more important for the initial attraction of the attention of the viewer than for visibility *Font* • [25_At_4] Risk communication research has found little evidence supporting trends in danger symbols, font choices, and other typographical features, including caps, bold, colors, exclamation marks, and spacing, that significantly affect attention, perception, or behavioral response. *Placement* • [93_At_1] Placement is a key factor; signs in an airport should be placed where passengers must make choices on where to go, at a height above eye level to maximize the number of people seeing them.

At	Summary Findings - Attentiveness
Visual Warnings (cont.)	• [150_At_1] Warnings presented in task instructions for a chemistry experiment resulted in greater compliance than warnings posted on the wall. *Lighting – Static* • [104_At_1] In the absence of smoke, very bright signs bothered the participants. An upper limit was suggested under these conditions by 92 % of the participants, with a range from 70 cd/m² to 700 cd/m². *Lighting – Dynamic* • [26_At_1]Flashing lights increased drivers' awareness of road hazard warning signs and increased the interpretation of the sign as signaling a hazard. • [71_At_1] Conspicuousness of an exit sign can be improved by adding a flashing light. • [102_At_1] Flashing lights attract attention. • [102_At_3] Green flashing lights are able to overcome participant proximity; i.e., exit signs with green flashing lights attracted people who were closer to other exits. • [102_At_4] Green flashing lights are able to overcome a less advantageous angle of observation; i.e., exit signs with green flashing lights on a side exit attracted people who could see an exit at the end of the corridor. • [103_At_1] Green flashing lights placed next to some exits in an office building were not mentioned by evacuating occupants as a factor influencing exit choice. Although more people evacuated through those exits, the difference was not significant. • [103_At_2] In a movie theater, all those evacuating used the exits with green flashing lights in preference to those with normal exit signs. Flashing lights may be more effective in influencing the choice of exits in less familiar places. *Lighting –Wayguidance systems* • [12_At_1] During an emergency, PLM lighting with a one inch strip along the front edge of each step in a stairwell is preferred to emergency lighting at an average of 37 lx. • [62_At_1] The highest probability of making correct route decisions (98 %) was found for a wayguidance system consisting of a combination of tactile and green light cold cathode lighting systems. • [62_At_2] Electrically powered and photoluminescent low location lighting wayguidance systems resulted in the lowest percentage of evacuees (69 % to 79 %) making correct escape route decisions. • [72_At_1]Traveling flashing lights provide effective wayguidance in both light and thick smoke, although smoke decreases the effectiveness. • [102_At_2] Because flashing lights attract attention and green is associated with safety, green flashing lights can be used to overcome familiarity issues and direct people to an emergency exit. *Individual and group characteristics* • [88_At_1] People younger than 40 are more likely to look for warnings on a package than those over 40.

At	Summary Findings - Attentiveness
Visual Warnings (cont.)	*Sleep arousal* • [14_At_1] Even with light intensities on the face exceeding those from commercially available visual alarms, strobe lights cannot be relied upon to wake a sleeping person. Tactile devices should be used with the flashing strobes to wake up hard of hearing people during a fire.
Auditory Warnings	*Design* • [41_At_1] Short form and staccato messages are easier to recall. The first statement in an alert message should be an attention statement to alert the population that a message is about to be given. The second statement should describe the problem. The final statement should present the action that is required to react to the problem. For maximum recall, people must hear the message at least twice. • [46_At_1] An auditory warning needs to not only sound at a suitable volume for the location; it also needs to be effective and appropriate for the particular population being warned. For instance, in special care baby units in hospitals, only the staff should be informed. • [74_At_1] A voice alarm system consisting of an alarm tone followed by a female voice to attract attention and a male voice to deliver instructions resulted in a successful area evacuation, with personnel vacating to receiving floors either above or below their own. • [74_At_2] Change voices during the emergency broadcast message to attract attention. • [122_At_1] Messages that violate expectations attract attention. Warnings should therefore be distinct from normal ambient sounds and given in a different voice than that used for normal announcements. *Alarm tones* • [109_At_1] To reduce the delay time before evacuation, a recognizable fire alarm signal (the standard T-3 evacuation signal as described in ISO 8201) should be installed. *Synthesized speech* • [68_At_1] Synthesized voice messages can both alert and inform simultaneously. *Message length* • [32_At_1] An analysis of ten years of print and media coverage of emergencies and crises in the U.S. shows that a sound bite in the print media has an average length of 27 words, a sound bite in the broadcast media lasts for nine seconds, and each sound bite (print and broadcast) consists of three messages on average. • [32_At_2] People in high-stress situations are able to process only three messages at a time, much lower than the seven messages that can be processed under normal circumstances. • [32_At_6] Organize key messages into sound bites three bullets long that contain a maximum of 27 words total and that can be spoken in nine seconds. • [65_At_5] An eight-unit set of auditory instructions (four turns and four street names) was completed successfully by 36 % of participants, while a six-unit set

At	Summary Findings - Attentiveness

Auditory Warnings (cont.)

(three turns and three street names) was completed successfully by 53 %.

- [65_At_6] Eight units represent the retention limit of auditory messages for design purposes.
- [65_At_9] For routes that require more than four turns and recall of more than four street names, a trailblazer system is recommended.

Complexity

- [65_At_1] For auditory messages about driving route changes, staccato (terse and telegraphic) and short-form (brief and concise but grammatically complete) styles were preferred to a conversational style, which was additionally associated with the poorest recall.
- [65_At_7] Avoid a conversational style when presenting a complex message over the radio.

Urgency

- [7_At_2] An urgent message delivered by an emotional female voice is more effecting in arousing young adults than either sounds of a house fire with breaking glass (2.5 dBA higher) or a shift between these two signals (3.75 dBA higher).
- [10_At_1] Signal words spoken in a female voice with an emotional tone result in a higher reported intent to be careful than words spoken in a male voice or in a monotone or whisper.
- [68_At_2] An audible warning must be loud enough to attract attention and be heard over ambient noise, but not so loud that it is painful and distracting. This may leave a narrow range for manipulation to affect perceived urgency.

Pitch

- [18_At_1] Low frequency (400 Hz and 520 Hz) square wave alarms in T-3 patterns are significantly more effective at waking young adults than square waves at higher frequencies, white noise, whoops, a rising sequence of three pure tones or a rising sequence of three square waves.
- [18_At_2] Evidence suggests that the low frequency square wave is both the most detectable signal while awake and the most alerting while asleep. This is consistent with the concept of critical bandwidths.
- [19_At_1] A mixed frequency T-3 signal, including the fundamental tone at about 520 Hz and several harmonics, is significantly more effective than high or low frequency T-3 signals or a male voice for waking older adults.
- [27_At_2] To draw attention away from a conversation, a change in voice pitch (e.g., from male to female) or notification by a tone other than speech is needed.
- [41_At_2] Greater speech intensity and greater pitch variability are desirable.
- [41_At_4] The most common attention-demanding signal uses a frequency of modulation between one Hz and three Hz. Pitch modulation works better than changes in loudness, with the frequency varying by at least 100 Hz.

Sound level

- [2_At_3] The sound pressure during the 'on' phase of the ANSI standard audible emergency evacuation signal must exceed the highest level of the background noise averaged over 60 seconds and be at least 65 dBA;
- [2_At_4] For arousing people from sleep, the sound pressure of the ANSI

At	**Summary Findings - Attentiveness**
Auditory Warnings (cont.)	standard audible emergency evacuation signal must be at least 75 dBA at the normal head position of the sleeping person, with all doors closed. • [5_At_1] Under a high cognitive workload, singular, instructive auditory messages were processed more accurately at higher sound levels (65 dB compared to 45 dB), even though both sound levels were in the range considered optimal for speech stimuli in quiet environments. • [5_At_2] Higher sound levels for auditory warnings may be especially helpful in enhancing the performance of older people with some hearing loss. • [19_At _2] Significantly higher volumes (about 10 dBA) are required to wake young adults than older adults for a mixed frequency T-3 signal. • [38_At_2] To attract attention and get people to act, add to discomfort through loud alarms (for kinesthetic and auditory communicators). • [41_At_6] The alarm signal should sound at 10 dB above the sound level of the message. • [46_At_2] In factories and other noisy workplaces, the alarm needs to compensate for the use of hearing protectors. • [47_At_1] Alarms should attract attention but not startle. • [140_At_2] The ADA regulations require verbal messages in elevators to be provided at sound levels 10 dB minimum above ambient, without exceeding 80 dB measured at the source. *Timing and delivery rate* • [18_At_3] The insertion of a 10 second period of silence into a 520 Hz square wave T-3 signal tends to reduce the time to awaken young adults. • [41_At_3] A speed under 110 words per minute sounds too slow, and the message may be perceived as less important. • [41_At_5] The signal should be no longer than five seconds in duration, followed by 0.5 seconds of silence before the message begins. • [90_At_6] Slower rates of speech always result in higher recall. • [90_At_8] Auditory messages should be delivered slowly for maximum recall. • [146_At_1] More items in a list are recalled at a slower speech rate (four seconds per word) than at a faster speech rate (one second per word). • [146_At_2] At a faster speech rate (one second per word), the probability of recall of a word on a list is linearly dependent on the number of times it appears for at least seven repetitions. • [146_At_3] At a speech rate of one second per word, consecutive repetition leads to better recall for two or fewer repetitions, and distributed repetitions are better for four or more repetitions • [146_At_4] At a slower speech rate (four seconds per word), recall is improved by repeating the information at distributed intervals. • [146_At_5] Word recall is independent of the pause time between words. *Ambient noise* • [2_At_2] The ANSI standard audible emergency evacuation signal assumes the existence of an audible alarm signal that can penetrate a particular background noise pattern in a given building. • [47_At_2] An emergency alarm or alert should differ significantly from the ambient sounds. In environments where there are many natural sounds, a more

At	Summary Findings - Attentiveness
Auditory Warnings (cont.)	traditional alarm may be a better choice. - [49_At_1] Voice acoustics matter more than the sex of the speaker. Voices for warning messages should be selected based on minimal overlap with the ambient noise spectrum. - [49_At_2] In a complex noise environment, two voices with different characteristics (e.g., male and female) can be alternated to attract attention. - [49_At_3] The choice of speaker should depend on the acoustic qualities of the voice relative to the noise spectrum, not the speaker's sex. *Bilingual and non-native English speakers* - [19_At _3] Voice alarms with English text may be unsuitable for populations that include non-English speakers. *Sleep arousal* - [6_At_1] Female voice and T-3 alarms are equally effective at arousing young adults from sleep while sober and after alcohol consumption. The standard Australian home alarm is significantly less effective, requiring a sound level more than 20 dBA over the background noise to awaken sober individuals. - [6_At_2] Alcohol consumption significantly worsens the ability to awaken to auditory alarms. The most effective alarms require about 80 dBA to arouse young adults with 0.05 BAC, compared to 60 dBA while sober. - [6_At_3] Young adults are more likely to sleep through an auditory alarm as alcohol level increases, with 36 % of trials at 0.05 BAC resulting in no response below 95 dBA. - [6_At_4] Alcohol effects when arousing young adults from sleep with auditory alarms do not depend on gender. - [7_At_1] Children of ages six to 17 years and sleep-deprived young adults do not reliably awaken to alarm signals.

A.3 Comprehension (Co)

Co	Summary Findings - Comprehension
General Warnings	*Design* • [47_Co_2] Standardizing the meaning of alarms within similar settings (e.g., hospitals and other health care facilities) would reduce confusion and increase the understanding of medical alarms. • [75_Co_6] Comprehension can be studied through analysis of propositions, units of thought that are connected by a relationship. • [91_Co_1] An extensive literature is available on the communication of hazard information to the public and the public response during disasters. • [108_Co_4] Messages should be targeted for the social networks of vulnerable populations and developed by people that understand those communities. • [114_Co_3] To reduce the age related differences, older adults benefit from environments or contexts that provide cues for remembering information. • [114_Co_5] Older people have difficulty remembering time-based tasks, so it is better to present instructions in terms of event-based tasks (e.g., eat after breakfast vs. eat every eight hours). • [134_Co_2] Comprehension is improved when DMS messages are brief, use simple words and well-understood abbreviations, standardize the word order, and standardize the order of message lines. • [139_Co_2] Effective risk communication requires the engagement of communication representatives in designing and evaluating educational materials, emergency messages, and dissemination practices, including language resources and accessibility. *Validation* • [75_Co_5] Before a message is used, there is a need to "walk through" the message to identify every place where misconception can occur. • [89_Co_3] Testing should be done before deployment of the warning to make sure that it is comprehensible by the intended audience. • [135_Co_12] The use of graphics to display traffic information requires testing on drivers to determine the optimal amount of information that can be displayed without overloading the driver. *Content* • [25_Co_8] Tell people how and where to get further information. • [42_Co_1] Drivers expect DMS to provide information on the problem or reason for the sign first, followed by the location of the problem. Using standardized ordering of words and messages lines can also enhance understanding. • [43_Co_1] Information on the cause of an incident was appreciated by those affected. • [43_Co_2] Real-time information is appreciated. • [43_Co_4] Drivers prefer to receive information before experiencing the conditions. Information on the nature and impact of the situation is appreciated.

Co	Summary Findings - Comprehension

General Warnings (cont.)

- [43_Co_5] Information on the location of an incident is desired.
- [43_Co_7] Fifty percent of motorists regularly rely on the information provided by CMS. They favor messages that are simple, reliable and useful.
- [43_Co_9] Reports on upcoming conditions had a direct impact on behavior (reduction in speed). There were also some negative effects, including refocusing attention, encouraging the drivers to look for evidence of the conditions, testing the conditions, and more conservative passing behavior.
- [43_Co_10] The level of detail provided in a message influenced willingness to modify behavior.
- [100_Co_1] More research is needed on effective content for alerts, because alert technologies, such as sirens and weather radio alerts, can only convey a small amount of information.
- [152_Co_2] The description of the hazard should be specific and complete.
- [152_Co_3] The explanation of the consequences of the hazard and the directions for avoiding the hazard should be explicit.

Word choice

- [40_Co_3] The meaning of keywords used in the message should be clear to the intended audience, and their use should be consistent.
- [84_Co_3] The signal words WARNING and CAUTION are not perceived as denoting significantly different levels of hazard.
- [123_Co_1] Elderly people, non-native English speakers, college age students, and grade school children generally rate hazard words in the same order by implied meaning.
- [123_Co_2] While the word CAREFUL implies a higher level of carefulness than ATTENTION for college students (and in current standards), the two words were rated in the opposite order by elderly and non-native english populations.
- [123_Co_3] While the word DEADLY implies a higher level of carefulness than DANGER for college students (and in current standards), the two words were rated in the opposite order by non-native English speakers.
- [123_Co_4] Use more familiar words like STOP and DON'T on dangerous products that are likely to be used by non-english speakers.
- [148_Co_3] Words and phrases should be familiar and short, without multiple interpretations.

Complexity

- [25_Co_4] Verbal comprehension has been found to drop an average of approximately four grade levels at high levels of stimulation, stress, and distraction; emergency notification messages should therefore be written for no more than a sixth grade reading level.
- [25_Co_6] Scientific concepts, probability analyses, and variables/info on how the warning was prepared are too complex to include in emergency messages.
- [25_Co_7] Provide complex information and instructions using simple, brief and concrete language.
- [32_Co_1] Construct messages with a sixth to eighth grade education (average grade level of the audience minus four).
- [89_Co_2] Messages should be simple and straightforward. The text of the messages should only include simple words (no jargon) and sentence structure

Co	Summary Findings - Comprehension
General Warnings (cont.)	to avoid confusion. • [99_Co_1] Essential emergency information should be frequently repeated in a simple message to accommodate individuals with cognitive disabilities. • [111_Co_3] Messages should use simple words that are easily understood by non-native speakers, and the messages should avoid memory overload. • [113_Co_2] Warnings need to be understandable and in language with a common meaning. Messages should use simple, everyday language with declarative sentences. Officials who speak other languages should be available to help non-native speakers. • [114_Co_2] Cognitive abilities generally affect the comprehension and compliance stages of warning signals. Working memory, which is the ability to keep information in active in memory, declines with age. This is important in warnings or signs that require multiple steps. • [114_Co_4] Warning designs should have simple sentences, with the information given explicitly and not left to the reader to infer its meaning. *Timeliness* • [109_Co_2] Voice alarms should be given as soon as the emergency has been identified in order to reduce the delay time. • [127_Co_2] Captioning of news broadcasts should begin immediately as soon as audio information begins. *Communication channels* • [86_Co_1] Emergency managers should attempt to obtain feedback from the public. • [109_Co_1] A voice communication system should be installed in large public buildings. • [139_Co_1] Different communities obtain information from different sources, including social networks, ethnic media, and places of worship. *Alternative communication modes* • [13_Co_1] Not all people may understand the meaning of an alarm bell; a spoken message is important. • [43_Co_3] Motorists prefer receiving information by changeable message sign (CMS) and radio rather than telephone and television. • [127_Co_1] Television broadcasters and internet providers should provide accurate, real-time, and verbatim captioning of all news broadcasts, with contrasting background to ensure readability and nothing blocking the message. • [127_Co_3] Important new emergency information should be identified in captions as well as in the news broadcast. • [127_Co_4] All audio announcements should be broadcast with a simultaneous display of text, which should provide appropriate information, including the nature of the emergency and instructions on what to do. *Multiple alarms* • [2_Co_2] The temporal pattern for the ANSI standard audible emergency evacuation signal can also be applied to visual and tactile signals to aid occupants with impaired hearing and occupants in zones with intense background noise.

Co	Summary Findings - Comprehension

General Warnings (cont.)

Public perception

- [23_Co_2] Messages must be simple, take emotional states into account, and make sure that images (e.g., from television reports) and words match.

Stress

- [25_Co_5] Individuals in high stress situations are only able to process three distinctive issues or pieces of information in a single message unit.
- [107_Co_1] Fear (not panic) is a normal human reaction to extreme circumstances, which reduces the ability to reason through complex problems.
- [126_Co_1] Students of English as a second language are more limited in their ability to communicate subjective information when they are anxious than when they are relaxed and comfortable.
- [126_Co_2] Higher proficiency non-native English speakers feel less anxiety than lower proficiency speakers when asked to speak in a high pressure environment.
- [132_Co_1] Psychological factors that affect the communications of first responders at large-scale disasters include sensory overload, cognitive bias, speech center deficit, and suppressed emotions.

Cognitive load

- [132_Co_2] Message senders can transpose numbers or words without realizing it when their attention is focused elsewhere and not on the message itself.

Repetition

- [41_Co_2] For a DMS with more than one frame, the key word should be repeated in both frames to assist viewers in comprehending the sign.
- [44_Co_1] Better driver response (faster, correct action) was seen when each of multiple message phases was displayed for a relatively short duration and repeated.
- [98_Co_1] Multiple repetitions of warning messages may be needed.
- [113_Co_1] Repetition of a warning message that includes the actions that should be taken strengthens adaptive responses to the hazard.

Familiarity and training

- [9_Co_2] Accurate compliance to a set of commands increases with practice.
- [34_Co_1] Staff, occupants, and visitors should be trained on the meaning of distinctive safety sign colors, shapes, and graphical symbols.
- [40_Co_1] It is important to determine whether the audience for a message is familiar or not familiar with the surroundings or route.
- [43_Co_11] Familiarity with the available routes influenced willingness to modify behavior.
- [82_Co_1] Training is beneficial to the understanding of new symbols.
- [82_Co_2] Younger populations benefit more from training than older groups.
- [129_Co_2] The general public's ability to interpret evacuation plan diagrams should be improved through education and training.
- [132_Co_6] Training and experience promotes optimal radio use and teaches first responders about their own reactions to stress.
- [134_Co_1] Consistency in the design of DMS highway messages, over time and

Co	Summary Findings - Comprehension

General Warnings (cont.)

from region to region, is important for driver comprehension and traffic flow.

- [134_Co_3] Message elements in an unexpected location or order can result in driver confusion and increase the time to read the message.
- [142_Co_1] Familiarity with a symbol results in more accurate identification of its meaning.

Individual and group characteristics

- [25_Co_2] The varied backgrounds of individuals in the target audience affect the process by which they make sense of and respond to this information.
- [52_Co_2] Individual native speakers differ significantly in their ability to discriminate speech in high noise levels.
- [75_Co_2] Emergency messages that can be comprehended by lay people must take background knowledge into account.
- [75_Co_3] Due to differences in background knowledge, the health concepts described by physicians are not interpreted in the same way by their patients, who may therefore discount the explanation and devise an alternative, resulting in poor compliance.
- [75_Co_4] Prior knowledge provides context in which meaning is understood and assumptions from which to make inferences. It also affects what details are remembered and how they are organized for reasoning and decision-making.
- [92_Co_1] People have inherently different perceptions of the world around them – which they bring to any emergency – predisposing them to different responses, especially when situational information provided by the environment is conflicting.
- [92_Co_2] Four factors affect individual perception of and response to an emergency: 1) ability to process risk information, 2) access to social networks, 3) incentives to take the warning seriously, and 4) constraints on action.
- [108_Co_2] The interpretation of warning messages depends on the sociocultural context of the receiver, which is not well understood by those outside of the community.
- [108_Co_3] Tourists are unfamiliar with the environment, available resources, and possibly the language.
- [139_Co_4] A database of trusted sources, interpreters, and effective communication modes for diverse populations should be prepared.

Non-native English speakers

- [108_Co_1] Non-native English speakers may lack access to information in their native language or a good translation of the message.
- [128_Co_1] Government agencies are in general not properly staffed to produce and disseminate messages in multiple languages. Not all information on websites is translated, and content in languages other than English may be difficult to find.
- [128_Co_2] A training program is recommended for government representatives dealing with ethnic media, including crisis communication plans.
- [128_Co_3] Adequate multilingual staffing, including volunteers, is needed for governmental agencies to handle any emergency at any time.
- [128_Co_4] Arrangements should be made for English-language broadcast

Co	Summary Findings - Comprehension
General Warnings (cont.)	stations to air multilingual messages if ethnic stations are knocked off the air. • [128_Co_5] A multilanguage emergency alert system is being developed by a national nonprofit organization for ethnic news organizations. • [128_Co_6] Multicultural and multilanguage course material should be included in the training of emergency and media professionals and students. • [128_Co_7] Government agencies and first responder organizations need to develop bilingual preparedness materials. • [135_Co_8] Graphical information on DMS roadway signs provided faster and more accurate comprehension for non-native English speakers. • [139_Co_3] Messages must be culturally appropriate and accurately translated. • [148_Co_4] Messages should be in multiple languages when appropriate.
Visual Warnings	*Design* • [11_Co_1] Signs that follow the universal ergonomic principles of sign-content compatibility, familiarity and standardization are better understood by people from different cultural backgrounds. • [61_Co_1] The shape of a sign can simplify a message; for example, instead of having a rectangular-shaped sign with an arrow pointing in the right direction, the sign itself can be shaped like an arrow. • [93_Co_1] A sign's design should indicate its importance; important signs should be large, lighted, and positioned perpendicular to the walking direction. • [140_Co_1] The ADA regulations require text descriptors for pictograms to be located directly below the pictogram field. • [158_Co_27] In an instruction manual, conspicuous print (18-point Times font covered with transparent orange florescent highlighting) accompanied by icons was better understood and remembered than either text without the icons or plain text with icons. *Word choice* • [138_Co_1] Driver comprehension of dates on work zone PCMS was improved by giving the month as an abbreviation (AUG 25) rather than in numerical form (8/25). *Symbols* • [11_Co_2] Symbols on signs should be related to the message content. • [20_Co_1] Incorporated text did not increase the level of understanding of a symbol. • [35_Co_2] Concrete (real-world) icons are more readily understood by users than more symbolic icons. • [54_Co_1] Circles with a solid slash covering the symbol and a solid slash under the symbol were preferred to translucent or partial slashes for negative pictorials. • [73_Co_1] Use international symbols in signage to both simplify the information and inform occupants not familiar with the language. • [88_Co_1] A pictogram showing specific body parts is more effective than one that shows the entire body. Specific information is better than general information.

Co	Summary Findings - Comprehension
Visual Warnings (cont.)	• [88_Co_2] Pictograms depicting protective equipment and those with simple messages are easily recognized. • [95_Co_1] Graphical signs that indicate the location of an exit or that exits are located to the left/ right are well understood. Comprehension is about the same for text and graphical signs. • [95_Co_2] Graphical and text signs identifying that an exit is located further up the aisle were unsuccessful. • [115_Co_1] A solid shape is preferable to a line boundary for the boundary of a figure. • [115_Co_2] A closed figure should be used unless there is a reason for the figure to be discontinuous. • [115_Co_4] Symbols in a set of related signs should be as unified as possible. • [121_Co_2] For prohibitive symbols, the pictorial size should be at least 75 % of the inner diameter of the circle. • [121_Co_3] For prohibitive symbols, a slash that comprised 25 % of the area within the circle was rated the highest. Slash thicknesses that exceeded this value, reducing the available space for the pictorial and covering more of it, resulted in lower ratings. • [124_Co_1] A pedestrian signal with countdown only (no flashing hand) resulted in the least confusion about whether the pedestrian was allowed to be in the crosswalk. The standard orange flashing hand signal resulted in the most confusion. • [135_Co_2] Well-designed symbols can be advantageous because they are quicker to read and comprehend than words, can be recognized from farther away, and are more recognizable under poor environmental conditions such as darkness and weather. • [135_Co_6] An accident symbol showing two cars was better understood than a single car overturned. • [135_Co_11] Arrows on a traffic lane indicate that the lane is open beyond the point of congestion. • [148_Co_5] Clearly comprehensible symbols can augment or replace text in the message panel. Symbols attract attention and are beneficial for poor readers and non-English speakers. However, the meaning of the symbols must be clear. • [148_Co_6] A warning symbol should be bold and avoid irrelevant graphical details. It should be legible from a distance and after moderate degradation. • [151_Co_1] The meaning of prohibition signs was more rapidly comprehended when the slash in the red circle symbol was translucent or located under the pictorial (i.e., more of the pictorial could be seen) than when the slash covered the pictorial or was broken. • [152_Co_4] Pictorials facilitate warning comprehension and improve comprehensibility for people who do not understand the language. *Diagrams* • [20_Co_2] A series of symbols showing sequential steps is the most easily understood way to provide instructions for certain procedures, such as lifting a box. • [22_Co_1] Guidelines for pictorial airline safety instructions recommend using a

Co	**Summary Findings - Comprehension**
Visual Warnings (cont.)	minimum number of words. • [22_Co_3] Concrete drawings are more easily understood than abstract drawings. An ordered sequence of pictorials should be numbered. • [23_Co_1] People tend to rely more on images than words • [118_Co_1] Larger, less wordy, and more colorful airplane passenger safety cards are preferred. • [118_Co_2]Airplane passenger safety cards that display words integrated with diagrams or present pictures in pictogram sequence with little text are preferred. • [118_Co_3] Illustrations with words are remembered better than photographs with words. • [129_Co_3] Evacuation plan diagrams should have the following characteristics: comprehensibility, ease of being conventionally understood, clarity, ability to provide directions, and safety concepts. • [135_Co_1] Evaluation of the use of graphics on DMS roadway signs must take into account that drivers must be able to read and comprehend a message in only a few seconds (eight seconds or less) at highway speeds, possibly while engaged in a complex driving task. • [135_Co_3] New hardware technologies and the increasing use of symbols on static traffic signs encourage the use of graphic features on DMS roadway signs. • [135_Co_5] A three dimensional map perspective did not improve comprehension of a graphical DMS roadway sign. • [135_Co_7] Although a graphical route information panel (GRIP) was able to communicate a large amount of information, study results showed that viewing times were significantly higher than for text or other types of graphics. *Message length* • [34_Co_2] Adding supplementary text increases the size of safety signs and improves their comprehensibility and legibility. • [40_Co_2] The maximum length of a DMS message is controlled by the reading time available for the sign. • [40_Co_4]When splitting a message for a dynamic message sign (DMS), no more than two phases should be used. Each phase should be understood by itself. • [41_Co_1] People can remember an eight-word message better when it is broken up rather than displayed in full. • [41_Co_3] Only three units of information should be displayed at once. • [81_Co_2] Variable message signs (VMS) with a double message line were better understood and resulted in shorter response times than those with single or triple lines. • [134_Co_4] The average driver can process no more than eight words of four to eight characters each when driving at high speeds. For drivers to be able to recall all information when making a decision, there should be no more than two informational units on a single line, no more than three units on a single frame, and no more than four units in the full message. • [134_Co_5] There should be no more than four DMS message lines total, which should be divided into two frames of two lines each. • [136_Co_1] Safety messages (PCMS) should be simple, brief, and clear. Messages

Co	Summary Findings - Comprehension
Visual Warnings (cont.)	are limited to one or phases, each with up to three lines with no more than eight characters per line. • [137_Co_2] If cars are traveling under 40 mph, DMS with 12 inch characters can be used to display up to four units of information during the day, but only up to three units at night. • [138_Co_2] For four units of information delivered on work zone message boards, comprehension rates for two sequential PCMS were comparable to those for a large single-phase DMS and acceptable for highway use. • [138_Co_3] Significantly lower comprehension and recall rates were found for sequential PCMS displaying five units of information compared with a large DMS. • [138_Co_4] Neither portable changeable message signs (PCMS) nor dynamic message signs (DMS) were able to convey five units of information with comprehension rates that are acceptable for highway applications. • [144_Co_2] The response time was shorter for variable message signs (VMS) when there were fewer lines in the message. *Complexity* • [35_Co_1] Concrete (real-world) icons do not increase the complexity of a sign over more symbolic icons. • [35_Co_3] Complexity did not affect the user's ability to understand the meaning of a concrete (real-world) icon, and adding visual detail did not appear to make icons more concrete. • [75_Co_1] Message readability alone does not explain comprehension problems. • [114_Co_1] Signs or labels should avoid clutter or irrelevant information to increase the ability for people to select and process the key relevant information. • [115_Co_3] Figures should be as simple as possible, with only necessary features. • [120_Co_2] An increased number of map details results in poorer performance in cognitive mapping. • [121_Co_1] Solid pictorials are preferred to outline pictorials, likely because outline figures contain more lines, angles, and segments than solid figures and are thus more complicated visually. • [129_Co_1] The number of technical symbols in evacuation plan diagrams should be reduced. Metaphorical and abstract symbols should be eliminated. • [135_Co_9] The use of graphics makes it easier to communicate unusual scenarios on a roadway, such as high occupancy lanes or cash-only toll lanes, that would require many words to explain using text. • [136_Co_2] Present one thought per phase for PCMS roadway signs. • [136_Co_3] For messages that can be displayed on a single phase of a PCMS roadway sign, present the problem on top, the location on the second line, and the recommended action on the bottom • [136_Co_4] If a message can't be displayed in two phases, additional PCMS should be used. • [151_Co_2] Comprehension of prohibition signs was most rapid when the symbol was simple.

Co	Summary Findings - Comprehension

Visual Warnings (cont.)

Color

- [22_Co_2] Colored drawings or pictures are preferable to black and white for airline safety pictorials.
- [76_Co_1] Color has been shown to positively influence visual search tasks and counting, and to influence memory.
- [81_Co_1] A two-color scheme for VMS road signs resulted in the shortest response time and was preferred by drivers.
- [102_Co_1] Green is more strongly associated with emergency exits than other colors. This association translated into action during evacuation experiments; i.e., exits marked by green flashing lights were used preferentially.
- [102_Co_2] Red is more strongly associated with danger than with safety.
- [104_Co_1] Interpretation of the color of a sign (i.e., the meaning associated with it) is important for comprehension under a variety of clear and smoky conditions. The meanings of many colors are culturally dependent
- [135_Co_10] Congestion in lanes can be communicated using red for stop-and-go conditions, yellow for slow, and green for normal.
- [152_Co_1] Colored warning labels are perceived as more readable.

Contrast

- [76_Co_2] Colored signs must have high contrast between the background and the text.
- [140_Co_3] For two-way communication systems, a light can be used to indicate visually that assistance is on the way. Signs are needed to indicate the meaning of all visual signals.

Font

- [143_Co_1] Opinions on the meaning of font styles (e.g., formality, freshness, directness) may vary by cultural background.
- [148_Co_1] Text should be legible and high contrast, with a plain font, good use of white space, and bullet points to separate statements.
- [152_Co_5] For older people, print must be large enough to compensate for reduced visual capabilities. To accommodate the large print, clearer layouts using spacing or bullets can be used.

Angle

- [94_Co_1] If the same sign is used in multiple locations, all signs should be oriented in the same way to avoid confusion.
- [156_Co_1] The number of exit signs required to ensure sufficient coverage may currently be underestimated in situations where the angle of approach of some evacuees may be at an oblique angle, which reduces access to the information.

Message display rate

- [41_Co_4] For words between four and eight characters long, the exposure time should be at least one second per word, excluding prepositions. This is true for messages of up to eight words.
- [43_Co_12] A display rate of four seconds per frame and one of two seconds per frame with one repetition did not make a difference in the comprehension of a two-frame message by motorists.

Co	Summary Findings - Comprehension

- [43_Co_13] Flashing words increase reading times.
- [134_Co_6] Message comprehension increases as viewing time increases.
- [136_Co_6] Adjustable display rates (three seconds per phase minimum) are needed to allow each driver to read a PCMS message twice.
- [137_Co_1] Drivers require two seconds to read each unit of information on a DMS and can process no more than four units per message at highway speeds or five units at low speed. The required DMS viewing time is therefore up to eight seconds or 10 seconds respectively
- [144_Co_1] The response time was shorter for simultaneously displayed (discrete) messages than for the same message displayed sequentially in multiple phases. Drivers preferred the discrete displays.

Lighting – Static

- [43_Co_6] Under most conditions, fiber-optic signs performed better than LED signs for target value, legibility distance and viewing comfort.

Lighting – Dynamic

- [26_Co_1] Flashing lights increased drivers' awareness of road hazard warning signs and increased the interpretation of the sign as signaling a hazard.
- [102_Co_3] Green flashing lights are more strongly associated with safety than green strobe lights. However, this did not necessarily influence egress behavior.

Lighting –Wayguidance systems

- [15_Co_1] The nature of the target population should be considered when selecting the wayguidance/lighting system.
- [72_Co_2] When the traveling speed is fast and the spacing between lights is small, flashing wayguidance lights can appear to point in the opposite direction.
- [102_Co_4] Because flashing lights attract attention and green is associated with safety, green flashing lights can be used to overcome familiarity issues and direct people to an emergency exit.
- [133_Co_1] The UL 1971 Standard for Signaling Devices for the Hearing Impaired specifies the mechanical and electrical properties of these devices but does not specify how the device is to be used; for example, the ways in which the lights must flash to provide a specific message to occupants.

Surroundings and context

- [67_Co_1] The process of perceiving and comprehending road signs can be made more efficient by limiting the length of the message, presenting it in an uncluttered environment, and providing context.
- [97_Co_1] Care must be taken by television broadcasters not to block the visual emergency information, including with other captioning.
- [135_Co_4] A study of pictograms on DMS roadway signs showed that correct interpretation deteriorated rapidly with distance and in darkness,
- [142_Co_2] Context greatly increases the comprehension of a symbol.

Individual and group characteristics

- [57_Co_1] Older adults report that warning symbols helped them to understand the hazard, but when asked to identify the meaning of the symbols, older adults showed poorer comprehension than younger adults for a number of commonly

Co	Summary Findings - Comprehension
Visual Warnings (cont.)	used symbols. • [58_Co_1] People older than 52 years old generated fewer phrases associated with safety symbols than younger participants. • [58_Co_2] The first phrase that was written by younger participants was a more accurate description of the meaning of safety symbols than those written by older participants. • [119_Co_1] Younger adults comprehend standard safety symbols better than older adults. • [120_Co_1] Older adults create less accurate cognitive maps than younger adults. *Bilingual messages* • [3_Co_1] Although slightly longer gaze duration was found for a VMS displaying an alternating bilingual message than for a VMS displaying the messages in both languages simultaneously, the effect was not statistically significant. • [3_Co_2] For alternating bilingual messages on a VMS, 90 % of the drivers considered 2.0 seconds an appropriate length of time to read the message in their native language. • [3_Co_3] Gaze duration and longest fixation duration were significantly longer for bilingual VMSs giving a safety warning than for a monolingual display of air and road surface temperatures. • [43_Co_8] Alternating bilingual messages were no more physically demanding on the viewer than simultaneous displays. • [67_Co_2] The overall reading time for bilingual signs may be improved by better sequencing of the two languages or by differentiating them by colors or fonts.
Auditory Warnings	*Design* • [46_Co_1] Auditory messages are more likely to be confused if they consist of only single continuous tones, have the same temporal pattern, or share a similar on/off pattern. Warning systems should consist of a wide range of different sound types to reduce confusion. • [140_Co_2] Alarm signals should be distinguishable for different scenarios (e.g., fire, tornado) that require different responses. • [148_Co_7] A warning message needs to consider the audience and be understandable by an audience member with the lowest level of skills, including consideration of sensory-perceptual difficulties, cognitive limitations, and literacy levels. *Alarm tones* • [47_Co_3] Traditional continuous alarms are inappropriate because they can easily confuse, are hard to localize, and stay on until they are manually deactivated. They also distract from the task at hand and interrupt communication. • [47_Co_5] In addition to traditional alarms such as horns, bells, and buzzers, more natural sounds that signal a specific action may be advantageous, especially if they reduce the need to learn a set of alarms.

Co	**Summary Findings - Comprehension**

Auditory Warnings (cont.)

Content

- [31_Co_1] In spoken messages, letters are more difficult to identify correctly than numbers, which are more difficult than colors. The difference in difficulty is increased with noise.
- [160_Co_3] Numbers in an auditory message were repeated less correctly than other parts of speech.
- [160_Co_4] Nouns were the most easily comprehended in an auditory message, followed in order by numerals, general sentences, and numeral sentences.

Word choice

- [52_Co_3] The ability to discriminate words in speech depends on the commonality of their use.
- [64_Co_1] Common words are intelligible at a lower sound threshold in noise. Under test conditions, the required threshold for intelligibility was found to decrease by about 4.5 dBA per logarithmic unit of word frequency.

Message length

- [9_Co_1] Performance was found to deteriorate when the number of commands in a set exceeded three for native speakers and two for low proficiency non-native speakers.
- [9_Co_4] Accuracy in carrying out longer sets of commands improved when redundant words were eliminated.
- [9_Co_5) No more than two or three commands should be given in a single message.
- [25_Co_1] All messages should be brief and precise.
- [64_Co_2] Shorter words are intelligible at a lower sound threshold in noise. Under test conditions, the intelligibility threshold was found to decrease by approximately one dBA per letter.
- [148_Co_2] Instructions should be explicit but brief.
- [160_Co_2] Single words in an auditory message were more accurately understood and repeated than entire sentences.

Complexity

- [24_Co_1] Comprehension of an auditory message by non-native English speakers is improved by lowering the syntactic complexity of the message and repeating it.
- [41_Co_7] The language of an audio message should be concise rather than conversational. If the format of the message is standardized, then people can anticipate what piece of information will come next.
- [41_Co_8] The vocabulary of a message should be limited so that people can anticipate which words will be used and can process the information more quickly.
- [89_Co_1] Older adults are less able to draw inferences from novel warnings. Warnings that are specific, clear, and accurate reduce the need to make inferences and thus improve comprehension.
- [132_Co_3] The messages that were responded to with the most accuracy were stated clearly and avoided multiple interpretations.
- [132_Co_4] Overly wordy messages could cause confusion.

Co	Summary Findings - Comprehension
Auditory Warnings (cont.)	*Urgency* • [47_Co_1] The meaning of the alarm should be clear to the staff, and its characteristics should convey the level of urgency. • [132_Co_5] Calm, controlled voices were most easily understood and promoted calmness in others at a disaster scene. *Pitch* • [31_Co_5] Female speakers are more intelligible than males when the sound level of noise is equivalent to or greater than that of the message. • [41_Co_9] No differences in intelligibility have been found between high and low pitched voices, although low-pitched female voices are commonly used in audible warning systems with good results. • [49_Co_3] The choice of speaker should depend on the acoustic qualities of the voice relative to the noise spectrum, not the speaker's sex. • [74_Co_1] A higher pitched male voice is both authoritative and more easily understood. • [160_Co_5] Female voices are generally easier to understand than male voices. *Sound level* • [5_Co_1] Under a high cognitive workload, singular, instructive auditory messages were processed more accurately at higher sound levels (65 dB compared to 45 dB), even though both sound levels were in the range considered optimal for speech stimuli in quiet environments. • [5_Co_2] Higher sound levels for auditory warnings may be especially helpful in enhancing the performance of older people with some hearing loss. *Timing and delivery rate* • [2_Co_1] A standard temporal pattern for an audible emergency evacuation signal has been developed by ANSI, with the sole meaning of "evacuate the building immediately." • [2_Co_3] The ANSI standard audible emergency evacuation signal consists of an 'on' phase for 0.5 seconds followed by an 'off' phase for 0.5 seconds, repeated for a total of three successive 'on' periods and followed by an 'off' phase for 1.5 seconds. • [9_Co_3] Accurate execution of a set of commands did not depend on the pause length between words, word speed, or elimination of extraneous words. • [28_Co_1] The delay time for people to recognize words in continuous speech is about one-sixth seconds. • [30_Co_3] Comprehension of auditory messages by non-native speakers requires messages to be delivered slowly, repeatedly, and without distortion. • [31_Co_3] Comprehension is significantly better under all noise conditions when the message is delivered at a slower speech rate. • [31_Co_10] To improve comprehension, the voice used to deliver a message should not be heavily accented, and the speaking rate should be slow. • [41_Co_8] A speaker should spent more time on syllables and less on pauses. • [47_Co_4] A key factor that causes confusion in alarms is sharing similar temporal patterns like the same rhythm or repetition rate. Alarms for different purposes should have different temporal patterns.

Co	Summary Findings - Comprehension

Auditory Warnings (cont.)

- [160_Co_1] Comprehension of auditory messages deteriorates as speech rate increases.

Clarity

- [30_Co_1] Native English speakers are able to use their knowledge of the language to get the maximum amount of information from a severely distorted message.
- [30_Co_2] Non-native speakers are unable to make maximum use of the information in a severely distorted message, regardless of ability.
- [41_Co_7] The person delivering the message should have clear enunciation wtihout an obvious dialect.
- [49_Co_2] Rare and specific messages need to be more intelligible than common messages to be understood.

Cocktail party effect

- [27_Co_1] People find it difficult to distinguish and understand a spoken message when two voices of similar quality and volume are speaking simultaneously. When the competing messages are random phrases rather than a coherent passage, they cannot be distinguished.
- [27_Co_2] People who are deaf in one ear may find it especially difficult to understand a verbal message in the presence of competing speech.
- [31_ Co_6] When trying to follow one of two speakers, the message is easier to understand when the fundamental frequency of the competing speech is significantly different, such as of the opposite sex.
- [31_ Co_7] It is most difficult to follow a conversation when a competing voice is of the same sex and at a slightly higher volume.
- [31_ Co_8] Non-native listeners have a significant disadvantage in understanding speech when two people are speaking at once.
- [31_ Co_9] When two people are speaking at once, the disadvantage for non-native listeners is worse when the competing voice is louder and the speech rate is faster.
- [31_Co_12] A message delivered in English to a non-native English speaking population is less likely to be understood if there are competing voices speaking in the native language.

Ambient noise

- [31_ Co_2] Comprehension worsens faster for non-native listeners than for native listeners as background noise increases.
- [31_ Co_11] Masking noise, especially speech, should be eliminated whenever possible.
- [49_Co_1] Intelligibility should be tested under the ambient noise conditions.
- [52_Co_1] The ability to discriminate words in noise deteriorates about twice as fast for non-native speakers as for native speakers.
- [52_Co_4] Masking sounds need to be reduced as much as possible to improve the comprehension of auditory warnings by both non-native and native speakers.
- [160_Co_6] Participants repeated voice messages most accurately in the absence of noise.

Co	Summary Findings - Comprehension
Auditory Warnings (cont.)	• [160_Co_7] The effects of speech rate and voice type (male and female) on the comprehension of spoken messages were more pronounced in the presence of background noise. • [160_Co_8] For good comprehension of a spoken message, background noise should be decreased as much as possible. *Bilingual and non-native English speakers* • [13_Co_2] Spoken messages will not directly help those who don't understand the message. • [16_Co_1] Bilingual speakers of closely related languages may communicate with considerable transference, in which the syntax and word meaning from one language influences the other. • [16_Co_2] Mental processing of cognates by multilingual speakers differs from processing of non-cognates. • [16_Co_3] To reduce confusion for non-native English speakers, messages should minimize the number of false cognates, words that have similar form but different meanings, in multiple languages that are common in the target population. • [17_Co_1] Bilingual speakers may switch from one language to the other when a clause contains a cognate, a word similar in form and meaning to a word in the other language. • [25_Co_3] Multilingual or foreign language alerts may be necessary, because in crisis situations people have a tendency to revert back to thinking and understanding in their native language. • [31_ Co_4] Speaker voices with robust speech cues (without a heavy accent, resistant to noise masking) significantly improve comprehension for non-native listeners and are also helpful for native listeners. • [84_Co_1] Common warning signal words should be used when non-native speakers may be present. • [84_Co_2] Populations of non-native English speakers should be tested on their comprehension of words used in warning messages. • [111_Co_1] When facing memory overload or when unsure of the proper word, a non-native speaker with a very high level of English proficiency reverted to speaking in her native language. • [111_Co_2] When asked to carry out difficult tasks, a non-native speaker with a very high level of English proficiency reverted to thinking in her native language.

A.4 Credibility (Cr)

Cr	**Summary Findings - Credibility**
General Warnings	*Design* • [23_Cr_3] The CDC's five key steps in successful communication are: 1) execute a solid communication plan, 2) be the first source of information, 3) express empathy early, 4) show competence and expertise, and 5) remain honest and open throughout the crisis or emergency. • [23_Cr_4] Factors that lead to communication failure include mixed messages from multiple experts, late release of information, paternalistic attitudes, failure to counter rumors and myths as they come up, and public power struggles. • [25_Cr_2] The five communication steps for success are: 1) execute a solid communication plan, 2) be the first source for information (especially since the first message carries more weight), 3) express empathy early, 4) show competence and expertise, and 5) remain honest and open. • [91_Cr_1] An extensive literature is available on the communication of hazard information to the public and the public response during disasters. • [108_Cr_2] Messages should be targeted for the social networks of vulnerable populations and developed by people that understand those communities. • [136_Cr_2] Research-based guidelines for message design are not helpful unless they are in a form that can be used by field personnel that need to make quick decisions. • [139_Cr_5] Effective risk communication requires the engagement of communication representatives in designing and evaluating educational materials, emergency messages, and dissemination practices, including language resources and accessibility. *Positive messages* • [25_Cr_5] Reduce the number of negatively-dominated messages. • [32_Cr_1] Give three positive messages for every negative message. *Content* • [23_Cr_7] Effective communication includes acknowledging uncertainty, avoiding over-reassurance, giving people things to do, acknowledging fears and providing information to put the fear into context, addressing "what if" questions, and asking more of people. • [43_Cr_1] Information on the cause of an incident was appreciated by those affected. • [49_Cr_1] The message content (including the choice of signal words to convey importance, an explicit description of the risks, and the use of personal pronouns) is more influential than the sex of the speaker in promoting compliance. • [75_Cr_2] Science concepts taught to lay people are not easily integrated into practical knowledge, but may instead compete with other explanations more deeply rooted in tradition or experience. Communication with the public needs to address possible discrepancies between common lay theories and science.

Cr	Summary Findings - Credibility
General Warnings (cont.)	• [107_Cr_1] People are more reluctant to comply with instructions when the warning is vague or incomplete. Information should not be withheld out of a fear of panic. In fact, the lack of communication is a factor that contributes to panic. • [107_Cr_3] After an event, communication that explains the disaster and the disaster experience helps to provide closure and a transition to normal life. • [152_Cr_1] The description of the hazard should be specific and complete. • [152_Cr_2] The explanation of the consequences of the hazard and the directions for avoiding the hazard should be explicit. *Timeliness* • [4_Cr_5] Because "cry wolf" syndrome is not supported by the evidence from real natural disasters, evacuation notices should be given as early as possible. *Communication channels* • [4_Cr_2] Door-to-door communication by authority figures resulted in 97 % evacuation rates from beaches in two hurricanes. • [23_Cr_6] The public needs information about the crisis and about the plans for response and recovery, and officials need to listen and respond to feedback from the public. • [86_Cr_1] During an emergency, people seek confirmation of warnings from multiple sources. These can include emergency management agencies, internal web sites, electronic and print news media, web sites, and friends and neighbors. • [86_Cr_2] During an emergency, people develop their own independent emergency assessments (risk perception) and pass information on to others. • [92_Cr_1] Messages that are disseminated frequently and through the correct channels are more likely to achieve an appropriate public response. • [108_Cr_1] Cultural groups outside of the mainstream may be reluctant to trust warnings until they have confirmed the information with members of their own communities, thus delaying their response. • [113_Cr_5] Encouraging people to contact their neighbors with information will reduce the overloading of telephone systems, warn individuals that might not have received the warning, and help to confirm the warning for those who have received the message. People are less likely to seek confirmation if with family. • [128_Cr_6] Community leaders recommended neighborhood alert systems to organize members of the community to reach isolated people and those distrustful of government-based alerts • [139_Cr_2] Different communities obtain information from different sources, including social networks, ethnic media, and places of worship. *Alarm reliability* • [4_Cr_4] The "cry wolf" syndrome, in which unnecessary evacuations result in unwillingness to leave during the next event, is not supported by the evidence. • [8_Cr_1] Women and older residents living in a flood plain were found to be more likely than men and younger residents to prefer more warnings and accept the possibility of false alarms. • [8_Cr_2] The "cry wolf" effect is not supported by evidence. Overall, 78 % of

Cr	**Summary Findings - Credibility**

General Warnings (cont.)

residents living in a flood plain would prefer more warnings even if some of them turned out to not be accurate. A large majority (77 %) agreed that one or two false alarms would not reduce their confidence in future alarms.

- [8_Cr _3] Emergency planners should not be reluctant to issue warnings given the general preference of the public for more warnings and the acceptance of the possibility of false alarms.
- [37_Cr_1] With an unreliable alarm system (either false alarms or missed failures), operators are distracted from other tasks by monitoring the information that the alarm should be handling.
- [37_Cr_2] An unreliable alarm system causes more task disruption than no alarm system at all.
- [37_Cr_3] In a system prone to false alarms, the time to respond to the alarm is significantly longer, suggesting that the information is double-checked first.
- [47_Cr_1] A good alarm sounds only in the presence of danger, with no false alarms.
- [109_Cr_1] The credibility of the fire alarm system is reduced by false alarms, test alarms and prank alarms.
- [122_Cr_1] Attitudes toward previous false alarms and emergency drills affect behavior during the next emergency.
- [145_Cr_1] In a set of repeated emergencies requiring evacuation (four hurricanes in two months), public complacency decreased after the first two and then increased.
- [145_Cr_3] Too many unnecessary evacuations in a short period of time may result in increased public complacency.
- [145_Cr_4] Emergency response managers need to have an evidence-based understanding of the "cry wolf" effect in order to properly manage the possibility of public complacency.

Messenger

- [4_Cr_1] Evacuation notices from public officials are effective in several ways: they convince people that they are in danger, they represent the voice of authority, and some (15 % to 25 %) believe that there will be a penalty for noncompliance.
- [4_Cr_3] People hearing official evacuation notices are more than twice as likely to leave.
- [23_Cr_5] Credibility needs to be established from the beginning. Trusted, credible sources should be used to get communication out quickly, first expressing empathy and caring and then establishing competence, expertise, honesty and openness, commitment, and dedication to the audience.
- [25_Cr_1] An audience is likely to pay more attention to those who are perceived as most trustworthy, honest and without motives to deceive them. The best sources are those to whom the public can relate, especially if personally known by the public.
- [32_Cr_2] Spokespersons should practice their answers and only cite sources that are credible to the audience.
- [38_Cr_1] Provide a live warning announcement by a familiar authoritative source (for visual and auditory communicators).

Cr	Summary Findings - Credibility

General Warnings (cont.)

- [85_Cr_1] In evaluating trustworthiness in face-to-face communication, behavior was found to be more important than prior expectations.
- [85_Cr_2] Officials delivering messages should behave in accordance with social norms to be more credible.
- [86_Cr_4] The most effective warnings (highest level of compliance) are from a credible source and sufficiently specific that individuals can determine whether the threat pertains to them personally.
- [98_Cr_1] Members of the vulnerable communities should be involved in planning and disseminating messages in ways that are amenable to how the vulnerable populations use the technology.
- [98_Cr_2] Warning messages should be delivered by people trusted by the vulnerable population community, preferably those that have regular contact with them.
- [100_Cr_1] The source of an alert or warning influences a person's response. If the source of the information is not identified in the message, the message is unlikely to be deemed trustworthy.
- [107_Cr_2] People tend to be compliant and trustful of authorities during a rapid-onset disaster. This provides a window of opportunity (and a responsibility) for emergency managers to impact the outcome of the disaster.
- [109_Cr_2] A live voice should be used since recorded messages can be ineffective or dangerous. A live voice can be updated with new information, can convey the appropriate urgency, and is perceived as more credible and reliable.
- [113_Cr_1] Because they have access to early warning systems, a Federal authority should alert Americans to the nature of the threat in a national emergency.
- [113_Cr_2] Because they have greater credibility with local residents than Federal authorities, local officials should notify the public of specific actions to be taken.
- [122_Cr_2] The response to communication depends on the relationship of the communicator to the listener.
- [122_Cr_3] Credible sources have a wider range of language strategies to choose from, including more intense language that may attract more attention.
- [139_Cr_6] Trusted members of the community and channels of communication (formal and informal) are critical to disseminate and reinforce the messages, making sure that messages are received, understood, and followed.
- [139_Cr_7] A database of trusted sources, interpreters, and effective communication modes for diverse populations should be prepared.

Public perception

- [25_Cr_3] Be honest about the event – there is little evidence that honesty causes panic.
- [25_Cr_4] Describe what you know and don't know and what you are doing about it.
- [32_Cr_3] The EPA's seven rules of risk communication are: 1) accept and involve the public as a legitimate partner, 2) listen to people, 3) be truthful, honest, frank, and open, 4) coordinate, collaborate, and partner with other credible sources, 5) meet the needs of the media, 6) speak clearly and with

Cr	**Summary Findings - Credibility**

General Warnings (cont.)

compassion, and 7) plan thoroughly and carefully and evaluate performance.

- [43_Cr_2] Confidence in the information influenced willingness to modify behavior.
- [99_Cr_1] All official and media emergency information sources need to provide the same emergency phone numbers.
- [113_Cr_4] People will seek confirmation from multiple sources, so consistency is important.
- [135_Cr_1] Maintaining the credibility of DMS roadway signs requires drivers to be confident that they can obtain from them the information they need.
- [139_Cr_1] Partnerships are needed with community representatives in order to build trust and to obtain insight into the design of risk communication strategies and other emergency preparedness tools.
- [139_Cr_3] The level of trust in various messengers and communication channels and the degree to which information is sought from these sources must be assessed for each diverse community.
- [139_Cr_4] The attitudes of community members to emergency response practices must be understood.
- [139_Cr_8] Building and maintaining trust is of critical importance.

Stress

- [23_Cr_1] Expressing empathy to the public early in any crisis or emergency situation helps to establish and maintain a connection.
- [23_Cr_2] Communicators must be respectful of people's emotions and coping mechanisms and not attempt to interject humor.

Repetition

- [113_Cr_3] Repetition of a warning message that includes the actions that should be taken strengthens adaptive responses to the hazard.

Familiarity and training

- [47_Cr_2] As people become familiar with an alarm's reliability, they will match their response rates to the reliability of the alarm; if the alarm is 90 % reliable, they will respond 90 % of the time.
- [89_Cr_1] Presenting information in a familiar way to older adults both counters memory limitations and improves credibility.
- [109_Cr_3] Staff must be well-trained, since they will be perceived as reliable sources of information.
- [134_Cr_1] Consistency in the design of DMS highway messages is important to maintain the credibility of the agency responsible for the signs.
- [145_Cr_2] Even if they are ultimately deemed to be unnecessary, the first one or two evacuations in response to a repeated threat may decrease complacency at the time of the subsequent evacuation.

Individual and group characteristics

- [75_Cr_1] Due to differences in background knowledge, the health concepts described by physicians are not interpreted in the same way by their patients, who may therefore discount the explanation and devise an alternative, resulting in poor compliance.
- [86_Cr_3] Population segments differ in their access to warning messages and in

Cr	Summary Findings - Credibility
General Warnings (cont.)	their perception of the credibility of information sources. *Non-native English speakers* • [128_Cr_1] Government agencies are in general not properly staffed to produce and disseminate messages in multiple languages. Not all information on websites is translated, and content in languages other than English may be difficult to find. • [128_Cr_2] A training program is recommended for government representatives dealing with ethnic media, including crisis communication plans. • [128_Cr_3] Multicultural and multilanguage course material should be included in the training of emergency and media professionals and students. • [128_Cr_4] Government agencies and first responder organizations need to develop bilingual preparedness materials. • [128_Cr_5] Government agencies and first responder organizations need to develop trust and partner with organizations within non-English speaking communities.
Visual Warnings	*Design* • [136_Cr_1] Improper use of highway signage (PCMS) can destroy their credibility, create confusion, and promote improper driving behavior. • [136_Cr_4] The PCMS format should not have the appearance of an advertising display and should display only traffic-related information. *Word choice* • [136_Cr_6] The words "DANGER" and "CAUTION" should not be used in PCMS for road work (although there are exceptions). *Complexity* • [136_Cr_3] To maintain credibility of the PCMS communication mode, messages should not be vague or overly simple. *Placement* • [136_Cr_5] Roadway signs need to be positioned far enough upstream of the decision point to provide adequate warning but close enough that the message is relevant.
Auditory Warnings	*Pitch* • [74_Cr_1] A higher pitched male voice is both authoritative and more easily understood. • [122_Cr_4] A male voice is more persuasive for presenting specific instructions that need to be followed. • [122_Cr_5] Fear arousal makes listeners more receptive to low intensity messages, which are most effectively delivered by women.

A.5 Personalization (Ps)

Ps	Summary Findings - Personalization
General Warnings	*Design* • [40_Ps_1] The intended audience must be known when designing a message. • [45_Ps_1] Real-time information systems in transportation systems cause travelers to perceive that their wait time is reduced and reduces their anxiety. • [46_Ps_1] A user-centered approach to warning design should be used. • [87_Ps_2] Messages need to be communicate the level of risk clearly and precisely, encouraging as little inference as possible. • [91_Ps_1] An extensive literature is available on the communication of hazard information to the public and the public response during disasters. • [139_Ps_3] Effective risk communication requires the engagement of communication representatives in designing and evaluating educational materials, emergency messages, and dissemination practices, including language resources and accessibility. • [152_Ps_2] Effective warnings consist of four message components: a signal word to attract attention, identification of the hazard, explanation of the consequences if exposed to the hazard, and directives for avoiding the hazard. *Message ordering* • [136_Ps_2] For messages that can be displayed on a single phase of a PCMS roadway sign, present the problem on top, the location on the second line, and the recommended action on the bottom *Validation* • [25_Ps_5] Test and validate the effectiveness of messages. • [139_Ps_5] The design, execution, and evaluation of drills and exercises must include members of diverse populations. *Content* • [23_Ps_1] It is important to understand the specific audience (e.g., emergency or public health officials, victims, medical personnel, or the public) and provide the information that is needed. • [25_Ps_5] The four main functions of an emergency message are to provide vital information about the hazard, communicate the implications of the event and its urgency or seriousness, instruct on what to do next, and explain how recipients can respond. • [25_Ps_4] Use a "danger-action" formula for messages – keywords and phrases describing dangers (e.g., 'active shooter' or 'HAZMAT warning') and actions (e.g., 'evacuation now' or 'take shelter now') can be vetted in advance of the emergency. • [32_Ps_1] Address risk perceptions • [40_Ps_2] An effective message provides relevant information that can be applied by the intended audience. • [49_Ps_4] The message content (including the choice of signal words to convey

Ps	**Summary Findings - Personalization**

General Warnings (cont.)

importance, an explicit description of the risks, and the use of personal pronouns) is more influential than the sex of the speaker in promoting compliance.

- [75_Ps_1] Emergency communication needs to help lay people sort relevant from irrelevant information.
- [75_Ps_2] Recall of news items is best for names and places, lower for events, and very low for specific numbers. Recall is improved when the information is relevant, repeated, local, sensational or tragic, or matches previous knowledge.
- [92_Ps_1] An effective message includes information on what (people need to do), when (they need to do it), where (the danger is), why (the action must be taken) and who (the warning is from).
- [98_Ps_3] Messages should include instructions on how to shelter in place for those people that are unable to evacuate.
- [98_Ps_4] Showing members of the vulnerable populations taking protective actions such as using the shelters encourages people to believe that their needs will be met if they comply with the instructions.
- [100_Ps_1] A warning usually follows an alert, providing detailed information indicating who is at risk, where the risk resides, who is sending the warning message, and what protective actions should be taken.
- [113_Ps_1] Make emergency warnings as specific as possible, including the nature of the threat, the area and likelihood of impact, and what simple actions can be taken alone and with others to ameliorate the situation. Nonspecific warnings encourage nonchalance in the presence of a real threat. Since people seek to reduce uncertainty, standard generic warnings only lead people to seek more information.
- [113_Ps_2] Warning messages should be followed with a short list of easily recognized indicators of danger.
- [122_Ps_1] Unwarranted fear can be avoided by including specific details about the risks for each individual in a warning message.
- [128_Ps_2] Regulations on emergency message content for broadcast media are lacking.
- [134_Ps_2] In the case of a terrorist attack, people identified what to do, evacuation routes, the areas affected, safe areas, and the type of attack as critical information to display on the DMS.
- [136_Ps_4] The base message for roadway signs consists of seven elements: the type of roadwork, the location, lanes closed, the effect on travel, the action to take, the audience, and the reason to take the action. If not all information fits on the sign, methods are provided for condensing or prioritizing the message.
- [149_Ps_1] People exposed to personalized signs (that included their names) demonstrated more compliance than people exposed to an impersonal sign.
- [152_Ps_4] The description of the hazard should be specific and complete.
- [152_Ps_5] The explanation of the consequences of the hazard and the directions for avoiding the hazard should be explicit.

Word choice

- [83_Ps_2] Signal words suggest different levels of hazard.
- [98_Ps_2] The use of words such as 'immediate' and threatens' convey a sense of

Ps	**Summary Findings - Personalization**
General Warnings (cont.)	urgency. • [152_Ps_3] The signal word to attract attention should indicate the level of hazard present, such as DANGER or WARNING. *Communication channels* • [4_Ps_2] Door-to-door communication by authority figures resulted in 97 % evacuation rates from beaches in two hurricanes. • [25_Ps_1] A variety of communication channels should be explored for providing specific warning information to specific audiences. • [25_Ps_2] Automated notification can be used to address some of the challenges of contacting a large number of people in a small amount of time. • [128_Ps_4] Community leaders recommended neighborhood alert systems to organize members of the community to reach isolated people and those distrustful of government-based alerts • [128_Ps_5] Strategies for social media include sending multiple language audio and/or text messages to cell phones with brief guidance on actions to be taken. • [128_Ps_6] Women and children are good targets for disseminating emergency information to households. • [148_Ps_1] People receive warnings from indirect channels such as colleagues and friends in addition to direct channels. *Messenger* • [4_Ps_1] Evacuation notices from public officials are effective in several ways: they convince people that they are in danger, they represent the voice of authority, and some (15 % to 25 %) believe that there will be a penalty for noncompliance. • [86_Ps_1] The most effective warnings (highest level of compliance) are from a credible source and sufficiently specific that individuals can determine whether the threat pertains to them personally. • [109_Ps_2] A live voice should be used since recorded messages can be ineffective or dangerous. A live voice can be updated with new information, can convey the appropriate urgency, and is perceived as more credible and reliable. *Public perception* • [113_Ps_3] Emergency officials must seek to affect the perception of the hazard in the minds of the public, since perception and not the actual hazard is what determines people's response. Older people must be convinced that they are able to respond. Individual responses to any message will vary based on past experience, source of warning, and interpretation. • [139_Ps_2] The level of trust in various messengers and communication channels and the degree to which information is sought from these sources must be assessed for each diverse community. • [139_Ps_4] Trusted members of the community and channels of communication (formal and informal) are critical to disseminate and reinforce the messages, making sure that messages are received, understood, and followed. *Stress* • [87_Ps_1] With higher anxiety levels, people are more likely to assign threatening meaning to ambiguous information. Messages therefore need to be

Ps	**Summary Findings - Personalization**

<table>
<tr>
<td>General Warnings (cont.)</td>
<td>

communicate the level of risk clearly and precisely, encouraging as little inference as possible.

Cognitive load

- [138_Ps_1] Regardless of the format for the date, only about two-thirds to three-quarter of drivers were able to correctly interpret whether the PCMS message was applicable to them on their current trip or a later one.

Familiarity and training

- [51_Ps_1] The more hazardous the product, the more likely the label warning was read.
- [108_Ps_1] Those who have survived previous events such as hurricanes may be more reluctant to evacuate.
- [128_Ps_3] A training program in emergency communications for journalists is recommended.
- [132_Ps_1] Training and experience promotes optimal radio use and teaches first responders about their own reactions to stress.
- [134_Ps_1] Testing identified the most effective words and phrases for describing the type of incident, its location, and what action to take.
- [145_Ps_1] In a set of repeated emergencies requiring evacuation (four hurricanes in two months), public complacency decreased after the first two and then increased.
- [145_Ps_2] The most common reason given for not evacuating was the belief that the individual was not at risk.

Individual and group characteristics

- [98_Ps_1] Messages and practices must consider the needs, means, and abilities of vulnerable populations. There is often significant overlap among different vulnerable groups, and accommodating people with disabilities often improves response for everyone.
- [139_Ps_1] Different communities obtain information from different sources, including social networks, ethnic media, and places of worship.

Non-native English speakers

- [83_Ps_3] The interpretation of hazard words and colors is not consistent across cultures. United States students ranked colors from low to high hazard in the order Blue, Green, Yellow, Black, Orange, and Red; while Chinese students ranked them as Yellow, Blue, Green, Red, Black and Orange. The Chinese students rated signal words lower in hazard than the United States students.
- [125_Ps_1] Red, black, orange, and yellow connote hazard across different cultures (Spanish-speakers and English-speakers) in the United States.
- [125_Ps_2] The SKULL, SHOCK, and PROHIBITION symbols communicate higher hazard across both Spanish-speaking and English-speaking populations in the United States.
- [128_Ps_1] It would be useful for Spanish-language media websites to contain local news, information about developing emergency situations, and up-to-date lists of emergency contact persons.

</td>
</tr>
</table>

Ps	**Summary Findings - Personalization**
General Warnings (cont.)	*Effects on behavior* • [100_Ps_2] People take steps to verify that the information they are receiving about the threat is reliable and that a threat could materialize
Visual Warnings	*Design* • [136_Ps_1] The PCMS format should not have the appearance of an advertising display and should display only traffic-related information. *Content* • [41_Ps_1] Advisory sign messages should contain a problem statement, an effect statement, an attention statement (the audience of the message) and an action statement. • [135_Ps_1] DMS signs are particularly important for communicating unexpected circumstances, such as traffic incidents, roadwork, and special events, that require decisions to be made by drivers. • [135_Ps_2] Graphics on DMS roadway signs may be able to communicate more complex information regarding the type of problem, its location, the action to take, lane closures, consequences of the problem, the audience for the message, and the reason for taking the action. *Diagrams* • [135_Ps_4] Graphical interchange information was able to communicate congestion on specific lanes. *Urgency* • [51_Ps_2] The use of warning symbols on consumer products increases the perception of hazard. • [135_Ps_3] Pictogram symbols on DMS roadway signs may be (rightly or wrongly) associated with potential danger. • [136_Ps_5] The words "DANGER" and "CAUTION" should not be used in PCMS for road work (although there are exceptions). *Color* • [41_Ps_2] Presenting a keyword or phrase in a different color makes it more likely to be read first and increases its urgency. • [76_Ps_1] Colored signs are perceived to indicate more hazard than black and white signs. • [83_Ps_1] Colors suggest different levels of hazard. • [152_Ps_1] Colored warning labels are perceived as more likely to indicate a hazard. *Placement* • [136_Ps_3] Roadway signs need to be positioned far enough upstream of the decision point to provide adequate warning but close enough that the message is relevant. • [152_Ps_6] The visual warning should be located close to the hazard. *Surroundings and context* • [152_Ps_7] Visual clutter in the neighborhood of the warning should be reduced.

Ps	**Summary Findings - Personalization**

Auditory Warnings

Design

- [41_Ps_3] Short form and staccato messages are easier to recall. The first statement in an alert message should be an attention statement to alert the population that a message is about to be given. The second statement should describe the problem. The final statement should present the action that is required to react to the problem. For maximum recall, people must hear the message at least twice.
- [68_Ps_1] Message type, message format, sex of the speaker, average fundamental frequency, speech rate, and interval between messages all have a significant effect on perceived urgency for synthesized voice messages. Perceived urgency increases for keyword context (compared to syntax context), higher average frequency, faster speech rate, and shorter message interval.

Alarm tones

- [47_Ps_4] In addition to traditional alarms such as horns, bells, and buzzers, more natural sounds that signal a specific action may be advantageous, especially if they reduce the need to learn a set of alarms.
- [55_Ps_1] For pulse-based signals in background noise consistent with continuously-operating machinery, perceived urgency increases with increasing pulse level and with decreasing inter-pulse interval, and also depends on pulse format.
- [56_Ps_1] Auditory warnings consisting of signal pulses can built different levels of urgency into the warning itself, through selection of parameters including the fundamental frequency, the pulse sound level, and the inter-pulse interval.

Synthesized speech

- [48_Ps_4] Synthetic voices can be manipulated to increase their level of urgency, but it is more difficult to change the perceived appropriateness and believability.
- [59_Ps_4] Urgency can be imbued into synthesized speech through the use of three acoustic parameters: amplitude, fundamental frequency, and pitch range.
- [59_Ps_5] Synthesized speech messages need to be realistic, intelligible, and believable.
- [68_Ps_2] Synthesized male voices were perceived as more urgent than synthesized female voices with the same fundamental frequency, possibly because the male voice sounded more unnatural.
- [68_Ps_5] For synthesized voice messages, the most economical way to change the perceived urgency using quantitative variables is to modify the speech rate, followed by the fundamental frequency and finally the message interval.

Content

- [109_Ps_1] The voice message alarm in an emergency should be simple, direct, and truthful. It should give the location of the fire and what is expected of occupants, including specific directions to the exit.

Word choice

- [2_Ps_2] A key word or phrase (e.g., FIRE!) may be inserted during the 'off' phase between the three pulses of the ANSI standard audible emergency

Ps	**Summary Findings - Personalization**
Auditory Warnings (cont.)	evacuation signal.

- [48_Ps_3] Words with a highly negative connotation are considered more urgent, appropriate, and believable.

Urgency

- [47_Ps_1] The meaning of the alarm should be clear to the staff, and its characteristics should convey the level of urgency.
- [48_Ps_1] A message delivered in a higher pitched voice and at a faster speech rate is perceived as being more urgent.
- [48_Ps_2] Although an "appropriate" delivery of warning words by human speakers gave them the perception of being more believable and appropriate than words spoken "inappropriately," this was not found to be related to pitch or speech rate.
- [55_Ps_2] Certain signal characteristics can be modified to match the urgency perceived by the listener with the urgency required by the situation.
- [55_Ps_3] Through variation of parameters, proper design of pulse-based signals can assist listeners in deciding how urgently they need to respond.
- [56_Ps_2] Auditory signals with high frequency, fast speed, and high sound levels result in the highest ratings of perceived urgency.
- [59_Ps_1] Speakers are reliably able to convey different levels of urgency to listeners.
- [59_Ps_2] Urgency is transmitted through both the acoustic characteristics of the message and its meaning.
- [60_Ps_1] Words that have a stronger negative connotation and are spoken in an urgent, female voice are more effective in conveying urgency.
- [68_Ps_3] The relative effects of multiple variables in audible warnings on the perception of urgency can be quantified using Stephens' power law.
- [68_Ps_4] Quantification of the relative effects of multiple variables in audible warnings enables the transmission of equivalent levels of urgency through different parameters, as well as providing a technique to appropriately scale suggested prioritization of tasks.

Pitch

- [49_Ps_1] Urgency is better perceived in messages that are delivered urgently and in a higher pitched voice.
- [49_Ps_3] The higher perception of urgency in unambiguous female voices was based on acoustic characteristics rather than on the perceived sex of the speaker.
- [59_Ps_3] A female voice transmits a greater range of urgency than a male voice over styles from monotone to nonurgent to urgent.

Sound level

- [47_Ps_3] Alarms need to be audible within all environments of a building, such as in operating rooms. The sound level should allow people to communicate.
- [68_Ps_6] An audible warning must be loud enough to attract attention and be heard over ambient noise, but not so loud that it is painful and distracting. This may leave a narrow range for manipulation to affect perceived urgency.

Ps	Summary Findings - Personalization
Auditory Warnings (cont.)	*Timing and delivery rate* • [2_Ps_1] A standard temporal pattern for an audible emergency evacuation signal has been developed by ANSI, with the sole meaning of "evacuate the building immediately." • [41_Ps_4] The speaker should speak loudly at a moderately fast rate without speaking in a monotone. The message should be read in a "calm, matter of fact, and dignified manner," at a speed of about 175 words per minute • [41_Ps_5] A speed under 110 words per minute sounds too slow, and the message may be perceived as less important. • [47_Ps_3] Temporal patterning can help to communicate the level of urgency. *Ambient noise* • [47_Ps_2] In environments with unpredictable and fluctuating noise, the masking of alarms can be minimized by including several harmonics within the 500 Hz-4000 Hz band. Alarms with rich harmonic content within this frequency range can be played at a lower signal-to-noise ratio than more simple tones, allowing the noise levels to be minimized. These alarms are also easier to localize. • [49_Ps_2] Female voices may retain more emotional content (urgency) after acoustic manipulation to improve intelligibility in noise.

A.6 Action (Ac)

Ac	Summary Findings - Action
General Warnings	*Design* • [91_Ac_1] An extensive literature is available on the communication of hazard information to the public and the public response during disasters. • [92_Ac_4] Follow-up messages should be provided if the hazard and disaster risks change. • [92_Ac_6] Warnings for rapid-onset events, which materialize less than 15 to 30 minutes from detection, may be more expensive and require a higher level of planning. • [113_Ac_3] An all-clear signal is needed to prevent premature return to a hazardous area. *Validation* • [148_Ac_2] Warning system effectiveness should be tested using participants from the target population. *Content* • [23_Ac_1] People need constructive actions to take, which can be provided as a set of "must do," "should do," and "can do" choices. • [25_Ac_3] An 'all clear' message is important to indicate that there is no longer a threat. This type of message can be longer and more complex than initial alerts. • [38_Ac_3] Provide specific information about the emergency situation, including what is happening, what is being done, and what actions need to be taken (for digital communicators). • [43_Ac_1] Information on the cause of an incident was appreciated by those affected. • [43_Ac_2] Drivers prefer to receive information before experiencing the conditions. Information on the nature and impact of the situation is appreciated. • [43_Ac_7] The severity of an incident influences the need and desire for relevant information. • [43_Ac_8] Suggestions for alternative responses are required. • [43_Ac_10] Motorists were willing to pay for in-car information (men more than women). • [43_Ac_13] The level of detail provided in a message influenced willingness to modify behavior. • [73_Ac_1] Give occupants all needed information, such as directions on where to go in addition to what to do. • [86_Ac_3] Traffic management, transportation alternatives, and shelters must be arranged and communicated to the public. • [89_Ac_8] Protective action recommendations should include options that allow everyone to be protected regardless of age, ability, or financial means. • [92_Ac_1] An effective message includes information on what (people need to do), when (they need to do it), where (the danger is), why (the action must be

Ac	Summary Findings - Action

General Warnings (cont.)

taken) and who (the warning is from).

- [92_Ac_2] An effective message is specific, consistent, certain, clear, and accurate.
- [97_Ac_1] Examples of critical information to be communicated to the public during an emergency are: specific details about the affected areas; evacuation orders, including the areas to be evacuated and evacuation routes; approved shelter locations and/or instructions on sheltering in place; instructions on how to protect property; road closures; and instructions on obtaining relief assistance.
- [98_Ac_4] Messages should include instructions on how to shelter in place for those people that are unable to evacuate.
- [98_Ac_5] Showing members of the vulnerable populations taking protective actions such as using the shelters encourages people to believe that their needs will be met if they comply with the instructions.
- [100_Ac_1] A warning usually follows an alert, providing detailed information indicating who is at risk, where the risk resides, who is sending the warning message, and what protective actions should be taken.
- [107_Ac_3] Emergency managers should address the fears of unfamiliar hazards by disseminating information about the hazard, its potential consequences, and appropriate actions to minimize negative effects. These include both constructive actions that citizens can do themselves (and why) and what is being done by experts on their behalf. Such messages will both increase compliance and discourage actions that are not effective.
- [109_Ac_3] The voice message alarm in an emergency should be simple, direct, and truthful. It should give the location of the fire and what is expected of occupants, including specific directions to the exit.
- [152_Ac_1] Effective warnings consist of four message components: a signal word to attract attention, identification of the hazard, explanation of the consequences if exposed to the hazard, and directives for avoiding the hazard.

Complexity

- [89_Ac_2] Age affects working memory, which can be taxed by complex tasks such as safety procedures. Older adults may choose less protective procedures over more complex ones because they are aware of their limitations.
- [89_Ac_6] To avoid overtaxing working memory in older adults, warning messages should use simple sentence structure and break protective actions to be taken into a few, simple steps.
- [113_Ac_1] Warning messages should focus on simple protective or avoidance measures, since people naturally behave in familiar patterns during emergencies.

Timeliness

- [23_Ac_3] Different types of messages are effective at different times during the crisis.
- [109_Ac_2] Voice alarms should be given as soon as the emergency has been identified in order to reduce the delay time.
- [113_Ac_4] Issue warning messages as early as possible, instructing local officials to transmit the warning without delay.

Ac	Summary Findings - Action
General Warnings (cont.)	*Communication channels* • [25_Ac_1] Two-way communication can be used to monitor public knowledge and beliefs, misunderstandings, behavioral compliance, and the progress of the response. This is an effective way to correct rumors and misinformation. • [86_Ac_2] Emergency managers should monitor news media to assess the situation and to make sure that messages being given to the public are accurate. • [89_Ac_7] Feedback from the community is required to assess what messages are most effective. • [89_Ac_9] New technology may be very useful, but should be tested before deployment. • [92_Ac_3] Monitoring the public response after message dissemination, such as through social networks and video or in-person observation, provides information on whether the warning system is guiding behavior in a manner consistent with current knowledge of the hazard and disaster risks. If the warning is not effective, adjustments should be made. • [99_Ac_1] A well-designed and frequently practiced phone tree is an effective way to reach members of the disability community during an emergency. A reverse calling system allows the community to alert emergency professionals to issues of which they may not be aware. • [109_Ac_4] Closed-circuit TVs installed for security purposes can be used to monitor the situation. • [128_Ac_1] Strategies for social media include sending multiple language audio and/or text messages to cell phones with brief guidance on actions to be taken. *Alternative communication modes* • [21_Ac_3] An auditory warning resulted in faster pilot response than a pictorial warning, although the difference was not found to be significant. *Multiple alarms* • [106_Ac_2] The route decision-making of a visual wayfinding system was improved through the addition of a tactile component. • [106_Ac_3] The slow evacuation time for a tactile wayfinding system was improved by adding a visual component. *Messenger* • [4_Ac_1] Official evacuation notices are most effective when they are specific about the area affected, are delivered in a personalized way, describe the specific actions to be taken, and are worded aggressively and delivered in an urgent manner. • [4_Ac_2] People hearing official evacuation notices are more than twice as likely to leave. *Public perception* • [43_Ac_15] Confidence in the information influenced willingness to modify behavior. • [139_Ac_2] Effective emergency preparedness and response depends on the engagement of diverse communities.

Ac	Summary Findings - Action

General Warnings (cont.)

Stress

- [23_Ac_2] People seldom panic, but they can refuse to take good advice, become isolated alone or as a group, or become paralyzed and unable to act. Mental stresses can be anticipated and managed using appropriate risk communication strategies.
- [66_Ac_1] The response of a group to a dangerous situation varies, and may be related to the structure of the group and to the quality of its leadership.
- [66_Ac_4] The performance of divers on manual dexterity tasks was impaired with both increasing depth and worse diving conditions.
- [66_Ac_7] Low to moderate degrees of stress tend to improve performance, while high levels impair it.
- [66_Ac_8] Fear during dangerous situations is heightened by lack of communications and isolation.
- [66_Ac_9] Deterioration in performance, including manual dexterity, sensory-motor tasks, and secondary tasks that require divided attention, can be expected when physical danger is perceived.
- [66_Ac_10] Threatening conditions have been shown to inhibit correct responses to assigned tasks for subjects in military planes and dive chambers who were told they were in danger. Reducing the perception of danger may improve performance.
- [75_Ac_2] When physical danger is imminent, reactions are quick, with little information processing. Knowledge is critical for medium urgency decisions.
- [75_Ac_3] The effects of physical danger on cognition include impairment in the ability to learn new tasks and deterioration in memory retrieval, visual recognition of numbers, and performance on secondary tasks.
- [75_Ac_4] Increased fear can decrease accuracy, as demonstrated by an increase in navigation calculation errors by bombers as the aircraft approaches the target in wartime.
- [107_Ac_1] Disaster studies indicate that people tend to act in what they believe to be their best interest, given the limited information available. Shock reactions, panic flight, and anti-social behavior such as looting are relatively rare.
- [107_Ac_2] Citizens confronted with disaster are rational, responsible, and capable of action. Emergency authorities should expect fear (not panic), action, and compliance from the victims and structure messages accordingly.

Cognitive load

- [21_Ac_2] Low redundancy warnings were beneficial in improving response time for pilots under a heavy cognitive load.
- [67_Ac_4] Under the low workload condition, four-line signs resulted in increased minimum headway between the driver's car and the car in front, consistent with decreasing speed. Under the high workload condition, however, minimum headway decreased by an average of one second while reading four-line signs.

Repetition

- [44_Ac_1] Better driver response (faster, correct action) was seen when each of multiple message phases was displayed for a relatively short duration and

Ac	Summary Findings - Action

General Warnings (cont.)

repeated.

- [65_Ac_1] Internally redundant audio messages, with repetition only within the message, resulted in significantly less success in completing a set of instructions (31% of participants) than successively redundant messages, with repetition only of the entire message (57 % of participants).

Familiarity and training

- [29_Ac_1] The evacuees from a fire in a recreational facility were more likely to use routes with which they were familiar, rather than other potentially safer routes.
- [38_Ac_2] Use fire evacuation drills to teach and practice the proper actions (for kinesthetic communicators).
- [41_Ac_1] To direct people to diverge from the normal path, the message in its simplest form should contain the incident type, the location of the incident and the action that is required to be taken.
- [43_Ac_6] Drivers unfamiliar with the route requested information.
- [43_Ac_14] Familiarity with the available routes influenced willingness to modify behavior.
- [66_Ac_3] On the day of the jump, novice parachutists had a significantly higher auditory threshold and higher reactivity to anxiety-provoking words.
- [66_Ac_5] Visual recognition of numbers composed of dots against a distracting background deteriorated as novice parachutists approached the jump point.
- [66_Ac_6] Physiological effects of anxiety are reduced by experience.
- [74_Ac_2] A voice alarm system is capable of directing an evacuation in which speed and proper relocation of personnel do not depend on specific training.
- [75_Ac_5] Expertise and experience mitigate the effects of physical danger on performance.
- [89_Ac_1] Semantic memory, the store of information built by experience, is not affected by age and can be improved with training.
- [97_Ac_2] Interpreters for the deaf need to be part of emergency training and disaster relief assistance programs for the public.
- [98_Ac_6] Training and drills can be used to help familiarize people with emergency procedures.
- [110_Ac_2] Occupants are willing to enter stairwells even where the signage system is unfamiliar.
- [113_Ac_2] Emergency preparedness and planning combined with drills and practice associate emergency responses with more routine and familiar behavior. Personal history matters. Enhanced adaptive response is seen with survivors of previous disasters, are more likely to act or obey authorities quickly rather than seeking additional information.
- [129_Ac_1] The general public's ability to interpret evacuation plan diagrams should be improved through education and training.
- [134_Ac_1] Consistency in the design of DMS highway messages, over time and from region to region, is important for driver comprehension and traffic flow.
- [148_Ac_1] Training may be used in addition to warning messages.

Individual and group characteristics

- [66_Ac_2] The response of individuals to fear and anxiety varies considerably.

Ac	Summary Findings - Action
General Warnings (cont.)	Swedish airline pilots who were slow to perceive threats in pictures presented for very short times were more likely to have accidents.

- [75_Ac_1] Prior knowledge provides context in which meaning is understood and assumptions from which to make inferences. It also affects what details are remembered and how they are organized for reasoning and decision-making.
- [89_Ac_3] When making decisions, older adults consider fewer pieces of information than younger adults and are susceptible to biases.
- [89_Ac_4] Older adults may be more cautious in decision-making than younger adults, but the speed and quality of the decisions appears to be similar.
- [89_Ac_5] The number of feasible options available to older adults may be reduced due to physical disabilities, low financial means, and inadequate social networks.
- [92_Ac_5] Four factors affect individual perception of and response to an emergency: 1) ability to process risk information, 2) access to social networks, 3) incentives to take the warning seriously, and 4) constraints on action.
- [98_Ac_1] Research is needed on rapid onset evacuation of vulnerable populations in the workplace.
- [98_Ac_2] Conditions that affect emergency response include mobility disabilities, fatigue, respiratory conditions, emotional and cognitive difficulties, and vision or hearing loss. These conditions may be temporary or permanent.
- [108_Ac_1] People in vulnerable populations are often marginalized, and are less likely to receive warnings and to respond appropriately in emergency situations. Vulnerable populations include those based on age, race and ethnicity, socioeconomic status, gender, physical disability (e.g., mobility, hearing, and vision), language, work, and isolation.
- [108_Ac_2] Socioeconomically disadvantaged families may lack the ability to receive forecasts and warnings, a safe home environment and the transportation to evacuate.
- [108_Ac_3] Gender roles in family and work influence vulnerability. Women appear to be more responsive to warnings, but may find themselves and their children in more dangerous situations.
- [108_Ac_4] Elderly people may have less access to social networks and technology, and may find it more difficult to evacuate due to mobility problems.
- [108_Ac_5] Children may be on their own or in settings that are unprepared to handle the emergency. Conversely, schools may provide an avenue to provide information to other family members.
- [108_Ac_6] Tourists are unfamiliar with the environment, available resources, and possibly the language.
- [108_Ac_7] Interactions among disadvantages, such as age, poverty, disability, and isolations, multiply the risk.
- [139_Ac_1] Racially and ethnically diverse communities remain vulnerable to disproportionate losses of life and property in disasters.
- [139_Ac_3] The specific vulnerability of each community and the availability of resources to carry out various types of emergency response, such as evacuation, isolation, and quarantine, must be understood.

Ac	Summary Findings - Action
General Warnings (cont.)	*Effects on behavior* • [13_Ac_1] During an evacuation, occupants of a building preferentially leave through either a familiar exit or an exit open to the outside. • [13_Ac_2] In an emergency, staff may direct customers to safe exits by opening them to the outside. More will respond if the exit is visible from many points. • [43_Ac_3] When relevant information was provided prior to an incident, motorists acted upon it. • [43_Ac_4] Approximately 50 % modified behavior accordingly when provided with new information via a sign. • [43_Ac_11] Route choice can be influenced by information, assuming alternatives exist. • [45_Ac_1] Real-time information systems influence behavior by encouraging passengers to run when the time before the arrival of the next train is small. • [86_Ac_1] A significant time delay after an evacuation warning almost always occurs, due to psychological preparation (including information seeking) and logistical preparation. • [98_Ac_3] The most accessible exits from a building may be crowded by nondisabled evacuees. • [109_Ac_1] Voice communication informing passengers of the type of incident, its location, and what to do was as successful as staff in provoking action, with a delay time of only 15 seconds after the voice message.
Visual Warnings	*Design* • [136_Ac_1] Improper use of highway signage (PCMS) can destroy their credibility, create confusion, and promote improper driving behavior. *Content* • [40_Ac_1] Advice telling the readers what to do in the situation should be placed in the last line of the DMS message. The reason for the advice must also be supplied, or people will ignore it. • [124_Ac_2] People prefer countdown pedestrian signals to a flashing hand signal with no countdown because the additional information provides more control over the decision. *Word choice* • [43_Ac_12] Abbreviations can be used (with the understanding of 75 %), with proper nouns and familiarity aiding understanding. • [131_Ac_1] People are more likely to comply with a warning sign when the word "DANGER" is used compared to "STOP" or "CAUTION." *Symbols* • [69_Ac_1] Compliance with safety instructions was higher for a warning message that combined a pictograph with text than for either pictograph or text alone. • [141_Ac_1] Participants walked faster to their destination when symbols were used in wayfinding signs than when they were guided by multilingual signs. • [150_Ac_2] Adding a pictorial to a written warning in instructions for a chemistry experiment was not found to increase compliance.

Ac	Summary Findings - Action
Visual Warnings (cont.)	*Message length* • [67_Ac_1] One or two-line monolingual and two-line bilingual variable message signs (VMS) did not affect driving speed or distance to the car in front (headway). • [67_Ac_2] Both four-line monolingual and four-line bilingual signs affected driver performance by slowing travel speed (by about 11 kph) while approaching and passing the sign. • [67_Ac_5] The response time to carry out instructions from the VMS road sign increased with the number of lines of text. • [81_Ac_2] Variable message signs (VMS) with a double message line were better understood and resulted in shorter response times than those with single or triple lines. • [137_Ac_1] If less time than two seconds per informational unit is available for reading a DMS, drivers may slow down, focus less attention from their driving, and contribute to traffic problems. • [144_Ac_2] The response time was shorter for variable message signs (VMS) when there were fewer lines in the message. *Complexity* • [110_Ac_1] Simple sign designs are more effective than signs with complex designs, especially those presenting a lot of information. *Color* • [81_Ac_1] A two-color scheme for VMS road signs resulted in the shortest response time and was preferred by drivers. • [102_Ac_1] Green flashing lights attracted people to an exit over a standard emergency exit sign, even when the starting point was further away. A green or orange strobe light was less effective. • [102_Ac_2] Green is more strongly associated with emergency exits than other colors. This association translated into action during evacuation experiments; i.e., exits marked by green flashing lights were used preferentially. *Font* • [25_Ac_2] Risk communication research has found little evidence supporting trends in danger symbols, font choices, and other typographical features, including caps, bold, colors, exclamation marks, and spacing, that significantly affect attention, perception, or behavioral response. *Placement* • [94_Ac_2] In public spaces, signs should be placed out of people's reach if possible in order to avoid vandalism. • [136_Ac_2] Roadway signs need to be positioned far enough upstream of the decision point to provide adequate warning but close enough that the message is relevant. • [150_Ac_1] Warnings presented in task instructions for a chemistry experiment resulted in greater compliance than warnings posted on the wall. *Message display rate* • [144_Ac_1] The response time was shorter for simultaneously displayed

Ac	Summary Findings - Action

Visual Warnings (cont.)	(discrete) messages than for the same message displayed sequentially in multiple phases. Drivers preferred the discrete displays.
	Lighting – Static
	• [1_Ac_2] People traveled faster at higher lighting levels (3.0 lx compared to 1.0 lx), both with and without smoke.
	• [29_Ac_2] At low luminance levels, travel times and arithmetic errors increased.
	• [62_Ac_1] The movement speed through a smoky escape path is relatively insensitive to luminance.
	Lighting – Dynamic
	• [101_Ac_1] Static flashing lights were able to attract the attention of people unfamiliar with an escape route but were unsuccessful in directing movement in a particular direction. An arrow may have helped.
	• [102_Ac_3] Green flashing lights are more strongly associated with safety than green strobe lights. However, this did not necessarily influence egress behavior.
	• [102_Ac_5] Green flashing lights are able to overcome participant proximity; i.e., exit signs with green flashing lights attracted people who were closer to other exits.
	• [102_Ac_6] Green flashing lights are able to overcome a less advantageous angle of observation; i.e., exit signs with green flashing lights on a side exit attracted people who could see an exit at the end of the corridor.
	• [103_Ac_1] Green flashing lights placed next to some exits in an office building were not mentioned by evacuating occupants as a factor influencing exit choice. Although more people evacuated through those exits, the difference was not significant.
	• [103_Ac_2] In a movie theater, all those evacuating used the exits with green flashing lights in preference to those with normal exit signs. Flashing lights may be more effective in influencing the choice of exits in less familiar places.
	• [130_At_1] Strobe lights alone are not able to attract sufficient attention to overcome sleep.
	• [149_Ac_1] Safety signs using dynamic LED displays were not found to increase compliance over static LED displays.
	Lighting –Wayguidance systems
	• [62_Ac_2] The highest probability of making correct route decisions (98 %) was found for a wayguidance system consisting of a combination of tactile and green light cold cathode lighting systems.
	• [62_Ac_3] Electrically powered and photoluminescent low location lighting wayguidance systems resulted in the lowest percentage of evacuees (69 % to 79 %) making correct escape route decisions.
	• [72_Ac_1] Traveling flashing lights provide effective wayguidance in both light and thick smoke, although smoke decreases the effectiveness.
	• [72_Ac_2] The effectiveness of traveling flashing lights is more sensitive to the spacing of the lights than to the traveling speed.
	• [72_Ac_3] Decreasing the spacing improves the effectiveness of traveling flashing guidance lights.
	• [72_Ac_4] When the traveling speed is fast and the spacing between lights is

Ac	**Summary Findings - Action**

	small, flashing wayguidance lights can appear to point in the opposite direction.
Visual Warnings (cont.)	• [94_Ac_1] In the United States, there is a right hand bias, such that without other information people tend to turn to the right. This bias should be considered when designing wayfinding signage.
	• [102_Ac_4] Because flashing lights attract attention and green is associated with safety, green flashing lights can be used to overcome familiarity issues and direct people to an emergency exit.
	• [103_Ac_3] Green flashing lights at exits can be used to influence exit choice, but their effectiveness may depend on the setting.
	• [106_Ac_1] Two continuous visual wayfinding systems (incandescent and PLM) and a combination of visual semi-continuous and tactile systems outperformed other wayfinding systems in terms of route adoption and evacuation times.
	• [106_Ac_4] A wayfinding system with continuous markings is superior to traditional exit signs.
	• [154_Ac_1] Wayguidance systems with lighting tracks mounted on the wall resulted in significantly faster walking speeds in smoke than overhead lighting. This was true for stairwells, stair landings, and corridors.
	• [154_Ac_2] Walking speeds for a wayguidance system that illuminated at lower than recommended levels were found to exceed speeds under normal overhead lighting by more than a factor of three in a long corridor.
	• [154_Ac_3] Directional information incorporated into wayguidance systems may contribute significantly to their effectiveness and compensate for lower levels of illumination. Outlining the exit door may also improve performance.
	Smoke
	• [1_Ac_1] People walked slower in the presence of smoke.
	• [101_Ac_2] In smoke-filled conditions people often use walls for guidance, suggesting that a tactile system would be useful in wayfinding.
	• [106_Ac_5] Continuous tactile wayfinding systems are recommended for optical density exceeding 1.5, visual continuous systems are recommended for optical density between 0.1 and 1.5, and traditional exit signs are recommended only for optical density under 0.1.
	Surroundings and context
	• [149_Ac_2] The personalized sign that was located the farthest distance away but in the least cluttered area resulted in the greatest compliance.
	Individual and group characteristics
	• [1_Ac_3] Younger adults walked faster than older adults at lower levels of floor illuminance.
	• [43_Ac_5] The effectiveness of a sign on influencing response depended on whether the motorists had to be at a place at a particular time or had some flexibility.
	• [43_Ac_9] Women appeared to be less likely to be influenced by information than men.
	• [124_Ac_1] Older people and women are more likely to show caution when crossing the street.
	• [144_Ac_3] Older drivers showed slower response times to variable message

Ac	Summary Findings - Action
Visual Warnings (cont.)	signs (VMS) and less response accuracy than younger drivers. • [144_Ac_4] Female drivers showed slower response times to variable message signs (VMS) but higher response accuracy than male drivers. *Bilingual messages* • [67_Ac_3] Recovery of the original speed was slower for four-line bilingual VMS road signs, possibly reflecting a longer time to process the information.
Auditory Warnings	*Design* • [41_Ac_2] Short form and staccato messages are easier to recall. The first statement in an alert message should be an attention statement to alert the population that a message is about to be given. The second statement should describe the problem. The final statement should present the action that is required to react to the problem. For maximum recall, people must hear the message at least twice. • [74_Ac_1] A voice alarm system consisting of an alarm tone followed by a female voice to attract attention and a male voice to deliver instructions resulted in a successful area evacuation, with personnel vacating to receiving floors either above or below their own. *Synthesized speech* • [21_Ac_1] Pilots in a simulated flight task study responded faster to synthesized speech warnings. • [68_Ac_1] Synthesized voice messages can both alert and inform simultaneously. *Content* • [74_Ac_3] An effective voice alarm system includes the nature of the emergency and directions (what is happening, what to do, and why alternatives should not be used), assurance that someone is in charge, specifics, information to promote calmness, and repetition of essential components. *Message length* • [65_Ac_2] An eight-unit set of instructions (four turns and four street names) was completed successfully by 36 % of participants, while a six-unit set (three turns and three street names) was completed successfully by 53 %. • [65_Ac_3] For routes that require more than four turns and recall of more than four street names, a trailblazer system is recommended. *Urgency* • [56_Ac_1] Auditory signals with high frequency, fast speed, and high sound levels result in the fastest response times. *Pitch* • [122_Ac_1] A male voice is more persuasive for presenting specific instructions that need to be followed. *Sound level* • [38_Ac_1] To attract attention and get people to act, add to discomfort through loud alarms (for kinesthetic and auditory communicators).

Ac	Summary Findings - Action
Auditory Warnings (cont.)	• [55_Ac_1] For operators engaged in a task in a noisy environment, response time was shorter by 60 milliseconds for a higher pulse sound level compared to that for a lower pulse level. This is enough to improve the performance of a fighter jet pilot. *Timing and delivery rate* • [2_Ac_1] A standard temporal pattern for an audible emergency evacuation signal has been developed by ANSI, with the sole meaning of "evacuate the building immediately."

APPENDIX B   Source Reviews

 # 1. Akizuki Y, Tanaka T, Yamao K. (2009). Calculation model for travel speed and psychological state in escape routes considering luminous condition, smoke density and evacuee's visual acuity. *Fire Safety Science, 9*, 365-376.

This experiment was designed to find a way to calculate evacuation behavior based on a person's ability to see in smoke. The space for the experiment was divided into three parts: the adapting space, the travel space and the questionnaire space. In the travel space, 20 W florescent lights were installed in the ceiling at 3.6 meter intervals and infrared sensors were placed on the wall every three meters. The experiment tested people's evacuation performance in luminous conditions with and without smoke. Under smoky conditions, t.he smoke density (optical density per meter of path length) varied from 0.2 m^{-1} to 1.8 m^{-1}. The visibility of the testing environment was measured using visual angle, background luminance, contrast and visual acuity. The 30 young adults and 30 older adults participating in the study were each given a visual acuity test beforehand. After being exposed to the illuminance of the adapting space for 2.5 minutes, participants entered the travel space and walked until they arrived at the questionnaire space. There they were asked about the visibility of the travel space, the ease of walking, and their feelings of anxiety while walking.

No significant change of speed was observed as participants walked from the beginning of the travel space to the end. Subjects walked slower in smoke than without smoke. The speed of the participants increased with higher illuminance levels on the floor. In the presence of smoke, speed was higher for 3.0 lx lighting than for 1.0 lx. An interaction was found between age and illuminance, with the younger age group travelling faster at lower levels of floor illuminance.

Key findings:
- [1_Ac_1] People walked slower in the presence of smoke.
- [1_Ac_2] People traveled faster at higher lighting levels (3.0 lx compared to 1.0 lx), both with and without smoke.
- [1_Ac_3] Younger adults walked faster than older adults at lower levels of floor illuminance.

2. American National Standard ANSI ASA S3.41-1990. (R2008). *Audible Emergency Evacuation Signal.*

S		Pc	At	Co	Cr	Ps	Ac

An international standard for an audible signal that means "evacuate the building immediately" was published by the International Standards Organization (ISO) in December 1987 as ISO-8201. Such a signal, if understood universally, may reduce confusion and reduce the time for evacuation. The American National Standards Institute (ANSI) developed a standard signal that conforms to this ISO standard. Since no sound can reliably penetrate all the conditions of background noise, the approach taken by the ANSI standard is to specify a recognizable temporal pattern for the 'on' and 'off' periods, assuming the existence of an audible alarm signal that can penetrate a particular background noise pattern in a given building. An advantage of this approach is that the standardized temporal pattern can also be applied to visual and tactile signals to aid those who have impaired hearing. Visual and tactile signals are also useful in cases in which the background noise is so intense that no signal is capable of reliable penetration.

According to the standard, this emergency evacuation signal shall "indicate imminent danger and signify unambiguously that evacuation from the building is immediately necessary." It is only for use in the zones to be evacuated immediately, and must never be used in zones with different prescribed emergency action, e.g., relocation to a safe place or defense in current location. The standard specifies two parameters for the audible emergency evacuation signal: 1) the temporal pattern and 2) the required minimum sound pressure level at all places within the zone intended to receive the signal. The specific requirements are:

- A 'three-pulse' temporal pattern consisting of an 'on' phase sounding for 0.5 seconds ± 10 % followed by an 'off' phase for 0.5 seconds ± 10 %, repeated for a total of three successive 'on' periods and followed by an 'off' phase for a duration of 1.5 seconds ± 10 %;
- Repetition of the signal for at least 180 seconds, or for the length of time appropriate to carry out the evacuation;
- Sound pressure during the 'on' phase that exceeds the highest level of the background noise averaged over 60 seconds and is at least 65 dBA;
- Visual and tactile signals to be provided if the averaged background noise exceeds 110 dBA;
- For arousing people from sleep, sound pressure of at least 75 dBA at the normal head position of the sleeping person, with all doors closed.

A key word or phrase (e.g., FIRE!) may be inserted during the 'off' phase between the three pulses. Examples of the application of the temporal pattern are included in the appendix of the ANSI standard.

Key findings:
- [2_Co_1] [2_Ps_1] [2_Ac_1] A standard temporal pattern for an audible emergency evacuation signal has been developed by ANSI, with the sole meaning of "evacuate the building immediately."
- [2_Pc_1] [2_At_1] [2_Co_2] The temporal pattern for the ANSI standard audible emergency evacuation signal can also be applied to visual and tactile signals to aid occupants with impaired hearing and occupants in zones with intense background noise.

- [2_Co_3] The ANSI standard audible emergency evacuation signal consists of an 'on' phase for 0.5 seconds followed by an 'off' phase for 0.5 seconds, repeated for a total of three successive 'on' periods and followed by an 'off' phase for 1.5 seconds.
- [2_Pc_2] [2_At_2] The ANSI standard audible emergency evacuation signal assumes the existence of an audible alarm signal that can penetrate a particular background noise pattern in a given building.
- [2_Pc_3] [2_At_3] The sound pressure during the 'on' phase of the ANSI standard audible emergency evacuation signal must exceed the highest level of the background noise averaged over 60 seconds and be at least 65 dBA;
- [2_Pc_4] [2_At_4] For arousing people from sleep, the sound pressure of the ANSI standard audible emergency evacuation signal must be at least 75 dBA at the normal head position of the sleeping person, with all doors closed.
- [2_Ps_2] A key word or phrase (e.g., FIRE!) may be inserted during the 'off' phase between the three pulses of the ANSI standard audible emergency evacuation signal.

3. Anttila V, Luoma J, Rämä P. (2000). Visual demand of bilingual message signs displaying alternating text messages. *Transportation Research Part F, 3*, 65-74.

T		Pc	At	Co	Cr	Ps	Ac

The Finnish National Road Administration started testing bilingual variable message signs (VMS) in 1997, with text message road signs alternating between Finnish and Swedish. An evaluation of this approach sought to determine whether alternating bilingual messages on a single message board are as easily read and understood as a separate sign for each language.

The first phase of the evaluation considered driver acceptance. In this study, 350 drivers were interviewed after passing a road sign. Of these, 76 % recalled the sign correctly. Of the drivers that recalled the sign correctly, 90 % considered the time provided for reading the message in their native language (2.0 seconds) to be appropriate. However, potential issues were raised for older drivers. In a field study of driver eye movement, 44 people were recruited to evaluate three VMS types. The first was a sign saying 'LOOSE GRAVEL" alternately in Finnish and Swedish for 2.0 seconds in each language, separated by a blank display for 0.5 seconds. The second sign displayed the same message simultaneously in both languages. The third sign displayed air and road surface temperatures in Finnish. A general warning symbol was also displayed for the alternating and simultaneous bilingual VMS. The independent variables included sign type and subject age. The dependent variables consisted of the number of eye fixations, total gaze duration and duration of the longest fixation.

The results showed no significant difference between the three signs in the percentage of subjects fixated on the sign, with 89 % to 100 % of drivers fixated on the sign in each case. No significant difference was found between the two bilingual signs for gaze duration, although a slight tendency toward longer gaze duration was found for the alternating message sign. Gaze duration and longest fixation duration were significantly longer for both bilingual VMSs than for the monolingual air and road surface temperature display. No visual response effect was found that suggested that alternating bilingual messages on this particular VMS were more demanding the same message shown in both languages simultaneously. However, the number of gaze fixations was higher than that found for simple signs in previous work, which indicates that the visual demand of VMS may be higher. Although older drivers tended to show longer gaze durations than younger drivers, the effect was not significant (and, according to the authors, was also confounded by a higher level of bilingual knowledge among the younger drivers).

Key findings:
- [3_Co_1] Although slightly longer gaze duration was found for a VMS displaying an alternating bilingual message than for a VMS displaying the messages in both languages simultaneously, the effect was not statistically significant.
- [3_Co_2] For alternating bilingual messages on a VMS, 90 % of the drivers considered 2.0 seconds an appropriate length of time to read the message in their native language.
- [3_Co_3] Gaze duration and longest fixation duration were significantly longer for bilingual VMSs giving a safety warning than for a monolingual display of air and road surface temperatures.

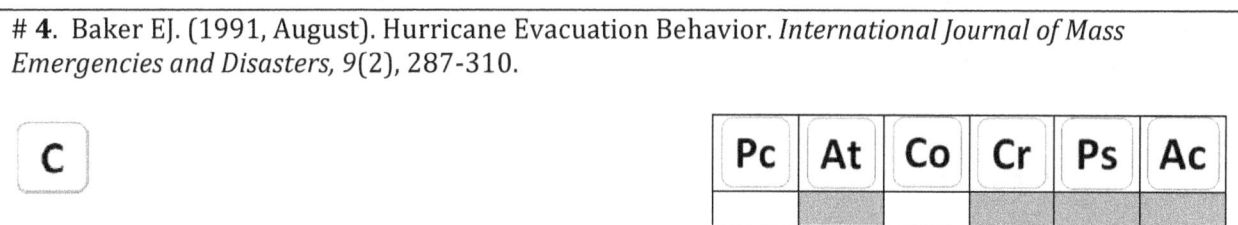

 # 4. Baker EJ. (1991, August). Hurricane Evacuation Behavior. *International Journal of Mass Emergencies and Disasters, 9*(2), 287-310.

C

Pc	At	Co	Cr	Ps	Ac

Public response to hurricane warnings has been studied through sample surveys of U.S. coastal residents involved in twelve hurricanes between 1961 and 1989. People tend to stay in their residential area during a storm when they feel safe and to leave when they feel unsafe, so it is important to determine the factors that influence the perception of safety. Evacuation notices from public officials are effective in several ways: they convince people that they are in danger, they represent the voice of authority, and some (15 % to 25 %) believe that there will be a penalty for noncompliance. Informal networks of friends and acquaintances may reinforce safety messages and help to convince people to leave, or may pressure people to stay. The survey database has been analyzed to determine the effects of the physical hazard, warning notices, hurricane experience, hurricane awareness, length of local residence, past evacuation behavior, and demographics.

Evacuation rates from high-risk areas are usually high (an average of 83 % in seven hurricanes). In moderate-risk areas, evacuation rates are usually between 55 % and 65 %, which may indicate a need for warning messages to target these areas specifically. People evacuating from low-risk areas (20 % to 40 % from locations near the coast), known as the "shadow" evacuation, contribute to traffic congestion and increase the time to evacuate those in danger. Messages for these people to stay in their homes are problematic because they carry legal risk. Residents of mobile homes are more likely to evacuate, as are residents who believe their own house is in danger of flooding or wind damage.

Warning messages from public officials are the strongest factor in evacuation, with the possible exception of risk area. Official evacuation notices are most effective when they are specific about the area affected (convincing people the message applies to them), are delivered in a personalized way (including door-to-door), describe the specific actions to be taken (what to do and where to go), and are worded aggressively and delivered in an urgent manner. Dissemination by the news media is not sufficient. Door-to-door communication by authority figures resulted in 97 % evacuation rates from beaches in two hurricanes. People hearing official evacuation notices are more than twice as likely to leave (the difference in evacuation rate between those who heard and those who didn't was 84 % to 20 % in one hurricane and 88 % to 8 % in another). The first or primary source of information about the storm, knowledge about hurricanes and hurricane safety, and demographics are not strong predictors of response.

In hurricane evacuations, past exposure to hurricanes, including false experiences and unnecessary evacuations (the "cry wolf" effect), is generally not a determining factor in whether individuals choose to evacuate or not. People that have evacuated before are more likely to evacuate again. In 1980, the city of Galveston evacuated even before the National Weather Service had issued a warning. Even though the city was not in the eventual path of the hurricane, 80 % of people that evacuated said that they would do the same thing in the future under similar conditions and 10 % said that they would leave even sooner. After hurricane Diana, only 5 % of evacuees said they would not do so in the future. Fewer than 5 % of the people that did not evacuate cited previous unnecessary evacuations as the reason that they did not evacuate. In 1985, Panama City and

Panama City Beach were evacuated three times, with approximately the same percentage (78 %) evacuating from the beach area each time. While the mainland area saw a slight decline in the percentage of evacuees over the three evacuations, the differences were not statistically significant.

Key findings:
- [4_Cr_1] [4_Ps_1] Evacuation notices from public officials are effective in several ways: they convince people that they are in danger, they represent the voice of authority, and some (15 % to 25 %) believe that there will be a penalty for noncompliance.
- [4_At_1] [4_Ac_1] Official evacuation notices are most effective when they are specific about the area affected, are delivered in a personalized way, describe the specific actions to be taken, and are worded aggressively and delivered in an urgent manner.
- [4_At_2] [4_Cr_2] [4_Ps_2] Door-to-door communication by authority figures resulted in 97 % evacuation rates from beaches in two hurricanes.
- [4_Cr_3] [4_Ac_2] People hearing official evacuation notices are more than twice as likely to leave.
- [4_Cr_4] The "cry wolf" syndrome, in which unnecessary evacuations result in unwillingness to leave during the next event, is not supported by the evidence.

Key recommendations:
- [4_Cr_5] Because "cry wolf" syndrome is not supported by the evidence from real natural disasters, evacuation notices should be given as early as possible.

5. Baldwin CL, Struckman-Johnson D, Galinsky AM, Williams EH. (1999). Presentation level and the attentional workload of comprehending verbal messages: Implications for design of verbal warnings. *Proceedings of the 43rd Annual Conference of the Human Factors and Ergonomics Society*, 893-897. Houston, Texas: Human Factors and Ergonomics Society.

E		Pc	At	Co	Cr	Ps	Ac

Auditory warnings are often used in high stress, multitasking environments. They are designed to alert and provide information to an operator who may be performing multiple tasks. The intent of this study was to look at how the sound level of a warning alters the attentional demands placed on an operator in a multi-task environment. By considering the volume at which a warning will gain the attention of a person concentrating on another task, the study addressed whether or not a warning is sufficiently effective as long as it is audible. The experiment tested the auditory processing of the truthfulness of simple sentences and of mental arithmetic while the operator was engaged in a driving task.

In the experiment, participants were given dual tasks consisting of simultaneously performing a simulated driving test and responding to an auditory prompt. The auditory stimuli were presented at sound levels of (65, 55, or 45) dB. In the sentence-processing experiment, participants heard a short sentence via headphones and had to verbally respond yes or no depending on whether the statement was true or false. In the auditory arithmetic experiment, participants heard a two digit number and had to subtract the smaller digit from the larger digit and respond with the result. The results showed that participants responded incorrectly more frequently in both tests as the sound level decreased, even though the auditory level was always in the range considered optimal for the presentation of speech stimuli in quiet environments. This indicates that results from studies of performance carried out under low workload conditions will not necessarily translate to high workload conditions. Consideration of the presentation level for auditory signals may enhance the performance of older people, who may have some hearing loss.

Key findings:
- [5_At_1] [5_Co_1] Under a high cognitive workload, singular, instructive auditory messages were processed more accurately at higher sound levels (65 dB compared to 45 dB), even though both sound levels were in the range considered optimal for speech stimuli in quiet environments.

Key recommendations:
- [5_At_2] [5_Co_2] Higher sound levels for auditory warnings may be especially helpful in enhancing the performance of older people with some hearing loss.

 # 6. Ball M, Bruck D. (2004, September). The Effect of Alcohol Upon Response to Fire Alarm Signals in Sleeping Young Adults. *Proceedings of the Third International Symposium on Human Behaviour in Fire.* Belfast, Northern Ireland. J Shields (Ed.), Interscience Communications, London, UK, 291–302.

F			Pc	At	Co	Cr	Ps	Ac

Alcohol consumption increases the fire fatality risk for people in all age groups, and is an important issue for young adults, who tend to be less experienced drinkers and deeper sleepers. The intent of this research was to investigate the effect of alcohol on arousal threshold for three alarm signals.

The participants in this study were 12 students (seven males, five females) in the age range of 18 to 25 years old from Victoria University. The minimum sound volume was sought at which students would awaken to any of three alarm signals when they were sober or had a blood alcohol concentration (BAC) of 0.05 or 0.08. The first alarm signal was a female voice warning of danger from fire and instructing the subject to wake up and investigate. The second alarm was a high frequency signal from a standard Australian home alarm. The third alarm was a low frequency Temporal-Three (T-3) pattern from ISO 8201. Subjects were tested on a different night for each level of intoxication with three to seven days between trials. The alarm signal was played during the deepest stage of sleep (stage four) in each participant's bedroom, with a background noise level of about 50 dBA. To familiarize participants with the alarms, all signals were played for them before they went to bed. Each alarm was initially sounded at 35 dBA for 30 seconds and then increased by 5 dBA until 95 dBA was reached, at which point the alarm was allowed to sound for 3.5 minutes. Results are shown in Table 11. The female voice and T-3 signals were equally effective, and significantly better than the standard alarm in waking the subjects. Subjects were more likely to sleep through the alarm as alcohol level increased, with 36 % of trials at 0.05 BAC resulting in no response below 95 dBA. The sex of the student was not a significant factor.

Table 11: Sound levels required to awaken people who are sober or under the influence of alcohol.

Alarm Signal	Mean Threshold Waking Level (dBA)		
	Sober	0.05 BAC	0.08 BAC
Female voice	59.58	79.33	80.50
Standard alarm	72.50	85.00	87.50
T-3 pattern	59.17	78.25	82.92

Key findings:
- [6_Pc_1] [6_At_1] Female voice and T-3 alarms are equally effective at arousing young adults from sleep while sober and after alcohol consumption. The standard Australian home alarm is significantly less effective, requiring a sound level more than 20 dBA over the background noise to awaken sober individuals.
- [6_Pc_2] [6_At_2] Alcohol consumption significantly worsens the ability to awaken to auditory alarms. The most effective alarms require about 80 dBA to arouse young adults with 0.05 BAC, compared to 60 dBA while sober.
- [6_Pc_3] [6_At_3] Young adults are more likely to sleep through an auditory alarm as alcohol level increases, with 36 % of trials at 0.05 BAC resulting in no response below 95 dBA.
- [6_Pc_4] [6_At_4] Alcohol effects when arousing young adults from sleep with auditory alarms do not depend on gender.

 # 7. Ball M, Bruck D. (2004, September). The salience of fire alarm signals for sleeping individuals: A novel approach to signal design. *3rd International Symposium on Human Behaviour in Fire.* Belfast, Northern Ireland. J Shields (Ed.), Interscience Communications, London, UK, 303-314.

F		Pc	At	Co	Cr	Ps	Ac

Previous research points to important individual differences in response to auditory signals during sleep, including: the auditory arousal threshold (AAT) of sleepy and sleep-deprived individuals does not decline during the night as is seen in alert individuals; children of ages six to 17 years and sleep-deprived young adults do not reliably awaken to alarm signals; increasing age is associated with increasing frequency of awakening and decreasing stimulus to awaken; and individual differences cause greater variability than sleep stage or age. Previous research also shows that emotional content is processed by the sleeping brain and reduces the AAT. A naturalistic sound has been found to be more effective than beeping, and the perception of urgency is higher for appropriate words and the human voice, especially female. The purpose of this pilot study was to select a meaningful signal with emotional content to be used in future alarm studies.

The first of two phases in this study was a survey for people associated with Victoria University asking them what sounds would 1) make them feel a negative emotion, 2) draw their attention while sleeping, and 3) feel the need to investigate upon awakening. Responses were given by 163 people, who provided 1447 answers to the three open-ended questions. For negative emotion, the top 15 most frequent responses were classified as: expressions of human emotion (28.25 % of responses), manufactured alarm sounds (12.3 %), and other sounds conveying potential emotional distress (6.9 %). For sounds that would draw attention while sleeping, the top 16 responses were classified as: expressions of human emotion (18.0 %), manufactured alarm sounds (17.2 %), and other potentially unusual or naturalistic alerting sounds (18.9 %). For sounds that would make the person want to investigate after awakening, the top 15 responses were classified as: expressions of human emotion (18.4 %), manufactured alarm sounds (9.6 %), and naturalistic alerting sounds (35.6%).

For the second phase, three new alerting signals were developed and tested: a naturalistic fire sound that included glass breaking, a female voice urgently reporting danger from a fire and telling the subject to get up and investigate, and a combination that shifted between these two signals. The duration of all three alarms was 30 seconds. The subjects were four male and four female young adults in the age range 18 to 25 years, all of whom self-reported as deep sleepers. Each alarm tone was started at 35 dBA followed by 30 seconds of silence and repeated with 5 dBA increments until 95 dBA, at which point the sound was played continuously for 3.5 minutes. Results are shown in Table 12 below. The female voice was found to be most successful (lower EEG response time and AAT) at awakening the subjects, and no advantage was found for shifting between the two signals. Standard deviations were large, which may reflect the small sample size and large individual variability.

Table 12: Sound levels required for three types of alarm signal to wake young adults.

Alarm Signal	AAT (dBA)	EEG Response Time (s)
Naturalistic house fire	50.00	198.00
Female voice	47.50	167.00
Combined fire/voice	51.25	203.13

Key findings:
- [7_Pc_1] [7_At_1] Children of ages six to 17 years and sleep-deprived young adults do not reliably awaken to alarm signals [from previous research].
- [7_Pc_2] [7_At_2] An urgent message delivered by an emotional female voice is more effecting in arousing young adults than either sounds of a house fire with breaking glass (2.5 dBA higher) or a shift between these two signals (3.75 dBA higher).

 # 8. Barnes LR. (2006, July). *Public perceptions of flash flood false alarms: A Denver, Colorado case study* (Undergraduate Paper). Colorado Springs, Colorado: Natural Hazards Center, University of Colorado at Colorado Springs.
http://colorado.edu/hazards/awards/paper-competition/Barnes2006.pdf

 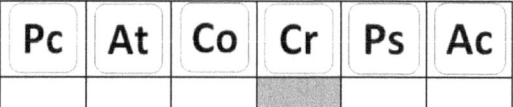

The goal of this case study was to examine the effect of false alarms on public response to natural hazard warnings, in order to improve the effectiveness of public warning systems. Predictions of the intensity and location of natural disasters such as blizzards, hurricanes, and floods are not always accurate, and evacuations and other protective actions taken in response to a warning may be carried out in vain. Although the conventional wisdom is that such false alarms reduce people's willingness to respond to the next such event (the "cry-wolf" effect), research indicates that this may not be the case if the event and the reason for the warning are understood. In fact, false alarms may provide a learning opportunity for both the public and emergency personnel. Previous research indicates that women have higher levels of risk perception (a strong predictor of evacuation compliance), are better prepared, and are more likely to believe warning messages. Although older people have been found to be less likely to receive warnings and more likely to be injured or die in a disaster, the research is inconclusive on whether they are less likely to take protective actions. This study surveyed residents of an area that floods periodically to examine their attitudes toward false alarms and the effects of demographics on these attitudes.

Surveys were mailed between September 2004 and January 2005 to 2800 Denver residents that lived in flood plains. The response rate was 16.5 %. Percentages of respondents in different demographic groups that agreed with two statements related to false alarms are shown in Table 13. Overall a large majority (78 %) said that they would prefer having more warnings even if some of them turned out to not be accurate. Statistically significant relationships were found with gender (more females than males agreed) and with age (agreement increased with age). A large majority (77 %) also said that one or two false alarms would not reduce their confidence in future alarms, with more women than men agreeing and a significant increase in agreement with age.

Table 13: Percentage agreeing with survey questions related to false alarms.

Survey Question	Gender (% Agree)		Age (% Agree)			
	Male	Female	18-35	36-55	56-75	75+
Prefer more warnings even if it means more false alarms or close calls	73	83	67	81	80	93
One or two false alarms or close calls would [not] reduce confidence in future warnings	73	81	68	80	76	90

Emergency planners should not be reluctant to issue warnings given the general preference of the public for more warnings and the acceptance of the possibility of false alarms. Planners should also take into account the differences in attitude between men and women and between old and young. Women and elderly individuals tend to be more vulnerable in a disaster, which may be a factor in why they also tend to prefer more warnings and take them more seriously.

Key findings:
- [8_Cr_1] Women and older residents living in a flood plain were found to be more likely than men and younger residents to prefer more warnings and accept the possibility of false alarms.
- [8_Cr_2] The "cry wolf" effect is not supported by evidence. Overall, 78 % of residents living in a flood plain would prefer more warnings even if some of them turned out to not be accurate. A large majority (77 %) agreed that one or two false alarms would not reduce their confidence in future alarms.

Key recommendations:
- [8_Cr_3] Emergency planners should not be reluctant to issue warnings given the general preference of the public for more warnings and the acceptance of the possibility of false alarms.

 # 9. Barshi I, Healy AF. (1998). Misunderstandings in voice communication: Effects of fluency in a second language. In AF Healy, LE Bourne Jr. (Eds.), *Foreign language learning: Psycholinguistic studies on training and retention* (pp. 161-192). Mahwah, NJ: Lawrence Erlbaum Associates.

L		Pc	At	Co	Cr	Ps	Ac

In this set of experiments, the difference in comprehension between native and non-native English speakers was studied by how well they could carry out a set of commands. The variables included the number of commands, the speed of delivery, the length of the pause between words, the word speed, and the presence of extraneous words in the commands.

The 18 participants in these studies were undergraduates at three levels of English proficiency: native speakers, high proficiency non-native speakers, and low proficiency non-native speakers. The proficiency of the latter two groups was determined by scores on a test similar to TOEFL listening comprehension scores. Each group was comprised of an equal number of subjects. The task required subjects to follow instructions to move through a 3-D grid. The instructions were presented as a set of commands to move a certain number of paces left/right, up/down, or forward/back. Each set of commands consisted of between one and six commands, which were delivered at a fast or slow speed, with or without a pause between each word. Subjects listened to the entire set of commands first before starting to execute them. A total of six sets of commands were carried out by each subject.

In the first two studies, each command consisted of the direction of movement, the number of paces, and two syntax words (e.g., "turn right two squares"). Experiment one varied the length of the pause between words, with either no pause or a pause of 750 milliseconds. Experiment two varied the word speed, compressing the words to either 50 % or 75 % of their original duration and providing no pause between words. Experiment three decreased the length of each command by eliminating the two syntax words. For example, "turn right two squares" became "right two," thus removing redundant information.

In each experiment, accuracy was found to increase with practice and decrease with the number of commands. Performance was found to deteriorate when the number of commands in a set exceeded three for native speakers and two for low proficiency non-native speakers. No statistically significant effect was observed for pause length, word speed, or elimination of redundant words in the commands. However, significance was found for the interaction of redundancy and command length, with accuracy on longer sets of commands increasing when redundant words were eliminated. Based on these findings, the authors recommend no more than two or three commands be given in a single message.

Key findings:
- [9_Co_1] Performance was found to deteriorate when the number of commands in a set exceeded three for native speakers and two for low proficiency non-native speakers.
- [9_Co_2] Accurate compliance to a set of commands increases with practice.
- [9_Co_3] Accurate execution of a set of commands did not depend on the pause length between words, word speed, or elimination of extraneous words.
- [9_Co_4] Accuracy in carrying out longer sets of commands improved when redundant

words were eliminated.

Key recommendations:

- [9_Co_5) No more than two or three commands should be given in a single message.

 # 10. Barzegar RS, Wogalter MS. (1998, April). Effects of auditorily-presented warning signal words on intended carefulness. In MA Hanson (Ed.), *Contemporary Ergonomics 1998: Proceedings of the Annual Conference of the Ergonomics Society* (pp. 311-315). London, UK: Taylor & Francis.

E		Pc	At	Co	Cr	Ps	Ac

This study investigated how strongly the choice of signal words affects compliance to instructions in a verbal warning message, relative to the sex, style, and volume of the voice.

Seventy two undergraduates (36 males and 36 females) at North Carolina State University each listened to three lists of 43 prerecorded signal words presented in different voice styles and rated them by how carefully they would intend to carry out instructions following these alerts. The words consisted of five signal words (DANGER, WARNING, CAUTION, NOTICE, and DEADLY) spoken in monotone, emotional, and whisper styles by three male and three female speakers. Words were presented eight seconds apart in random order. Male and female subjects were divided evenly into four groups to listen to either male or female speakers at a sound level of either 60 dBA or 90 dBA. After hearing each word, the subjects rated how carefully they would intend to behave.

Female speakers produced a statistically greater carefulness rating. The emotional voice was rated the highest in promoting carefulness, followed by the whisper, and with the monotone receiving the lowest ratings. The difference in ratings between the emotional and monotone styles was statistically significant. The ranking of tested signal words in order from highest carefulness rating to lowest was: DEADLY, DANGER, WARNING, CAUTION, and NOTICE. Differences between ratings for signal words were statistically significant, except for the difference between warning and caution.

Key findings:
- [10_At_1] Signal words spoken in a female voice with an emotional tone result in a higher reported intent to be careful than words spoken in a male voice or in a monotone or whisper.
- [10_At_2] The choice of signal word matters. Out of five words, the word DEADLY results in the highest reported intent to be careful, followed by DANGER, WARNING/CAUTION, and NOTICE.

11. Ben-Bassat T, Shinar D. (2006). Ergonomic guidelines for traffic sign design increase sign comprehension. *Human Factors, 48*(1), 182-195.

E

Pc	At	Co	Cr	Ps	Ac

This study looked at whether signs that comply with the ergonomic principles of sign-content compatibility, familiarity and standardization are more likely to be understood. A group of undergraduate students were asked to describe the meaning of 30 street signs, of which 15 were used internationally and 15 were unique to a specific country. The students also evaluated each sign in terms of three important ergonomic principles: compatibility, familiarity and standardization. Their ratings on standardization and compatibility compared well with evaluations of the same signs by human factors and ergonomic specialists.

Comprehension of the meaning of the 30 signs by the students varied widely. A high degree of correlation was found between comprehension and the expert assessment of the sign's compliance with ergonomic principles. Although international standards are difficult to establish, the study suggests that symbols on signs should be consistent and should resemble the real-life action that they are trying to portray. A good symbol design may incorporate humans or common modes of transportation. The findings suggest that designing signs that comply with spatial compatibility, conceptual compatibility, and physical representation and standardization guidelines will positively influence the understanding of street signs.

Key findings:
- [11_Co_1] Signs that follow the universal ergonomic principles of sign-content compatibility, familiarity and standardization are better understood by people from different cultural backgrounds.
- [11_Co_2] Symbols on signs should be related to the message content.

12. Bénichou N, Proulx G. (2007, September). Evaluation and comparison of different installations of photoluminescent marking in stairwells of a highrise building. *Proceedings of Interflam Conference 2007, 11th International Conference on Fire Science and Engineering,* London, UK, 1-12.

Bénichou, N, Proulx G. (2007). *Evaluation and comparison of different installations of photoluminescent marking in stairwells of a highrise building,* (Report No. NRCC-49230) National Research Council Canada.

F		Pc	At	Co	Cr	Ps	Ac

Photoluminescent material (PLM) is a combination of inorganic chemical compounds that has the ability to store light photons and emit light over time. After being charged by a light source, the PLM can emit light until the energy has been depleted. These materials can be used for exit signs, directional signage, path markings and other wayfinding systems. When there is a power loss, the photoluminescent material can help to safely guide building occupants to a safe location in the dark. The objectives of this experiment were threefold:

1. Assess the effectiveness of a PLM wayfinding system in three office building stairwells
2. Compare the effectiveness of a PLM wayfinding system with stairwell emergency lighting
3. Develop an installation guide for a PLM wayguidance system

The evacuation drill of an 11 story office building with four stairways was filmed. The subjects of this study were the approximately 4000 employees in the building. During the drill, three of the staircases were equipped with PLM lighting while the other stairwell was illuminated by emergency lighting at a level of 37 lx. One of the PLM staircases had L shaped markers on the sides of the steps, the second had a one inch wide marking along the front edge of each step and the third had both the front edge marking and the L shaped marker on each step. The cameras were located inside the stairwells to capture the behavior of the occupants. When the subjects exited the building, they were handed a questionnaire to fill out about their feelings of comfort and safety in the stairwells.

Questionnaires were returned by 421 people, of whom 65 % were female and 35 % were male. Among those that responded, 26 % had asthma and 23 % were overweight. These two conditions were the most common limiting conditions of the tested population. From the questionnaire, 22 % of the occupants experienced crowding in the stairwell. It was not difficult to find the handrail in the stairwells with PLM lighting or reduced emergency lighting; however, the largest complaint was not being able to see because of poor lighting. The staircases with the PLM marking along the front of each step received higher visibility ratings (62% and 67%) than either the emergency lighting staircase (56%) or the PLM staircase with only L shaped markers (50%). The questionnaire also asked whether stair features such as the handrail, first step, last step, and directional signs were easy to locate or visible. The results showed that for the stairwells with PLM lighting, the L-shaped markers alone made it harder for occupants to identify each steps. The stairwell with reduced emergency lighting was given comparable ratings to the PLM lighting with L shapes along the sides of the stairs. In both stairwells, almost half of the occupants commented that it was difficult to see.

Key findings:
- [12_Pc_1] Although it was not difficult to find the handrail in a stairwell illuminated by

either PLM lighting or reduced emergency lighting, almost half the people evacuating under either condition complained that it was difficult to see.

- [12_Pc_2] L-shaped markers along the sides of the stairwell alone made it hard for occupants to see the steps.
- [12_Pc_3] [12_At_1] During an emergency, PLM lighting with a one inch strip along the front edge of each step in a stairwell is preferred to emergency lighting at an average of 37 lx.

Key recommendations:

- [12_Pc_4] For PLM lighting in a stairwell, the front edge of each step should be delineated so that occupants can see the steps clearly.

13. Benthorn L, Frantzich H. (1999). Fire alarm in a public building: How do people evaluate information and choose an evacuation exit? *Fire and Materials 23*(6), 311-315.

			Pc	At	Co	Cr	Ps	Ac
F								

The first part of this study investigated the dependence of the choice of exit on distance and on whether the exit is open to the outside or closed. Emergency messages were transmitted over wireless headphones to 35 women and 29 men, ranging in age from 16 to 75, in an Ikea warehouse. An initial alarm ring signal sounded for 10 seconds, followed by a silent pause for 10 seconds and a 45-second prerecorded message to leave the building due to a "technical failure". The spoken message was repeated twice, and did not specify leaving by any particular exit. Since trials were not simultaneous, and since other customers not participating in the study were present, no social interaction effects were included. The ring signal was perceived as an unspecified alarm by 41 % of subjects, as a fire alarm by 19 %, and as indicating the need to evacuate by 6 %. The authors speculated that the low percentage that thought of the ring signal as a fire alarm may be due to the low sound level in the headphone and the use of headphones rather than the building PA system. Most subjects understood the voice alarm to mean a serious problem (44 %) or evacuation (38 %). However, five subjects did not understand the voice alarm message or did not interpret it as indicating a problem. In the absence of information about which exit to take, subjects were more likely to choose a familiar exit (by the cash registers). Exits that were open were more likely to be used than those that were closed. In an emergency, staff may direct customers by opening emergency exits to the outside.

The second part of the study looked at the identification of emergency signage. The subjects were asked to identify six standard symbols associated with fire, evacuation, or danger in the workplace. Almost all subjects identified the naked flames forbidden sign. The warning for radioactivity and the fire hose sign were the most difficult to identify.

Key findings:
- [13_Co_1] Not all people may understand the meaning of an alarm bell; a spoken message is important.
- [13_Co_2] Spoken messages will not directly help those who don't understand the message.
- [13_Ac_1] During an evacuation, occupants of a building preferentially leave through either a familiar exit or an exit open to the outside.

Key recommendations:
- [13_Ac_2] In an emergency, staff may direct customers to safe exits by opening them to the outside. More will respond if the exit is visible from many points.

14. Bowman S, Jamieson D, Ogilvie R. (1995). Waking effectiveness of visual alerting signals. *Journal of Rehabilitation Research and Development, 32*(1), 43-54.

D		Pc	At	Co	Cr	Ps	Ac

The three main modes by which a sleeping person may be warned of an emergency are through auditory, tactile and visual alarms. Most warnings are auditory; however, this is ineffective for the 8 % of Americans that are hard of hearing. Two experiments were conducted to determine whether visual alarms can reliably awaken people from deep sleep. The first study adjusted the strobe intensity of light sources on the market to find the point at which subjects were aroused from sleep. The second study looked at the relationship between visual signals, the sleep stage, and the effectiveness of waking up a participant. Experiments took place in a sleep lab so that the sleep stage could be monitored.

For experiment one, a strobe light with a one Hz pulse rate was turned on at a low setting after at least five minutes of Stage three or four sleep. The intensity was varied across different trials to seek the intensity that would reliably wake someone up after 2.5 minutes. The light level was increased every 20 seconds to 150 seconds until the subject woke up and demonstrated that they were awake by sitting up and putting their feet on the floor. Seven women with normal hearing participated in this experiment. The equipment was tested on the first subject. For this subject and four others the light was located on the ceiling at the foot of the bed. Another subject was tested with the light on the ceiling directly overhead, and the final subject was tested with the light directly over her face. The intensity of the strobe was increased in 30 seconds intervals until the subject woke up. In the morning, each subject filled out a questionnaire about their sleep. The results of this experiment showed that none of the visual signals consistently woke up the sleeping subjects at any stage of sleep.

Thirteen women participated in the second experiment. The strobe light was located close to the subject at a distance of 75 cm from the pillow. The light intensity on the sleeper's face was 7.6 lx for the first trial. If they did not wake up within five minutes, the light intensity was raised to 19.9 lx for subsequent trials. The highest level exceeded light intensities expected on the sleeper's face for all commercially available visual smoke alarms. Under all conditions, when the strobe was not able to arouse subjects shortly after onset of the alarm it was usually unsuccessful in awakening them at all. This experiment again found the flashing strobe unsuccessful in consistently awakening subjects at any sleep stage. Subjects were awakened from slow wave sleep in fewer than 30 % of the trials under the higher light intensity, and were awakened from REM sleep in about 50 % of the trials under both light intensities.

The study concludes that visual fire alerting devices should not be the only means to wake up a sleeping person, and that the additional use of tactile devices should be tested.

Key findings:
- [14_Pc_1] [14_At_1] Even with light intensities on the face exceeding those from commercially available visual alarms, strobe lights cannot be relied upon to wake a sleeping person. Tactile devices should be used with the flashing strobes to wake up hard of hearing people during a fire.

 # 15. Boyce KE, Shields TJ, Silcock GWH. (1999). Toward the characterization of building occupancies for fire safety engineering: Capability of people with disabilities to read and locate exit signs. *Fire Technology, 35*(1), 79-86.

F		Pc	At	Co	Cr	Ps	Ac

Boyce et al. studied the ability of a population of 118 people with a range of disabilities (25 with visual impairments) to locate and read code-compliant exit signs. Three types of exit signs (non-illuminated, internally illuminated, and light-emitting diode (LED) - all with white text on green) were tested at a vertical placement of 2.3 meters from the floor. All signs complied with BS 5499 Part 1, Specification of Fire Safety Signs.

Each participant was taken into a room that they had not seen before, and the distances were measured from where they could just read the signs. The lettering on familiar signs was changed; for example, letters in the word EXIT were switched from E to B or I to L. The participants were asked to indicate when they could read the different letters in the sign. The maximum distance from which the participants were asked to read the sign was 15 meters. Questionnaires were given to the participants after they located the signs.

The signs could be located from broadly comparable distances (approximately 14 meters). However, there were differences in the distance from which the sign could be read (see Table 14). As the level of vision impairment increased, the distance at which people could read the signs decreased, as expected. The performance of those with visual disabilities varied significantly in response to the type of sign being used, with the LED sign found to be the most visible and legible.

Table 14: Distance at which three types of signs can be read by people with and without visual disabilities.

Type of sign	Legibility distance (m) by population type		
	All participants	Visual disability	No visual disability
Non-illuminated sign	13.3	11.4	13.7
Illuminated sign	14.2	12.9	14.5
LED sign	14.6	14.0	14.7

Key findings:
- [15_Pc_1] LED exit signs are more visible and legible than illuminated or non-illuminated signs, especially for visually impaired people.

Key recommendations:
- [15_Co_1] The nature of the target population should be considered when selecting the wayguidance/lighting system.

 # 16. Broersma M. (2009). Triggered codeswitching between cognate languages. *Bilingualism: Language and Cognition, 12*(4), 447–462.

L		Pc	At	Co	Cr	Ps	Ac

This study looks at triggered codeswitching for bilinguals between two languages (Dutch and English) that share a large number of cognates, words with similar form and meaning in both languages. The interest is in codeswitching that arises in natural settings with other bilingual speakers and is not necessarily consciously chosen by the speaker. Previous research has shown that mental processing of cognates by multilingual speakers differs from processing of non-cognates, leading to the activation of both languages and potential codeswitching through selection of a lemma in the other language. This study addresses the question of whether triggered codeswitching only occurs following the cognate (trigger word), or whether preceding words and clauses may be involved.

A twenty four minute conversation between a 73 year old woman and a researcher was recorded. The woman was a native Dutch speaker that had been living in New Zealand for 34 years and had become fluent in English. The researcher was also bilingual in Dutch and English, and spoke only Dutch during the interview. Cognates (trigger words) accounted for 71.4 % of the words in the conversation, and the informant regularly codeswitched between Dutch and English. Her speech demonstrated transference, the influence of one language on another, in the word order of sentences (English ordering of Dutch words), the exchange of English meaning for Dutch words, mixed morphology of words, and pronunciation. The interview was coded for analysis by six judges, all native Dutch speakers, to reduce the level of subjectivity. A statistical analysis of the interview found that codeswitching was more likely to happen when the word preceded a trigger word, followed a trigger word, or was in the same basic clause as a trigger word. No dominance was found for a specific type of cognate; triggered codeswitching occurred with proper nouns, cognate content words with good overlap between languages, cognate content words with moderate overlap, and cognate function words.

Key findings:
- [16_Co_1] Bilingual speakers of closely related languages may communicate with considerable transference, in which the syntax and word meaning from one language influences the other.
- [16_Co_2] Mental processing of cognates by multilingual speakers differs from processing of non-cognates.

Key recommendations:
- [16_Co_3] To reduce confusion for non-native English speakers, messages should minimize the number of false cognates, words that have similar form but different meanings, in multiple languages that are common in the target population.

 # 17. Broersma M, de Bot K. (2006). Triggered codeswitching: A corpus-based evaluation of the original triggering hypothesis and a new alternative. *Bilingualism: Language and Cognition, 9*(1), 1–13.

L		Pc	At	Co	Cr	Ps	Ac

Codeswitching is the combining of two languages into a coherent whole for communication between multilingual speakers. This paper examines the claim that codeswitching is often observed to occur in the neighborhood of a cognate, a word that is similar in form and meaning between the two languages. The facilitation of codeswitching from one language to another by cognates is known as the triggering hypothesis. In its original form, this hypothesis is inconsistent with models of language production, since it suggests that the surface structure representing a sentence is formed before the choice of language. An adjustment of the triggering hypothesis proposes that trigger words that are sufficiently similar in form and meaning allow activation of the lemma in the other language for the next clause. Rather than codeswitching during the words immediately preceding or following the trigger word, the adjusted hypothesis predicts codeswitching based on clauses that contain a trigger word. This can be tested statistically.

The conversations of three young, male, Dutch-Moroccan Arabic speakers were recorded. These two languages are unrelated (4.7 % of the words in the exchange were cognates), and the lexical transfers were items that were common to both languages for the individual speaker. The tapes were then examined for instances where the primary language shifted, which were analyzed statistically to test the triggering hypothesis in its original and adjusted forms. The adjusted hypothesis was supported, with clauses containing a trigger word more likely to be switched from one language to the other.

Key findings:
- [17_Co_1] Bilingual speakers may switch from one language to the other when a clause contains a cognate, a word similar in form and meaning to a word in the other language.

 # 18. Bruck D, Ball M, Thomas I, Rouillard V. (2009, June). How does the pitch and pattern of a signal affect auditory arousal thresholds? *Journal of Sleep Research, 18*(2), 196–203.

P		Pc	At	Co	Cr	Ps	Ac

The purpose of this study was to determine the characteristics of pitch and timing of the most effective waking signal. This follows previous research that showed that a 520 Hz square wave signal is considerably more effective than the high-pitched (3000+ Hz) standard Australian smoke alarm signal at waking sleepers. For people who are awake, high-pitched sounds have been found to be more effective at attracting attention, but mixed sounds in the low to mid frequency range (within the range of human speech) are more easily heard. The phenomenon of critical bandwidths, in which a sound composed of multiple frequency peaks more than a critical bandwidth apart is perceived as louder than the level indicated by an objective measurement, may explain why square waves and other mixed tones have lower audibility thresholds than pure tones or human speech.

Thirty nine young adults (18 males and 21 females in the age range of 18 to 27 years) were tested over three nights. In Part A, the auditory arousal threshold (AAT) was determined for nine different alarm signals while each participant was in Stage four sleep. The signals consisted of short beeps (in a T-3 pattern) of square waves, pure tones, whoops, and white noise in the low to mid frequency range. Four tones were presented on each of two nights separated by at least a week. On a third night, the ninth tone was presented along with Part B tones, in which the temporal pattern was varied by inserting periods of silence. Each signal started at 35 dBA and increased by 5 dBA every 30 seconds (Phase A) or by 10 dBA every 66 seconds (Phase B) until the subject awakened or the signal reached 95 dBA, where it continued for an additional 3.5 minutes. As shown in Table 15, 400 Hz and 520 Hz square wave alarms were found to be the most effective for waking the participants. This is consistent with the concept of critical bandwidths and data that suggest that frequencies around 520 Hz are optimally audible while awake.

Table 15: Sound levels required to wake young adults for a variety of alarm signals.

Part A: Alarm Signals	Mean AAT (dBA)
400 Hz square wave	46.2
520 Hz square wave	45.5
800 Hz square wave	51.8
1600 Hz square wave	53.2
White noise	59.6
400 Hz to 1600 Hz whoop	61.3
400 Hz to 800 Hz whoop	66.3
Sequence of three pure tones (400 Hz, 800 Hz, 1600 Hz)	60.5
Sequence of three square waves (520 Hz, 800 Hz, 1200 Hz)	54.6

The insertion of 10 seconds or 21 seconds periods of silence into the 520 Hz square wave T-3 signal tended to cause subjects to awaken sooner, as shown in Table 16, although this effect was not statistically significant given the small sample size.

Table 16: Sound levels required to wake young adults for a T-3 alarm with pauses of varying duration.

Part B: Temporal Variations on 520 Hz square wave in T-3 pattern	Mean AAT (dBA)
On continuously	52.4
On 12 seconds / Off 10 seconds	47.3
On 12 seconds / Off 21 seconds	49.5

In this and other similar studies, the participants are primed to respond. The question that arises is whether the same sounds would be considered significant in an unprimed home situation, absent previous experience with fire alarms or regular education?

Key findings:
- [18_Pc_1] [18_At_1] Low frequency (400 Hz and 520 Hz) square wave alarms in T-3 patterns are significantly more effective at waking young adults than square waves at higher frequencies, white noise, whoops, a rising sequence of three pure tones or a rising sequence of three square waves.
- [18_Pc_2] [18_At_2] Evidence suggests that the low frequency square wave is both the most detectable signal while awake and the most alerting while asleep. This is consistent with the concept of critical bandwidths.
- [18_Pc_3] [18_At_3] The insertion of a 10 second period of silence into a 520 Hz square wave T-3 alarm signal tends to reduce the time to awaken young adults.

 # 19. Bruck D, Thomas I. (2008). Comparison of the Effectiveness of Different Fire Notification Signals in Sleeping Older Adults. *Fire Technology, 44*(1), 15-38.

F		Pc	At	Co	Cr	Ps	Ac

The subjects in this study were 42 adults 65 to 85 years old. Four different auditory signals were tested to determine which was the most effective at waking the subjects. The first was a mixed frequency T-3 signal (500 Hz to 2500 Hz) from a fire alarm audio demonstration CD. This signal includes odd harmonics in addition to the fundamental tone at about 520 Hz. The second was a male voice (200 Hz to 2500 Hz) that warned of danger from fire and instructed the person to get up and investigate. (A male voice was used rather than a female voice because pilot studies showed that a male voice was more effective at waking young adults). The third was a high pitched (3000 Hz) T-3 signal recorded from an actual fire alarm. The fourth was a low pitched (500 Hz) T-3 signal, generated synthetically. The auditory arousal threshold (AAT), the minimum volume level that results in awakening, was measured for each subject over two nights. Two different signals were tested during a single night, and the other two signals were tested three to seven days later. To ensure that all subjects were familiar with the different signals, the signals were played for the subjects before they went to sleep. In the test, the signal started at 35 dBA for 30 seconds and then increased by five dBA every 30 seconds until the subject awakened. At the loudest volume level of 95 dBA, the sound was played continuously for 3.5 minutes or until awakening. At the 75 dBA level, 4.6 % slept through the mixed T-3 alarm, while 14 % to 18.3 % slept through the other alarms. At 95 dBA, the mixed T-3 alarm woke all subjects, while not all subjects were awakened by the other alarms. Ranked by response time, the mixed T-3 performed the best and high T-3 was the worst. No gender differences were found, but older individuals were more likely to sleep through higher pitch alarms. When compared with young adults (18 to 26 years old) studied previously, the AAT was significantly less for the mixed T-3 signal (57.9 dBA for young adults, 48.0 dBA for older adults), whereas no significant differences were found for arousal by the male voice. The authors noted that language barriers could be a problem for people responding to voice alarms, based on the failure of the voice alarm to awaken two subjects that did not speak English.

Key findings:
- [19_Pc _1] [19_At _1] A mixed frequency T-3 signal, including the fundamental tone at about 520 Hz and several harmonics, is significantly more effective than high or low frequency T-3 signals or a male voice for waking older adults.
- [19_Pc _2] [19_At _2] Significantly higher volumes (about 10 dBA) are required to wake young adults than older adults for a mixed frequency T-3 signal.
- [19_Pc _3] [19_At _3] Voice alarms with English text may be unsuitable for populations that include non-English speakers.

 # 20. Burt CDB, Henningsen N, Consedine N. (1999). Prompting correct lifting posture using signs. *Applied Ergonomics, 30*(4), 353-359.

E		Pc	At	Co	Cr	Ps	Ac

This set of studies was carried out in Australia. Manual lifting is required in many occupations and can cause back injury if performed incorrectly. There have been many attempts to correct lifting techniques, but most training efforts have failed to demonstrate a reduction in lifting injuries. It was hypothesized that the inclusion of instructional signs on heavy packages might prompt the adoption of correct lifting techniques.

The signs designed for evaluation in this study were constructed with the aim of communicating the five steps to properly lift a heavy object. Four symbols already in use were used in this study, along with five new symbols specifically created for this research. Some of the symbols included text, while others did not.

The participants in the first study were 105 undergraduate students, who completed an experimental questionnaire before a lecture. The questionnaire showed the nine instructional symbols and explained that the symbols were designed to encourage people to use correct lifting techniques. Students were asked to estimate how well the general population would determine the correct meaning of each symbol and rank it from one (no one will understand) to seven (everyone will understand). In the second study, 22 City Council workers were shown various symbols and were asked to describe exactly what the symbol was trying to communicate. A third study was completed in which a researcher posed as a courier, who asked for assistance with a package. Displayed on the package were symbols informing people of the correct handling technique.

The results showed that incorporated text did not increase the level of understanding of the symbol. The symbol that gave step-by-step (narrative) pictures of lifting a box was the most easily understood. The procedure required a sequence of actions, each of which had to be performed correctly. Results indicate that introducing a series of sequential symbols is the clearest way to show how to follow certain steps in a manual handling situation.

Key findings:
- [20_Co_1] Incorporated text did not increase the level of understanding of a symbol.
- [20_Co_2] A series of symbols showing sequential steps is the most easily understood way to provide instructions for certain procedures, such as lifting a box.

21. Byblow WD. (1990). Effects of redundancy in the comparison of speech and pictorial displays in the cockpit environment. *Applied Ergonomics, 21*(2), 121-128.

| E |

Pc	At	Co	Cr	Ps	Ac

It is the responsibility of the design engineer to develop the most efficient warning system for a pilot. Pilots need to maintain their attention on the view from the cockpit and simultaneously monitor the state of the controls. In the case of a mechanical failure, a pilot should not be distracted from the flight task while he or she tries to determine the nature of the failure. Redundancy in a warning system may be beneficial or harmful, where redundancy is defined as "a reduction in information in a system from the maximum possible, owing to unequal probabilities of occurrence of signals." In this study, pilot response to messages with and without system redundancy were compared for speech and pictorial displays.

Sixteen right-handed participants with a mean age of 22.6 years old were recruited for the study. The subjects had to respond to warnings about problems or failures for the electrical, propulsion, or engine systems on the right side, left side, or both sides of the aircraft while engaging in simulated flight tasks with a Falcon F-16 fighter simulator. Participants were asked to bomb as many buildings as possible without crashing the aircraft. They controlled the flight task with their right hand and responded to warnings with their left hand. Subjects were randomly assigned to conditions with either high redundancy (the system was named or shown in addition to the component with the problem) or low redundancy (only the component with the problem was named or shown). The order of warnings (pictorial icons or voice) was varied across subjects. The results showed that subjects responded faster to synthesized speech warnings. The phrase structure was manipulated in low redundancy speech messages to maintain the quality of the content without increasing the message length. This resulted in quicker responses than high-redundancy speech messages. Subjects who listened to spoken warnings showed better performance than those who were shown pictorial icons, although the difference was not found to be significant.

Key findings:
- [21_Ac_1] Pilots in a simulated flight task study responded faster to synthesized speech warnings.
- [21_Ac_2] Low redundancy warnings were beneficial in improving response time for pilots under a heavy cognitive load.
- [21_Ac_3] An auditory warning resulted in faster pilot response than a pictorial warning, although the difference was not found to be significant.

 # 22. Caird JK, Wheat B, McIntosh KR, Dewar RE. (1997). The comprehensibility of airline safety card pictorials. *Proceedings of the Human Factors and Ergonomics Society 41st Annual Meeting Vol 2,* 801-805.

E		Pc	At	Co	Cr	Ps	Ac

Design guidelines have previously been studied to improve airline safety card pictorials. The current design guidelines for airline pictorials recommend using a minimum number of words. The use of colored drawings or photographs is preferable to black and white pictures, and concrete drawings, which directly represent real-world phenomena, are preferred to abstract representations. The ordered sequence of pictorials should be numbered. Pictorials were found to be preferable to words, and drawings are preferable to photographs. Larger, colorful graphics that are very organized and have fewer words were also found to be preferred.

The degree of comprehension of a variety of airline safety cards that describe evacuation was studied. Nine safety cards were chosen for the study based on uniqueness of design style, the presence of pictorials that were thought to either aid or interfere with comprehension, and adherence to the current design guidelines. After viewing the safety cards, 113 participants described the sequence of actions that they would take to exit the plane during an emergency. The results showed that none of the nine pictorials met the ANSI Z535.3 Standard for Criteria for Safety Symbols, which requires 85 % of the participants in a comprehension test to understand the meaning of a symbol.

Key findings:
- [22_Co_1] Guidelines for pictorial airline safety instructions recommend using a minimum number of words.
- [22_Co_2] Colored drawings or pictures are preferable to black and white for airline safety pictorials.
- [22_Co_3] Concrete drawings are more easily understood than abstract drawings. An ordered sequence of pictorials should be numbered.

 # 23. Centers for Disease Control and Prevention. (2002, September). *Crisis and emergency risk communication.* Washington, DC: CDC (Author).
Online course: http://emergency.cdc.gov/cerc/CERConline/index2.html.

C		Pc	At	Co	Cr	Ps	Ac

This 267-page document accompanies a free course offered by the Centers for Disease Control and Prevention on how to communicate risk to the public during crises and emergencies. The link to the course contains links to useful material on other websites, as well as worksheets and checklists that can be downloaded. Some highlights of the nine modules pertaining to the topic of this annotated bibliography are described here.

Good communication during a crisis is necessary to counter harmful human behaviors. Module 1 explains the core principles of crisis and emergency risk communication (CERC). Expressing empathy to the public early in any crisis or emergency situation helps to establish and maintain a connection. The CDC suggests using phrases such as, "I share your concern" or "I understanding how frightened you must be." People need constructive actions to take, which can be provided as a set of "must do," "should do," and "can do" choices. There are many different ways to react to crisis – communicators must be respectful of people's emotions and coping mechanisms and not attempt to interject humor. It is important to understand the audience for the information. For example, the health conditions and needs of older adults are different from other segments of the population. The five key steps in successful communication are: 1) execute a solid communication plan, 2) be the first source of information, 3) express empathy early, 4) show competence and expertise, and 5) remain honest and open throughout the crisis or emergency. Factors that lead to communication failure include mixed messages from multiple experts, late release of information, paternalistic attitudes, failure to counter rumors and myths as they come up, and public power struggles. Credibility needs to be established from the beginning. The public needs information about the crisis and about the plans for response and recovery, and officials need to listen and respond to feedback from the public.

In Module 2, CDC discusses the psychology of a crisis. Mental stresses can be anticipated and managed using appropriate risk communication strategies. People seldom panic, but they can refuse to take good advice, become isolated alone or as a group, or become paralyzed and unable to act. People tend to rely more on images than words and may seek out information that reinforces what they already believe to be the case. Messages must be simple, take emotional states into account, and make sure that images (e.g., from television reports) and words match. Over-reassurance must be avoided – the public will tolerate 'it's less serious than we thought" more so than "it's more serious than we thought." Effective communication includes acknowledging uncertainty, giving people things to do, acknowledging fears and providing information to put the fear into context, addressing "what if" questions, and asking more of people. Different types of messages are effective at different times during the crisis; recommendations for each phase of the crisis are provided.

Module 3 discusses the message and the audience, which could consist of emergency or public health officials, victims, medical personnel, and/or the public. Trusted, credible sources should be used to get communication out quickly, first expressing empathy and caring and then establishing competence, expertise, honesty and openness, commitment, and dedication to the audience. Messages should provide the information sought by the specific audience. A worksheet entitled "Message Development for Communication" provides help in designing the actual crisis or emergency message.

Module 4 provides information on crisis communication plans. The phases of the crisis for communication purposes are: Precrisis, Initial, Maintenance, Resolution, and Evaluation. Each phase requires its own set of considerations and planning. A link is provided to the Community Planning Toolkit for State Emergency Preparedness Managers from the Department of Health and Human Services, which includes guidance on effective crisis communication for people with disabilities: http://www.hhs.gov/od/disabilitytoolkit/index.html

Module 5 focuses on the role and best practices for public health spokespersons in the event of a crisis or emergency. Spokespersons need special training to perform this role. It is important for this person to connect with the audience. CDC suggests that spokespersons form a mental picture of the individuals to whom they are talking. This person must be perceived as trustworthy and credible, empathetic and caring.

The final three modules are less relevant to the purpose of this annotated bibliography. Module 6 discusses working with the media, Module 7 presents communication with stakeholders and partners, and Module 8 discusses the use of traditional and nontraditional channels to disseminate emergency communication. The final page of the document collects all the tools and checklists that were presented throughout the course to help plan and evaluate crisis and emergency communication programs.

Key recommendations:
- [23_Cr_1] Expressing empathy to the public early in any crisis or emergency situation helps to establish and maintain a connection.
- [23_Ac_1] People need constructive actions to take, which can be provided as a set of "must do," "should do," and "can do" choices.
- [23_Cr_2] Communicators must be respectful of people's emotions and coping mechanisms and not attempt to interject humor.
- [23_Ps_1] It is important to understand the specific audience (e.g., emergency or public health officials, victims, medical personnel, or the public) and provide the information that is needed.
- [23_Cr_3] The CDC's five key steps in successful communication are: 1) execute a solid communication plan, 2) be the first source of information, 3) express empathy early, 4) show competence and expertise, and 5) remain honest and open throughout the crisis or emergency.
- [23_Cr_4] Factors that lead to communication failure include mixed messages from multiple experts, late release of information, paternalistic attitudes, failure to counter rumors and myths as they come up, and public power struggles.
- [23_Cr_5] Credibility needs to be established from the beginning. Trusted, credible sources should be used to get communication out quickly, first expressing empathy and caring and then establishing competence, expertise, honesty and openness, commitment, and dedication to the audience.
- [23_Cr_6] The public needs information about the crisis and about the plans for response and recovery, and officials need to listen and respond to feedback from the public.
- [23_Ac_2] People seldom panic, but they can refuse to take good advice, become isolated

alone or as a group, or become paralyzed and unable to act. Mental stresses can be anticipated and managed using appropriate risk communication strategies.

- [23_Co_1] People tend to rely more on images than words
- [23_Co_2] Messages must be simple, take emotional states into account, and make sure that images (e.g., from television reports) and words match.
- [23_Cr_7] Effective communication includes acknowledging uncertainty, avoiding over-reassurance, giving people things to do, acknowledging fears and providing information to put the fear into context, addressing "what if" questions, and asking more of people.
- [23_Ac_3] Different types of messages are effective at different times during the crisis

 # 24. Cervantes R, Gainer, G. (1992, Winter). The effects of syntactic simplification and repetition on listening comprehension. *TESOL Quarterly, 24*(4), 767-770.

L			Pc	At	Co	Cr	Ps	Ac

The purpose of this paper is to investigate the absolute and relative effects of linguistic (syntactic) simplification and repetition on listening comprehension for speakers of English as a second language. Several previous studies have shown that repetition improves comprehension, but the effects of linguistic simplification are not as well understood. The authors of this study conducted two different experiments – first to look at the effects of syntactic simplification alone on listening comprehension, and then to compare the relative effects of syntactic simplification and repetition.

The subjects in the first experiment were 54 first year and 22 senior English majors at Fukuoka University, with Japanese as their native language. One group listened to a syntactically more complex passage, and the other group listened to a syntactically less complex passage. The group with the less complex lecture recalled significantly more information on a post-test. English proficiency level was not found to be a significant factor.

The second experiment was carried out with 54 first year and 28 senior English majors at Fukuoka University with Japanese as their native language, who had not participated in the first experiment. Three versions of a short lecture were recorded. The second version was of higher syntactic complexity than the first, and the third was identical to the second with each dictation segment repeated once. Comprehension was tested through a partial dictation test that required filling in blanks on a test paper with the deleted words or phrases. Comprehension was poorest for the group listening to the complex version without repetition, but scores for the other two groups (the lower syntactic complexity and higher syntatact complexity with repition) were not significantly different.

Key findings:
- [24_Co_1] Comprehension of an auditory message by non-native English speakers is improved by lowering the syntactic complexity of the message and repeating it.

 # 25. Chandler RC. (2010). *Emergency Notification.* Santa Barbara, CA: Praeger.

C		Pc	At	Co	Cr	Ps	Ac

The purpose of this book is to discuss the challenges and opportunities for communicating more effectively with the public during emergency situations The book begins with a review of legal requirements, industry standards and public expectations related to emergency communication, then describes the latest efforts by the Department of Homeland Security (DHS) to develop an effective, integrated, reliable, and comprehensive system to alert the American people, including efforts to alert and warn individuals via their personal mobile devices.

Chapter 2 presents the basics of emergency notification, beginning with an overview of public alert systems and communication channels that could provide more focused and specific warning information. The latter include telephones (e.g., through phone trees, reverse 911, hotlines, and call centers), digital signage, SMS and text messaging, pagers, email, instant messaging, web pages, and social networking services. A standardized alert message format known as the Common Alerting Protocol has been developed. The chapter concludes with a brief history of mass notification systems, including Emergency Broadcast Systems, Emergency Alert Systems, and call centers; as well as a discussion on specialized alert systems, including the Health Alert Network, J-ALERT System, action shooter warnings, AMBER alerts, and Silver alerts, and the benefits of automated notification.

Chapter 3 discusses the challenges associated with emergency communication, along with some techniques to overcome them. The challenges include potential breakdowns of the communication system and technologies, the compressed timescale and evolving information for rapidly occurring events, the difficulty in conveying accurate meaning, and collaboration issues. The first notification message is critical for effective emergency management. All messages should be brief and precise. Communicators should assume that the public will process only the first 30 seconds (about 30 words) of a message; because of this, it is important to front load each message with the most relevant and critical information at the time. There are several barriers to effective communication with the public. People are continuously faced with multiple cues during an emergency and must decide which most deserve their attention. The varied backgrounds of individuals in the target audience affect the process by which they make sense of and respond to this information. Specific issues that need to be considered for effective emergency communication include native language reversion, selective attention and perception, impaired mental functioning during stress, and the importance of the source of the message. Multilingual or foreign language alerts may be necessary, because in crisis situations people have a tendency to revert back to thinking and understanding in their native language. Emergency communications must recognize that the amount of message information that people can process decreases in a state of stress. Individuals may selectively notice and remember certain things during an emergency, and they may also experience impaired mental functioning. Verbal comprehension has been found to drop an average of approximately four grade levels at high levels of stimulation, stress, and distraction; emergency notification messages should therefore be written for no more than a sixth grade reading level to ensure the widest possible range of comprehension among the target audience. Additionally, individuals in high stress situations are only able to process three distinctive issues or pieces of information in a single message unit. The source of the message is important because an audience is likely to pay more attention to those who are perceived as most trustworthy, honest and without motives to deceive them. The best sources are those to whom the public can relate, especially if personally known by the public. The five communication steps for success are: 1) execute a solid communication plan, 2) be the first source for information (especially since the first message carries more weight), 3) express empathy early, 4) show competence and expertise, and 5) remain honest and open.

Chapter 4 describes the process of planning for emergency communication and urges the creation of a document that can guide emergency communication in the future. The important stages of emergency planning are: 1) delegate responsibility and authority for emergency communication, 2) manage the information provided to effectively utilize emergency communication strategies, 3) manage the message and get accurate, well-crafted information out as quickly as possible (and even develop some phrases, sentences and vocabulary words that are appropriate and approved by management before a disaster occurs), 4) acknowledge the range of notification systems and technology, 5) understand the audience that will be receiving information, 6) build a contact database of a variety of different sources you will need to contact in an emergency, 7) develop an information policy including what information will be needed at what times and for which people, 8) understand the legal review and issues that may affect your organization, 9) plan the message, 10) plan ways in which to communicate with your organizations' employees, 11) anticipate common hazards and needs, 12) educate and train personnel on the plan, and 13) test and validate the emergency communication plan. Scientific concepts, probability analyses, and variables/info on how the warning was prepared are too complex to include in emergency messages. Two-way communication can be used to monitor public knowledge and beliefs, misunderstandings, behavioral compliance, and the progress of the response. This is an effective way to correct rumors and misinformation.

Chapter 5 describes the six phases of an emergency crisis: warning, risk assessment, response, management, resolution, and recovery, of which the first three phases are of central interest in this document.

Chapter 6 discusses using automated notification to address some of the challenges of contacting a large number of people in a small amount of time. The four basic types of automated notification systems are: on-premise model systems, hosted model systems, hybrid systems, and the Software-as-a-Service (SaaS) model systems. The benefits of these systems include: 1) individuals can choose/prioritize methods of receiving information; 2) individuals can receive messages that are accurate and consistent with other information; 3) the sender can track whether messages were delivered successfully; 4) many systems allow for two-way communications; 5) the systems are redundant in that messages can be sent to the receiver via multiple pathways; and 6) reliability can be built in by using multiple data centers to store contact information, leasing dedicated phone lines from multiple national carriers, and allowing messages to be initiated from multiple places.

Chapter 7 deals with the development of the emergency messages themselves. The four main functions of an emergency message are to provide vital information about the hazard, communicate the implications of the event and its urgency or seriousness, instruct on what to do next, and explain how recipients can respond. Ways in which emergency message content can be improved include:
- Design a template to prepare warnings in advance of emergencies.
- Use a "danger-action" formula for messages – keywords and phrases describing dangers (e.g., 'active shooter' or 'HAZMAT warning') and actions (e.g., 'evacuation now' or 'take shelter now') can be vetted in advance of the emergency.

- Provide complex information and instructions using simple, brief and concrete language.
- Tell people how and where to get further information.
- Use the 3 & 30 principle, placing the most important elements of the message in the first three sentences and 30 words. The message may be an alert to get people's attention and direct them to a place with additional information.
- Use the 60 & 6 principle, writing the notification message at a sixth-grade or lower reading level and with a reading ease score of at least 60 on a scale of 100, to simplify the grammar, syntax, vocabulary, and reading level for people under stress.
- Risk communication research has found littler evidence supporting trends in danger symbols, font choices, and other typographical features, including caps, bold, colors, exclamation marks, and spacing, that significantly affect attention, perception, or behavioral response.
- An 'all clear' message is important to indicate that there is no longer a threat. This type of message can be longer and more complex than initial alerts.
- Be honest about the event – there is little evidence that honesty causes panic.
- Describe what you know and don't know and what you are doing about it.
- Reduce the number of negatively-dominated messages.
- Test and validate the effectiveness of messages.

The final chapter describes a method for message mapping that enables emergency messages for all audiences to be drafted in advance of the crisis.

Key recommendations:
- [25_Ps_1] A variety of communication channels should be explored for providing specific warning information to specific audiences.
- [25_At_1] The first notification message is critical for effective emergency management.
- [25_Co_1] All messages should be brief and precise.
- [25_At_2] People are continuously faced with multiple cues during an emergency and must decide which most deserve their attention.
- [25_At_3] Communicators should assume that the public will process (i.e., pay attention to) only the first 30 seconds (about three sentences or 30 words) of a message. Each message should be front-loaded with the most relevant and critical information at the time. (3 & 30 principle)
- [25_Co_2] The varied backgrounds of individuals in the target audience affect the process by which they make sense of and respond to this information.
- [25_Co_3] Multilingual or foreign language alerts may be necessary, because in crisis situations people have a tendency to revert back to thinking and understanding in their native language.
- [25_Co_4] Verbal comprehension has been found to drop an average of approximately four grade levels at high levels of stimulation, stress, and distraction; emergency notification messages should therefore be written for no more than a 6th grade reading level.
- [25_Co_5] Individuals in high stress situations are only able to process three distinctive issues or pieces of information in a single message unit.
- [25_Cr_1] An audience is likely to pay more attention to those who are perceived as most trustworthy, honest and without motives to deceive them. The best sources are those to whom the public can relate, especially if personally known by the public.
- [25_Cr_2] The five communication steps for success are: 1) execute a solid communication plan, 2) be the first source for information (especially since the first message carries more weight), 3) express empathy early, 4) show competence and expertise, and 5) remain

honest and open.

- [25_Co_6] Scientific concepts, probability analyses, and variables/info on how the warning was prepared are too complex to include in emergency messages.
- [25_Ac_1] Two-way communication can be used to monitor public knowledge and beliefs, misunderstandings, behavioral compliance, and the progress of the response. This is an effective way to correct rumors and misinformation.
- [25_Ps_2] Automated notification can be used to address some of the challenges of contacting a large number of people in a small amount of time.
- [25_Ps_3] The four main functions of an emergency message are to provide vital information about the hazard, communicate the implications of the event and its urgency or seriousness, instruct on what to do next, and explain how recipients can respond.
- [25_Ps_4] Use a "danger-action" formula for messages – keywords and phrases describing dangers (e.g., 'active shooter' or 'HAZMAT warning') and actions (e.g., 'evacuation now' or 'take shelter now') can be vetted in advance of the emergency.
- [25_Co_7] Provide complex information and instructions using simple, brief and concrete language.
- [25_Co_8] Tell people how and where to get further information.
- [25_At_4] [25_Ac_2] Risk communication research has found little evidence supporting trends in danger symbols, font choices, and other typographical features, including caps, bold, colors, exclamation marks, and spacing, that significantly affect attention, perception, or behavioral response.
- [25_Ac_3] An 'all clear' message is important to indicate that there is no longer a threat. This type of message can be longer and more complex than initial alerts.
- [25_Cr_3] Be honest about the event – there is little evidence that honesty causes panic.
- [25_Cr_4] Describe what you know and don't know and what you are doing about it.
- [25_Cr_5] Reduce the number of negatively-dominated messages.
- [25_Ps_3] Test and validate the effectiveness of messages.

26. Charlton SG. (2006). Conspicuity, memorability, comprehension, and priming in road hazard warning signs. *Accident Analysis & Prevention, 38*(3), 496-506.

| T | | Pc | At | Co | Cr | Ps | Ac |

The purpose of this study was to evaluate the conspicuity, memorability, and comprehensibility of several road hazard warning signs in current use, and to determine how these signs affect the actions of drivers.

Thirty-three participants were seated in a driving simulator with a high-resolution video screen. They were asked to drive at 50 km/h down a road along which they encountered multiple instances of the sixteen road warning signs being tested. The participants were instructed to press the pedal while driving to signal that something had attracted their attention. The number and type of reported items were recorded.

In the second part of the study, the participants were told to press the brake pedal whenever they noticed a hazard or hazard warning sign, and to name the sign aloud. The results showed that flashing variable message sign (VMS) warnings attracted more attention than other designs and were more likely to be interpreted as signaling a hazard. For road hazard warnings, the flashing VMS format was slightly more conspicuous than a large dimension sign format and was equal in comprehensibility. For school warnings, the VMS format instilled a greater sense of hazard.

Key findings:
- [26_At_1] [26_Co_1] Flashing lights increased drivers' awareness of road hazard warning signs and increased the interpretation of the sign as signaling a hazard.

 # 27. Cherry EC. (1953). Some experiments on the recognition of speech, with one and with two ears. *Journal of the Acoustical Society of America, 25*(5), 975-979.

A		Pc	At	Co	Cr	Ps	Ac

Two different groups were used in this study to determine how people hear and respond to messages when more than one message is being received at the same time. Subjects in one group were given two different messages simultaneously in both ears. Subjects in the second group were given different messages in each ear. Differences in pitch, accent, and cadence were removed by using the same speaker for both messages.

Individuals in the first group could separate out the key phrases from each message, but only after several repetitions. However, when the message was just a string of clichés rather than a coherent passage, the two messages could not be distinguished. The second group experienced far less difficulty. They could repeat one of the messages entirely while listening at the same time, although they could then not recall any part of the message from the other (rejected) ear, nor could they detect a change in the language being spoken. However, they could observe when the message in the rejected ear changed from a male voice to a female voice or to a 400 Hz pure tone. Change from normal speech to a message in a male voice played backwards in the rejected ear was noted as sounding strange by some but went unnoticed by others.

Key findings:
- [27_Co_1] People find it difficult to distinguish and understand a spoken message when two voices of similar quality and volume are speaking simultaneously. When the competing messages are random phrases rather than a coherent passage, they cannot be distinguished.
- [27_At_1] A message may be ignored if there is not a sufficient means of attracting attention to it.
- [27_At_2] To draw attention away from a conversation, a change in voice pitch (e.g., from male to female) or notification by a tone other than speech is needed.
- [27_Co_2] People who are deaf in one ear may find it especially difficult to understand a verbal message in the presence of competing speech.

28. Cherry EC, Taylor WK. (1954). Some further experiments upon the recognition of speech, with one and with two ears. *Journal of the Acoustical Society of America, 26*(4), 554-559.

A		Pc	At	Co	Cr	Ps	Ac

The two goals of this study were to measure the average delay time in recognizing messages in continuous speech and to determine the factors that affect the ability to distinguish one spoken message from another. In the first set of experiments, subjects heard a continuous reading from light fiction in alternating ears, which they were asked to repeat out loud. At a switching rate of 3 Hz to 5 Hz, the ability of subjects to repeat the test was greatly diminished. This provides a measure (about one-sixth seconds) of the time it takes a listener to switch attention from one ear to the other, which can also be interpreted as the word recognition delay time. At higher switch rates, or at lower speaking rates, the listener can obtain a number of samples of each syllable and improve the ability to articulate the message.

The second experiment found that delaying a message by about 15 milliseconds in one ear changes the perception of a single voice into the perception of two voices. This is about 20 times greater than the real difference in arrival time between ears for a speaker standing to one side of the listener, which is one method by which a listener may locate a speaker.

Key findings:
- [28_Co_1] The delay time for people to recognize words in continuous speech is about one-sixth seconds.

29. Collins BL, Dahir MS, Madrykowski D. (1993). Visibility of exit signs in clear and smoky conditions. *Fire Technology, 29*(2), 154–183.

Collins BL, Dahir MS, Madrykowski D. (1992). Visibility of exit signs in clear and smoky conditions. *Journal of the Illuminating Engineering Society, 21*(1), 69-84.

Collins BL, Dahir MS, Madrykowski D. (1990, August). *Evaluation of exit signs in clear and smoke condition.* (NISTIR 4399), National Institute of Standards and Technology.

Collins BL. (1991). Visibility of exit directional indicators. *Journal of the Illuminating Engineering Society, 20*(1), 117–33.

Collins BL, Treado, SJ, and Oullette, MJ. (1994). Performance of compact fluorescent lamps at different ambient temperatures, *Journal of the Illuminating Engineering Society, 23*(2), 72-85.

F	I

Pc	At	Co	Cr	Ps	Ac

A series of articles by Collins, Dahir, and Madrykowski reviewed existing material on the visibility of exit signs in clear and smoky conditions and reported on their measurements of visibility for twelve signage systems in clear and smoky conditions.

Source	Key Findings
Rea, Clark, and Ouellette (1985) Clark, Rea, and Ouellette (1985)	Sixteen participants evaluated the visibility of exit signs under smoke-filled and clear conditions. Sign luminance ranged from 14 cd/m² to 1277 cd/m² for incandescent and fluorescent signs and 0.18 cd/m² to 0.61 cd/m² for tritium signs. Detectability and readability increased with luminance, and higher smoke densities were required to mask signs with higher luminance. Visibility was best for three red/white signs with luminances of 170, 391 and 1272 cd/m². Scattered light from ambient illumination reduces the readability of low contrast signs more than high contrast signs. Red signs were slightly less affected by scattering by the smoke. Critical smoke density levels identified by subjects with difficulty identifying the color red were significantly lower than for other participants. Translucent green materials are therefore likely to appear brighter to more people than translucent red materials for the same light source. Recommendations include increasing the brightness of the sign, reducing ambient lighting, using stencil-faced rather than panel-faced signs, and choosing translucent green over translucent red signs. Aiming the ambient light source away from the sign reduces the negative impact on detectability, particularly when smoke is present. Smart fixtures that increase sign brightness and decrease ambient light in smoky conditions are a good idea.
Wilson (1990)	Evaluated the effectiveness of five green/white signs that were internally lit, externally lit, or edge lit, with luminances ranging from 0.9 cd/m² to 720 cd/m². The two internally lit signs performed better than the others.
Rasmussen, Garner, Blethrow, and Lowrey (1979)	Studied the visibility of a variety of exit signs in aircraft cabins. Signs with background luminances of 31, 89, 140 and 158 cd/m² were viewed from 77 inches away. Similar levels of smoke were required to obscure signs with the highest three luminances, with less smoke needed to obscure the sign with the lowest luminance.

Schooley and Reagan (1980a)	Visibility is affected by a number of factors, including smoke, dust, contrast, color, adaptation, and lighted/opaque backgrounds in addition to glare, veiling reflections, and ambient illumination. For legibility , calculations of the visual angle suggest a minimum object size of 1.25 inches based on $$\text{Max Visual Angle} = 3438 \times d \ / \ D$$ where d is the dimension of the object perpendicular to the line of sight and D is the distance to the object. Observer visual acuity, sign contrast, and overall illuminance must also be considered.
Schooley and Reagan (1980b)	Five subjects observed three signs in a corridor: an isotope self-powered sign, an unlit electrical sign, and an illuminated electrical sign. Visibility and legibility were studied in light (53.8 lx to 538 lx), in darkness, and in smoke. Under non-smoky lit conditions, the luminance of the internally lit sign was reduced to 17.1 cd/m^2 from the normal brightness of 166.2 cd/m^2. When smoke was introduced, the internally lit sign was also assessed at 85.7 cd/m^2 and 188.5 cd/m^2. When the corridor was lit, all three signs could be seen from 300 feet and were legible at 125 feet. In a dark corridor, the self-illuminated sign could be seen from 125 feet, the externally lit electric sign could be seen from 75 feet, and the unlit sign could not be seen. When smoke was added to the normally lit corridor, the three signs were legible from about 40 feet to 50 feet. Increasing the luminance of the self-powered electric sign increased its visibility to 125 feet and its legibility to 50 feet.
Cohn (1978)	Six subjects observed white, green, yellow, red and blue exit lights under lit and dark conditions. They assessed the readability of the sign and the visibility of the arrow symbol, given the color and the illumination of the sign. The yellow sign was found to bethe most effective, followed by the green and the red. Subjects had difficulty in identifying the word 'EXIT' from 80 feet, which may be due to the style and size of the lettering.
Beyreis and Castino (1974)	A number of observers evaluated three exit signs in clear and smoky conditions. Under clear conditions, an electrically illuminated sign was visible and legible from 300 feet, a tritium sign was legible from 75 feet, and a phosphorescent sign was legible from 25 feet. Observers were then positioned 12 feet from the sign, ambient luminance was measured at 53.8 lx,, and smoke was gradually introduced. The electrically powered sign was visible at a smoke optical density per meter path length of 0.152 m^{-1}, while the tritium sign was visible at 0.095/feet. When the brightness of the electric sign was reduced to the same level as the tritium sign, legibility and visibility were reduced to match those of the tritium sign.
Sime (1985)	The evacuees from a fire in a recreational facility were more likely to use routes with which they were familiar, rather than other potentially safer routes.
Jaschinski (1982)	Studied movement under normal and emergency lighting conditions. The independent variables were luminance level (0.24 lx to 7.7 lx) and age group (18 to 30 years old and 50 to 70 years old). Performance was assessed in terms of the time to reach safety, the participant's evaluation of their experience, and the ability to perform simple arithmetic tests while walking.The escape time and number of arithmetic errors were larger for the lower luminance level of 0.24 lx, which also received a less favorable subjective assessment. The importance of

| | | | | | | emergency lighting was reduced when luminous signs were employed. |

Collins, Dahir, and Madrykowski also reported on a study of the effectiveness of several exit signs under clear and smoke-filled conditions with 21 participants (14 males and seven females between 18 and 60 years old). Fourteen of the participants had corrective vision. Twelve different systems were evaluated: 10 stencil/panel-faced internally lit signs, one incandescent sign, and one fluorescent sign. Stencil-faced signs had illuminated letters, and panel-faced signs had an illuminated background. The internally lit signs were red, green, blue-green, and red/green, and the incandescent sign and fluorescent signs were red. The luminance of each sign was measured along with the luminance of the letters and background. The contrast C between the lettering and the background was calculated according to $C = (L_g - L_l) / L_g$, where L_l is the lesser luminance and L_g is the greater luminance. The visibility level of each sign was derived from participant reports based on a seven-point scale, with observers located at a fixed distance from the signs. After visibilities of the signs were assessed under clear conditions, a propane burner was ignited to generate smoke. Observers assessed sign visibility at four points during smoke accumulation (at four specific luminance values of one sign) and recorded the time at which each sign was no longer visible. Results are reproduced in Table 17.

Table 17: Relationship between sign design and visibility.

Sign Type	Color	Letter Luminance (cd/m²)	Contrast between letter and background	Average observer rating of sign visibility in clear environment (7-pt scale)	Time for sign to disappear given smoke (s)	Average observer rating of sign visibility during smoke accumulation (7-pt scale)
Incandescent	Green	21.5	0.94	4.7	221	3.3
	Red	80.6	0.77	5.2	433	5.7
Fluorescent	Green	140.9	0.99	5.6	469	5.4
	Red	324.9	0.99	6.2	498	6.6
	Red	173.8	0.87	5.6	572	6.5
Internally lit / Electrical	Red	5.7	0.99	6.2	263	3.7
	Green	23.5	0.99	6.0	282	3.8
	Red/Green	0.3	0.97	5.7	250	3.1
	Blue / Green	22.6	0.99	6.0	278	3.7
	Red/Green	0.9	0.99	2.9	206	1.1
	Red	3.5	0.99	6.0	299	3.4
	Green	8.5	0.99	5.9	300	3.7
	Green	0.8	0.94	5.3	298	4.0

The majority of the signs received high ratings under clear conditions, suggesting the existence of a luminance threshold above which the sign is reasonably visible. Signs with higher initial luminance took longer to disappear in smoke. Signs with luminance under 20 cd/m² were no longer visible after 300 seconds. Signs with luminance over 70 cd/m² were visible for a longer time of smoke accumulation (between 430 seconds and 570 seconds). Although sign luminance was the most important factor, the configuration of the sign also affected visibility. Stencil-faced signs (with illuminated letters and opaque backgrounds) were somewhat more visible than panel-faced signs. Although it appears that the red sign variants out-performed the other signs, it should be noted that participants were more familiar with this coloring, since red signs were employed in the participants' place of work.

Key findings:

- [29_Pc_1] Visibility was generally best for red/white exit signs with luminances of 170, 391, 1272 cd/m² relative to other colors and designs.
- [29_Pc_2] Red exit signs were slightly less affected by the scattering effect of smoke than other colors.
- [29_Pc_3] Detectability and readability increase with luminance.
- [29_Pc_4] Aiming the ambient light source away from an illuminated sign reduces the negative impact on detectability, particularly when smoke is present.
- [29_Pc_5] Similar smoke densities were required to obscure signs with background luminance at and above 89 cd/m² when viewed from 77 inches away, with less smoke needed to obscure a sign with luminance of 31 cd/m².
- [29_Pc_6] Internally lit green/white signs outperformed externally lit and edge lit signs.
- [29_Pc_7] Visibility is affected by a number of sign design factors, including smoke, dust, contrast, color, adaptation, lighted/opaque backgrounds, and luminance levels.
- [29_Pc_8] Visibility is affected by a number of environmental factors, including glare, veiling reflections, and ambient illumination.
- [29_Pc_9] Visibility for a tritium exit sign deteriorated much faster in smoke than that for an electrically illuminated sign.
- [29_Ac_1] The evacuees from a fire in a recreational facility were more likely to use routes with which they were familiar, rather than other potentially safer routes.
- [29_Ac_2] At low luminance levels, travel times and arithmetic errors increased.
- [29_Pc_10] Signs with luminance over 70 cd/m² are visible under much heavier smoke conditions than signs with luminance under 20 cd/m².
- [29_Pc_11] Stencil-faced signs (with illuminated letters and opaque backgrounds) are more visible than panel-faced signs.
- [29_Pc_12] Scattered light from ambient illumination reduces the readability of low contrast signs more than high contrast signs.
- [29_Pc_13] For signage legibility, calculations of visual angle suggest a minimum object size of 1.25 inches.
- [29_Pc_14] Adding smoke to a lit corridor reduced legibility of illuminated exit signs from 125 feet to between 40 feet and 50 feet.
- [29_Pc_15] Under lit and dark conditions, illuminated yellow exit signs were the most readable, followed by green and red signs.
- [29_Pc_16] Under clear conditions, an electrically illuminated sign was visible and legible from 300 feet, a tritium sign was legible from 75 feet, and a phosphorescent sign was legible from 25 feet.

Key recommendations:

- [29_Pc_17] Given that many people have difficulty seeing the color red, translucent green materials are likely to appear brighter to more people than translucent red materials for the same light source.
- [29_Pc_18] Smart fixtures that increase sign brightness and decrease ambient light in smoky conditions would improve visibility.
- [29_Pc_19] Increasing the brightness, reducing ambient lighting, using stencil-faced

---

rather than panel-faced signs, and choosing translucent green over translucent red are recommended for exit signs.

 # 30. Conrad L. (1989). The effects of time-compressed speech on native and EFL listening comprehension. *Studies in Second Language Acquisition, 11*, 1-16.

L		Pc	At	Co	Cr	Ps	Ac

Language comprehension depends on the cognitive capacity of short-tem memory and the processing strategy used by the listener or reader. The efficiency of the process depends on the ability to gain the maximum amount of meaning from as little information as possible from the text. Efficient language processing focuses on semantic information, which is the least redundant, and uses syntactic and graphophonic (sound-symbol) information to confirm understanding or notify of an error. Only semantic information, which is informed by past knowledge and expectations, is stored in long-term memory. With their limited knowledge of the language, non-native speakers are expected to require more time to process surface aspects of a message, leaving less time for learning and storing the semantic meaning. In this study, the differences in strategy for processing aural messages among native English speakers and high and medium proficiency non-native speakers were examined using time-compressed speech.

Twenty nine native English speakers, 17 high-level non-native English speakers, and 11 low-level non-native English speakers listened to one of two sets of 16 simple English sentences compressed to (450, 320, 253, 216, and 196) words per minute. Native speakers were freshmen at Michigan State University, and non-native speakers were native Polish speakers studying at the graduate level at the Technical University in Wroclaw, Poland. Polish has a more flexible word order than English. The sentences were nine to 11 words long and constructed with syntactic structures designed to illuminate memorization strategies. Presentation rates faster than 275 to 280 words per minute are known to affect comprehension. The subjects heard and were asked to write down each sentence five times, with sentences repeated at sequentially slower rates starting at the fastest rate. Native speakers had nearly complete recall (96 %) at 320 words per minute. High-level non-native speakers had a 64 % recall at 216 words per minute, and low-level non-native speakers only had a 44 % recall at 196 words per minute. For native speakers, key content words (nouns, verbs, and adjectives in positions where nouns were expected) were most likely to be recalled. For both non-native speaking groups, recall was significantly better for words at the start or end of sentences, regardless of the part of speech.

Key findings:
- [30_At_1] Recall was significantly better for words at the start or end of sentences for non-native speakers.
- [30_Co_1] Native English speakers are able to use their knowledge of the language to get the maximum amount of information from a severely distorted message.
- [30_Co_2] Non-native speakers are unable to make maximum use of the information in a severely distorted message, regardless of ability.

Key recommendations:
- [30_At_2] Critical components of auditory messages should be at the beginning or end of messages for non-native speakers.
- [30_Co_3] Comprehension of auditory messages by non-native speakers requires messages to be delivered slowly, repeatedly, and without distortion.

31. Cooke M, Garcia Lecumberri, ML, Barker J. (2008, January). The foreign language cocktail party problem: Energetic and informational masking effects in non-native speech perception. *The Journal of the Acoustical Society of America, 123*(1), 414-427.

A		Pc	At	Co	Cr	Ps	Ac

Previous research indicates that the ability to understand speech in the presence of noise is different for native and non-native language listeners. The purpose of this study was to assess whether the energetic signature of the noise (sound level, frequency, and other acoustic measures) or the informational content (e.g., in a cocktail party with many people speaking at once) is a more important factor. Energetic masking occurs when some part of the speech signal is inaudible due to noise with similar acoustic characteristics, and informational masking occurs when the partial speech signal that is not masked energetically is unintelligible. Informational masking can be due to misallocation (in which sounds from the competing speech are attributed to the speaker), the higher cognitive load from processing speech from multiple speakers, attention diverted from the speaker, and the presence of a better-known language in the noise. Understanding how informational masking affects speech comprehension for non-native language listeners would be helpful in improving communication with this population in noisy environments.

Two different studies were conducted using 42 native Spanish speakers studying English as a foreign language at the University of the Basque Country (mean age 21.2 years) and 18 native English speakers. All non-native English speakers had attained the level of the Cambridge Advanced Examination. The speech selections were extracted from a standard library of 1000 sentences consisting of six common words including a color, letter, and number, each spoken by 34 speakers (18 males, 16 females). The first experiment focused on energetic noise. Each non-native speaker listened to 60 sentences in random order under four noise conditions: silence, and stationary noise at signal-to-noise ratios (SNR) of 6, 0, and -6 dB. Results were compared to an earlier study with native speakers. As expected, the native speakers correctly identified the spoken colors, letters, and numbers more often under all test conditions. The order of keyword difficulty from easiest to hardest, as measured by the number of errors was color, then number, then letter. The native advantage, the difference between native and non-native performance, increased in the same order. For color and number keywords, the native advantage increased as the noise level increased. Both groups scored significantly better when the speech rate was slower, especially at higher noise levels. Non-native speakers scored lower when the speaker had a heavy accent. Both groups found female speakers to be more intelligible when the sound level of noise is equivalent to or greater than that of the message.

The second study focused on informational masking. Two sentences from the library were paired as target and masking sentences. The same participants as in Experiment one were asked to identify the letter and number spoken by the speaker of the target sentence, which always contained the keyword "white." The target-masking ratios (TMR), or the difference in sound level between the target sentence and the masking sentence, were 6, 3, 0, -3, -6, and -9 dB. Twenty sentence pairs were prepared for each of three conditions: both in the same voice, different voices of the same sex, and male/female mixed. A block of these 60 sentence pairs, in random order for each listener, was prepared for each of the six TMRs and presented to the listener in random order. Both native and non-native listeners found comprehension easiest when the competing speaker was of the opposite sex and most difficult when it was the same speaker. For the same speaker and a speaker of the same sex, the most difficult situation was when the target sentence was only slightly softer (by 3 dB) than the competing sentence. The non-native disadvantage increased when the masking sentence was louder than the target sentence. Native speakers scored 12 % to 15 % better under the most favorable conditions and up to 30% better in the least favorable condition. Differences in frequency were found to help both groups equally. Non-native speakers benefited from a slower speaking rate, and they were more likely to answer with a response from the masking sentence. Informational masking plays a significant role in the performance deficit of non-native listeners.

Key findings:
- [31_Co_1] In spoken messages, letters are more difficult to identify correctly than numbers, which are more difficult than colors. The difference in difficulty is increased with noise.
- [31_ Co_2] Comprehension worsens faster for non-native listeners than for native listeners as background noise increases.
- [31_Co_3] Comprehension is significantly better under all noise conditions when the message is delivered at a slower speech rate.
- [31_ Co_4] Speaker voices with robust speech cues (without a heavy accent, resistant to noise masking) significantly improve comprehension for non-native listeners and are also helpful for native listeners.
- [31_Co_5] Female speakers are more intelligible than males when the sound level of noise is equivalent to or greater than that of the message.
- [31_ Co_6] When trying to follow one of two speakers, the message is easier to understand when the fundamental frequency of the competing speech is significantly different, such as of the opposite sex.
- [31_ Co_7] It is most difficult to follow a conversation when a competing voice is of the same sex and at a slightly higher volume.
- [31_ Co_8] Non-native listeners have a significant disadvantage in understanding speech when two people are speaking at once.
- [31_ Co_9] When two people are speaking at once, the disadvantage for non-native listeners is worse when the competing voice is louder and the speech rate is faster.

Key recommendations:
- [31_ Co_10] To improve comprehension, the voice used to deliver a message should not be heavily accented, and the speaking rate should be slow.
- [31_ Co_11] Masking noise, especially speech, should be eliminated whenever possible.
- [31_Co_12] A message delivered in English to a non-native English speaking population is less likely to be understood if there are competing voices speaking in the native language.

32. Covello V, Minamyer S, Clayton K. (2007, March). *Effective risk and crisis communication during water security emergencies: Summary report of EPA sponsored message mapping workshops* (Report No. EPA/600/R-07/027). Washington, DC: U.S. Environmental Protection Agency. http://cfpub.epa.gov/si/si_public_record_report.cfm?address=nhsrc/&dirEntryId=165863.

C		Pc	At	Co	Cr	Ps	Ac

This report summarizes results from three EPA workshops on water security risk communication message mapping. Message mapping is a process that supports emergency planning by predicting questions that are likely to be asked by the media and preparing clear and concise answers in advance. This saves considerable time during the crisis, enables quick and effective communication, and allows emergency officials and spokespersons to focus their attention on the specific messages needed as the crisis evolves. The purpose of risk communication is to improve knowledge and understanding, build trust and credibility, encourage feedback, regulate the level of concern, and provide guidance on protective actions. Research indicates that mental noise, the tendency of people to lose cognitive ability during trauma, can reduce the ability to process information by 80 % or more. Risk communicators must provide brief, credible and clear messages that can overcome this barrier. A message map is a template that can be used to develop such messages in advance of a crisis situation. The message map requires seven steps: 1) identify potential stakeholders, 2) identify their questions, 3) analyze the questions for common sets of concerns, 4) develop key messages, 5) develop supporting information, 6) test and practice the messages, and 7) plan for the delivery through appropriate channels.

Step 4 of the message mapping process, which describes how to develop key messages based on what the target audience most needs or wants to know, relates most closely to the purpose and goals of this annotated bibliography. An analysis of ten years of print and media coverage of emergencies and crises in the U.S. shows that a sound bite in the print media has an average length of 27 words, that a sound bite in the broadcast media lasts for nine seconds, and that each sound bite (print and broadcast) consists of three messages on average. This suggests organizing key messages into sound bites three bullets long that contain a maximum of 27 words total and that can be spoken in nine seconds. The results of studies on mental noise are consistent with this recommendation – people in high-stress situations are able to process only three messages at a time, much lower than the seven messages that can be processed under normal circumstances. Other recommendations are to present the most important messages first and last (since listeners under high stress have less recall for items in the middle of a message), to construct messages with a sixth to eighth grade education (average grade level of the audience minus four), to state the three key messages three times, to give three positive messages for every negative message (since people focus more on negative information than positive), to address risk perceptions, and to use visual aids and narratives (which provide redundancy and can be heard above the mental noise). Spokespersons should practice their answers and only cite sources that are credible to the audience.

The EPA's seven rules of risk communication are: 1) accept and involve stakeholders (including the public) as legitimate partners, 2) listen to people, 3) be truthful, honest, frank, and open, 4) coordinate, collaborate, and partner with other credible sources, 5) meet the needs of the media, 6) speak clearly and with compassion, and 7) plan thoroughly and carefully, and evaluate performance to learn from mistakes.

Key findings:

- [32_At_1] An analysis of ten years of print and media coverage of emergencies and crises in the U.S. shows that a sound bite in the print media has an average length of 27 words, a sound bite in the broadcast media lasts for nine seconds, and each sound bite (print and broadcast) consists of three messages on average.
- [32_At_2] People in high-stress situations are able to process only three messages at a time, much lower than the seven messages that can be processed under normal circumstances.
- [32_At_3] Listeners under high stress have less recall for items in the middle of a message.
- [32_At_4] People focus more on negative information than positive
- [32_At_5] Visual aids and narratives provide redundancy and can be heard above the mental noise in a high stress situation.

Key recommendations:

- [32_At_6] Organize key messages into sound bites three bullets long that contain a maximum of 27 words total and that can be spoken in nine seconds.
- [32_At_7] Present the most important messages first and last
- [32_Co_1] Construct messages with a sixth to eighth grade education (average grade level of the audience minus four).
- [32_At_8] State the three key messages three times.
- [32_Cr_1] Give three positive messages for every negative message.
- [32_Ps_1] Address risk perceptions
- [32_Cr_2] Spokespersons should practice their answers and only cite sources that are credible to the audience.
- [32_Cr_3] The EPA's seven rules of risk communication are: 1) accept and involve the public as a legitimate partner, 2) listen to people, 3) be truthful, honest, frank, and open, 4) coordinate, collaborate, and partner with other credible sources, 5) meet the needs of the media, 6) speak clearly and with compassion, and 7) plan thoroughly and carefully and evaluate performance.

33. Craig CS, Sternthal B, Leavitt C. (1976, November). Advertising wearout: An experimental analysis. *Journal of Marketing Research 13*(4), 365-372.

M		Pc	At	Co	Cr	Ps	Ac

Advertisers have assumed that high levels of message repetition are required to instill the message in the minds of consumers. An alternate hypothesis asserts that the effectiveness of an ad depends on the response to the first few exposures. This hypothesis has been supported by research on wearout, which suggests that there is an optimum level of exposure, beyond which the recall of a message decreases. At the time of this paper, the wearout effect was considered to be paradoxical because it had only been demonstrated under field conditions and not under controlled conditions.

The effect of message repetition on brand name recognition was studied in two controlled experiments. In a pretest for the first experiment, exposure to 12 print ads for five seconds each seven times consecutively was found to result in immediate recall of all 12 ads by 50 % or more of the subjects. Since no improvement in immediate recall was observed for further exposure, this was defined as the 100 % learning level. For this experiment, 180 Ohio State undergraduates were divided into three groups, defined as the 100 %, 200 % and 300 % "learning groups," which were exposed to the same 12 ads for seven, 14 and 21 repetitions respectively. For some participants, recall of brand names was tested immediately after exposure. The others were called unexpectedly and asked to recall brand names one, seven, or 28 days after the experiment. As expected, recall declined with time. Statistically significant evidence for wearout was observed for the 28 day delay, in which recall was better for the 200 % learning group than for either the 100 % or 300 % group.

The second experiment looked at some possible explanations for wearout. A group of 70 undergrads from Ohio State University were shown a series of 28 print ads for five seconds each, with six repetitions. To provide multiple levels of repetition for individual ads, the series contained two ads that appeared only once, two ads that appeared twice, two ads that appeared three times, and two ads that appeared four times. The same two ads were placed at the beginning and end to test for order effects, and four "vigilance" ads were switched with similar ads for different repetitions to ensure the attention of students, who had been asked to look for variations in the ad sequence. Each participant was recontacted zero, two, seven, 14, or 28 days later to test their recall of brand names. The study found that the effect of repetition was only significant for the groups contacted after seven or 14 days, in which cases the use of four times the level of repetition to learn the brand names resulted in significantly higher recall. The authors concluded that it was important to keep people attentive to the message rather than just repeating it.

Key findings:
- [33_At_1] The phenomenon of wearout suggests the existence of an optimal level of repetition for messages, such as those used in an educational campaign.
- [33_At_2] Message recall over time is enhanced by attention and by motivation to retrieve the information.

34. Creak J. (1997). About Viewing Distances. *Means of Escape* (Article 549).
http://www.means-of-escape.com/articles/549/about-viewing-distances/

F		Pc	At	Co	Cr	Ps	Ac

Creak examined the empirical and regulatory guidance supporting the correct sizing, illumination, and viewing distances of safety signs. Guidance on fire precautions in the UK (BS 5499) recommends that letters appearing on safety signs be 100 mm in height in order to be read from a distance up to 25 meters. This was derived from the understanding that people with normal vision are able to resolve details that subtend an angle of one minute (visual angle), leading to the relationship $D = 3437.75\ d$ between observation distance D and dimension of the detail d. However, these guidelines were intended to apply to EXIT and FIRE EXIT signs only, and need to be revisited for sign applications with letters that are more difficult to resolve and for other font types and symbols.

The nature of the letters will influence the relationship of viewing distance to detail for good resolution; for example, B and D are more difficult to resolve than E. Using a safety margin of 33 % to ensure reliable resolution of all letters (in non-serif fonts) by people with normal vision, the viewing distance is approximately 500 times the letter height. An additional safety factor of two was included in the first version of BS 5499, implying that the signs were designed for people with vision half of that of individuals with normal sight. This vision level is met by 95 % of adults under 79 years old with corrected vision and 76 % if glasses are not worn.

Research on the legibility of graphical symbols and Japanese characters performed in the 1960's and 1970's suggested that the distances involved in resolving symbols are not significantly different from letters, even though the symbols may be more complex. Making some allowance for people with poorer vision and again using a safety factor of two, the relationships $D = 172\ s$, where s is symbol height, or $D = 130\ h$, where h is sign height, were adopted in the 2002 revision of the BS 5499 standard. These relationships assert that a 100 mm tall symbol can be reliably understood from 17 meters.

The ability to identify signage through peripheral vision is important. Contrast, color, and sign shape were found to be important considerations in making signs conspicuous. Research indicates that a sign that has been designed in accordance with standard practices will be visible within 15 degrees of the direct line of sight, and is legible up to a distance of 130 times the height of the letters.

The relationship of 130 times the letter height does not apply if the lighting conditions are abnormal. In poorly lit areas, signs may need to be 25 % larger. Note that the eye perceives the amount of light that is emitted from a surface, its luminance (measured in cd/m^2), rather than the amount of light that reaches the surface, its illuminance (measured in lx). Good lighting of 100 lx reflects off a white surface at 30 cd/m^2, and excellent lighting for detailed work provides 130 cd/m^2. This knowledge can be used to correct the necessary sign dimensions for the lighting level.

Key recommendations:
* [34_Co_1] Staff, occupants, and visitors should be trained on the meaning of distinctive

safety sign colors, shapes, and graphical symbols.

- [34_Co_2] Adding supplementary text increases the size of safety signs and improves their comprehensibility and legibility.
- [34_Pc_1] Signs should be designed to ensure that the expected observation/recognition distance does not exceed $D=172\,s$, where s is the symbol size, or $D=132\,h$, where h is the overall sign height.
- [34_Pc_2] Safety signs should not be oriented at a displacement angle greater than 15 degrees from the direct line of sight.
- [34_Pc_3] Safety signs should not be placed near other signs or objects of similar size and color, or where color contrast is poor.
- [34_Pc_4] For safety signs that indicate the escape route, multiple reinforcing directional signs are preferable to a single larger sign at a longer viewing distance.

35. Curry M, McDougall S, de Bruijn O. (1998). The effects of the visual metaphor in determining icon efficacy. *Proceedings of the Human Factors and Ergonomics Society 42nd Annual Meeting*, 1590-1594.

E			Pc	At	Co	Cr	Ps	Ac

A concrete icon means that an icon resembles its real world counterpart. There is a concern that concrete icons may increase the complexity of a sign over more symbolic icons. To test this, a group of 40 students were asked to rate of set of icons for their concreteness while another 40 students rated the same icons for their complexity (level of visual detail). Two hundred and forty icons were taken from existing examples used in signs for aircraft, electrical equipment, vehicles, computers and public information. The results showed that visual detail is not related to icon concreteness.

In the second part of this study, a search and match task was used to measure viewers' understanding of icon meaning. Participants were given a function name and asked to match the function to one of the nine icons in an array. Response time and accuracy were measured to determine the ease with which participants were able to match the meaning to an icon. The results showed that concrete icons appear to give users better access to meaning. Complexity did not affect the ability of participants to access icon meaning, and adding visual detail did not appear to make icons more concrete.

Key findings:
- [35_Co_1] Concrete (real-world) icons do not increase the complexity of a sign over more symbolic icons.
- [35_Co_2] Concrete (real-world) icons are more readily understood by users than more symbolic icons.
- [35_Co_3] Complexity did not affect the user's ability to understand the meaning of a concrete (real-world) icon, and adding visual detail did not appear to make icons more concrete.

36. de Fleur ML, Rainboth ED. (1952, December). Testing message diffusion in four communities: Some factors in the use of airborne leaflets as a communication medium. *American Sociological Review, 17*(6), 734-737.

		Pc	At	Co	Cr	Ps	Ac
M							

In the mid-twentieth century, airborne leaflets were sometimes used to deliver messages to a community. This study looked at the effectiveness of this delivery method as a function of the leaflet to population ratio and the distribution of drops over time.

In this study, leaflets were dropped over four similar towns and researchers conducted follow up interviews five days after the first drop. Leaflets were dropped on a single day or over three days (bad weather prevented a fourth day) at leaflet to population ratios that were low (four per person for a single day, one per person each day over multiple days) or high (16 per person for a single day, four per person each day over multiple days). In the high leaflet to population ratio communities, 64 % were aware of the message, compared to 34 % in the low ratio communities. In the single drop communities, 46 % of the population were aware of the message compared to 55 % in the multiple drop communities (these communities had 25 % fewer leaflets).

Key findings:
- [36_At_1] The improved effectiveness of messages that are repeated over internals (rather than massed at one time) is confirmed by a study of airborne leaflets dropped over communities.
- [36_At_2] Airborne leaflets delivered over three consecutive days resulted in a better informed community than 33 % more leaflets delivered in a single day.
- [36_At_3] Increasing the number of airborne leaflets delivered to a community by a factor of four resulted in almost twice as many people (64 % compared to 34 %) being aware of the message.

Key recommendations:
- [36_At_4] A large number of messages repeated at intervals rather than all at once improves community awareness of the message.

 # 37. Dixon SR, Wickens CD, McCarley JS. (2007, August). On the independence of compliance and reliance: Are automation false alarms worse than misses? *Human Factors: The Journal of the Human Factors and Ergonomics Society, 49*(4), 564–572.

Dixon SR, Wickens CD, McCarley JS. (2006, March). *On the independence of compliance and reliance: Are automation false alarms worse than misses?* (Technical Report AHFD-05-16/MAAD-05-04). Prepared for Micro Analysis and Design. Savoy, Illinois: Aviation Human Factors Division, Institute of Aviation, University of Illinois at Urbana-Champaign.

E

Pc	At	Co	Cr	Ps	Ac

An ideal alarm system is reliable. If an alarm on a piece of equipment tends to sound when there is no reason for the alert (false alarm-prone) or fails to sound when the alert condition exists (miss-prone), the operator will not be able to trust the system. It is difficult to make diagnostic aids perfectly reliable. A designer may choose to set the criteria for alarm with a liberal bias (tending toward false alarms) or a conservative bias (tending toward misses), or at a neutral point. The optimal setpoint may depend on the relative costs and benefits of correct or incorrect responses as well as on operator response.

It is theorized that operator behavior differs depending on whether there are more false alarms, expected to result in longer times to respond to the alarm (or no response at all – the "cry wolf" effect), or more missed alarms, expected to result in closer monitoring of the raw data behind the alarm, potentially disrupting concurrant or main tasks. Previous research indicates that these two behaviors (labeled compliance and reliance, respectively) may not be independent, and that the task at hand may also be disrupted by false alarms. This study looked at the effects of an unreliable alarm system on operator performance.

Thirty two undergraduate students at the University of Illinois participated in 100 trials (20 practice and 80 experimental) that lasted 30 seconds each. In each trial, participants had to carry out two simultaneous tasks. The first was to use a joystick to track a target, and the second was to press a button when a gauge exceeded a specified value, which changed trial-to-trial. For some participants, the gauge task was accompanied by an auditory alert. Participants were notified after each trial whether or not they had been successful. The four experimental conditions were: baseline (no audible alarm), alarm with perfect accuracy, 40 % false alarms, and 40 % misses. Participants were told in advance whether or not the alarm was reliable, and if not, in which direction the alarm failures would lie. Both the false alarm and missed alarm conditions resulted in significantly higher tracking errors than both no alarm and perfect alarm conditions, indicating that participants paid more attention to the gauges when systematic failures occurred, independently of the type of failure. The detection of an out-of-range gauge level and the response time were significantly worse for the false alarm condition than for the other conditions, supporting the hypothesis that compliance is an issue for false alarms only.

Key findings:
- [37_Cr_1] With an unreliable alarm system (either false alarms or missed failures), operators are distracted from other tasks by monitoring the information that the alarm should be handling.
- [37_Cr_2] An unreliable alarm system causes more task disruption than no alarm system at

all.
- [37_Cr_3] In a system prone to false alarms, the time to respond to the alarm is significantly longer, suggesting that the information is double-checked first.

 # 38. Dobbs M, Fung A. (2009). *Enhancing occupant response through neuro linguistics* (Poster Paper). Umow Lai Pty. Ltd., Victoria, Australia.
http://www.umowlai.com.au/pdf/publications/Fire/Enhancing%20Occupant%20Response%20T hrough%20Neuro%20Linguistics.pdf

M		Pc	At	Co	Cr	Ps	Ac

Rapid building evacuation in an emergency depends on a good warning system. The design of this system needs to expedite the decision-making process in order to reduce the procrastination time. The warning system needs to encourage evacuation, discourage delay, and counter the tendency to dismiss alarm cues when they become familiar. Factors that delay evacuation include the individual's role (such as bystander or visitor), commitment to a task, affiliation (assembling to form a group), and avoidance (denial or minimization of the situation). Factors that encourage evacuation include responsibility, desire for safety, and physical evidence like smoke. Case studies of occupant behavior in actual fire scenarios, such as the 1987 King's Cross station fire and the 1985 Bradford City Stadium fire, suggest that the factors that delay evacuation often outweigh factors that encourage evacuation. In accounts of fires in multi-story buildings, groups of people tend to evacuate much more slowly than individuals, delaying until fire cues were apparent and authority was recognized. Because of differences in prior experience and thinking patterns, individuals often respond differently to the same environmental cues. Understanding this may lead to more effective warning methods.

People can be categorized into four groups by communication preference: visual (40 % of the population), kinesthetic (instinct- and emotion-driven, 40 %), auditory (8 % to 12 %), and digital (data-driven, 5 % to 8 %). Good decision-making can be promoted by tailoring the warning message to each preference. For visual communicators, announcements should be made by a person in authority and visual distractions should be ended. Kinesthetic communicators are more likely to act when they are uncomfortable, which can include loud alarms, cutting TV and internet signals and ending other activities, and allowing the smell of smoke to pervade the space. Since kinesthetic communicators learn by doing, fire evacuation drills are beneficial. Auditory communicators alert to uncomfortable sounds such as the fire alarm signal. They may need to have their attention drawn from their iPods. A live voice message in an urgent tone is more effective than a prerecorded message, and a familiar (authoritative) voice is best. For digital communicators, information about the fire situation will prompt them to evacuate; they need to know what has happened and what actions need to be taken. A specific, live or dedicated prerecorded message and a crowd moving in a direction where they could learn more are beneficial.

Key findings:
- [38_At_1] People can be categorized into four groups by communication preference: visual, kinesthetic, auditory, and digital. Tailored warning methods can be used to move each group toward evacuation.

Key recommendations:
- [38_At_2] [38_Ac_1] To attract attention and get people to act, add to discomfort through loud alarms (for kinesthetic and auditory communicators).
- [38_At_3] Disrupt activities by ending visual distractions, turning up lights, cutting TV and internet signals, and ending group activities (for all - visual, kinesthetic, auditory, and digital

communicators).

- [38_Ac_2] Use fire evacuation drills to teach and practice the proper actions (for kinesthetic communicators).
- [38_At_4] [38_Cr_1] Provide a live warning announcement by a familiar authoritative source (for visual and auditory communicators).
- [38_At_5] [38_Ac_3] Provide specific information about the emergency situation, including what is happening, what is being done, and what actions need to be taken (for digital communicators).

 # 39. Driskell JE, Willis RP, Copper C. (1992). Effect of overlearning on retention. *Journal of Applied Psychology, 77*(5), 615-622.

P		Pc	At	Co	Cr	Ps	Ac

This paper describes a meta-analysis of past studies on overlearning and its effectiveness in enhancing performance. The effects of three factors that moderate the relationship between overlearning and retention were considered: the degree of overlearning, the retention interval, and the type of task (physical or cognitive). A typical study of overlearning sets a performance criterion that requires a measureable amount of practice to attain. The subjects then go through additional practice trials, and retention is measured after a specific time interval following the training session. The literature is consistent in showing that increased overlearning results in increased retention; the meta-analysis seeks to derive values for the magnitude of this effect and the relative contributions of various factors.

A search of the literature identified 15 studies with 3771 subjects and 88 separate hypothesis tests. Each hypothesis test was evaluated to determine the amount of overlearning, the retention interval, and whether the task was physical or cognitive. Overall, overlearning was found to improve retention moderately. With 50% overlearning there was a small effect on retention, with 100% overlearning there was a moderate effect on retention; and with 150 % overlearning there was a large effect on retention. Although effective for both physical and cognitive tasks, overlearning was more effective for cognitive tasks. For physical tasks, retention was actually enhanced by increasing time intervals between training and testing, possibly reflecting "cheating," or practicing of skills during the interim. For cognitive tasks, longer intervals decreased the effects of overlearning, with a half-life of 19 days. This suggested that refresher training at about three week intervals would be helpful for maximum cognitive performance. Differences in ability level of subjects affected the initial performance level but not the rate of retention.

Key findings:
- [39_At_1] Repetition of key parts of a message (overlearning) helps with retention.
- [39_At_2] Refresher training at about three week intervals maintains optimal performance for cognitive tasks.

40. Dudek CL. (2006, February). *Dynamic message sign message design and display manual* (Texas Department of Transportation Report No. FHWA/TX-04/0-4023-P3). Texas Transportation Institute, College Station, TX.

T		Pc	At	Co	Cr	Ps	Ac
				▒		▒	▒

Dynamic message signs (DMS) are electric signs with an alternating display. They are also referred to as variable message signs (VMS) or (previously) as changeable message signs (CMS), and are widely used by transportation managers to repeat a message for drivers as they pass by. The DMS can be in the form of permanent or portable signs. This manual provides guidance for the operation and message design for new and experienced users, as well as examples for specific incidents.

The intended audience of a DMS must be known when creating a message, whether it is an entire population or a specific subset. It is important to determine whether the audience is familiar or not familiar with the surroundings or route. An effective message provides relevant information that can be applied by the intended audience. Advice telling the readers what to do in their current situation should be placed in the last line of the message. The reason for the advice must also be supplied, or people will ignore it. The reason should be given in the first line of the DMS while the advice should be the second.

The maximum length of a DMS message is controlled by the reading time available for the sign. The three action verbs USE, TAKE, and FOLLOW have an equivalent effect when used on signs. The word USE is more commonly used because it is shorter. On roads, the verb USE should be used on DMS signs when discussing a route that will take a person to their destination, TAKE should be used to insinuate that they are beginning the first leg of a route, and FOLLOW should be used when people will be guided by other signs along the way.

When splitting a message, no more than two phases should be used for the DMS sign. Each phase should be understood independently. When reducing a message, unimportant and redundant information can be omitted.

Key findings:
- [40_Ps_1] The intended audience must be known when designing a message.
- [40_Co_1] It is important to determine whether the audience for a message is familiar or not familiar with the surroundings or route.
- [40_Ps_2] An effective message provides relevant information that can be applied by the intended audience.
- [40_Ac_1] Advice telling the readers what to do in the situation should be placed in the last line of the message. The reason for the advice must also be supplied, or people will ignore it.
- [40_Co_2] The maximum length of a DMS message is controlled by the reading time available for the sign.
- [40_Co_3] The meaning of keywords used in the message should be clear to the intended audience, and their use should be consistent.
- [40_Co_4]When splitting a message for a dynamic message sign (DMS), no more than two phases should be used. Each phase should be understood by itself.

 # 41. Dudek CL, Huchingson RD, Stockton WR, Koppa RJ, Richards SH, Mast TM. (1978, September). *Human factors requirements for real-time motorist information displays, Volume 1 – Design Guide* (U.S. Department of Transportation Report No. FHWA-RD-78-5). Texas Transportation Institute, College Station, TX.

T

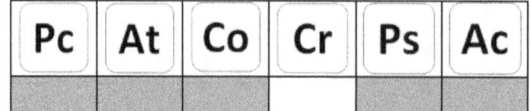

The purpose of this document was to provide guidelines for real-time driver displays used to manage traffic, including the development, design and operation of these displays. The information and recommendations are based on human factors laboratory and field studies, real-world experience, and expert opinion. A real-time control system is defined as "a control process with continuous monitoring of the system such that changes can be made in the system in ample time to improve the operation within the system." Specific recommendations for signage are provided for incident management of unusual conditions such as an accident.

During non-emergencies, there are two alternate philosophies regarding what to display on a sign:
- Always display a message to assure people that the sign is working and can provide assurance that everything is normal; or
- Display messages only when there is an emergency, because messages at other times are unnecessary, and unnecessary messages increase the likelihood of error.

The document favors the second philosophy. It is important for those operating the system to be specific about how the messaging is handled. The problem should be defined and the objectives of the sign should be discussed.

Advisory sign messages should contain a problem statement, an effect statement, an attention statement (the audience of the message), and an action statement. Research has shown that run-on or sequential messages are not good for drivers travelling at high speeds. However, a study showed that people can remember an eight-word message better when it is broken up rather than displayed in full. A message can be broken up by sequencing it or displaying it on multiple signs. When breaking up the message, it is necessary to assemble words into chunks that are compatible and memorable.

Repetition or redundancy helps viewers to remember the message. For a sign with more than one frame, the key word should be repeated in both frames to assist viewers in comprehending the sign. Presenting a keyword or phrase in a different color makes it more likely to be read first and increases its urgency.

Only three units of information should be displayed at one time if all three pieces of information need to be recalled by the viewer. Four units can be shown if one of the units is minor. A single word should be displayed for at least one second and a phrase for at least two seconds. If the message is in three or more phrases, the sign should be blank for one second between the phrases.

For a light-reflecting sign, the contrast ratio C between the brightness of the text or symbols (legend) and the background of the sign is defined as

$$C(\%) = (100)\frac{|R_B - R_L|}{R_B},$$

where R_B is the reflectance of the background and R_L is the reflectance of the legend. The contrast ratio should be at least 30 %, although recommended values are 40 % in daylight and 50 % at night. The required contrast also depends on the size of the letters and the viewing distance; the larger the sign, the lower the level of brightness needed to detect the sign from a farther distance. The color of the sign should contrast with its environment; a light background requires a dark sign and vice versa.

Figure 5 shows acceptable sign color combinations based on brightness contrast and the human eye sensitivity to color, shown in
Figure 6. Note that the eye is particularly sensitive to yellow and green, with a shift of the spectral response between day and night vision. If a sign is meant to be visible at night or with low lighting, colors in the middle of the spectrum should be used (light greens and yellows).

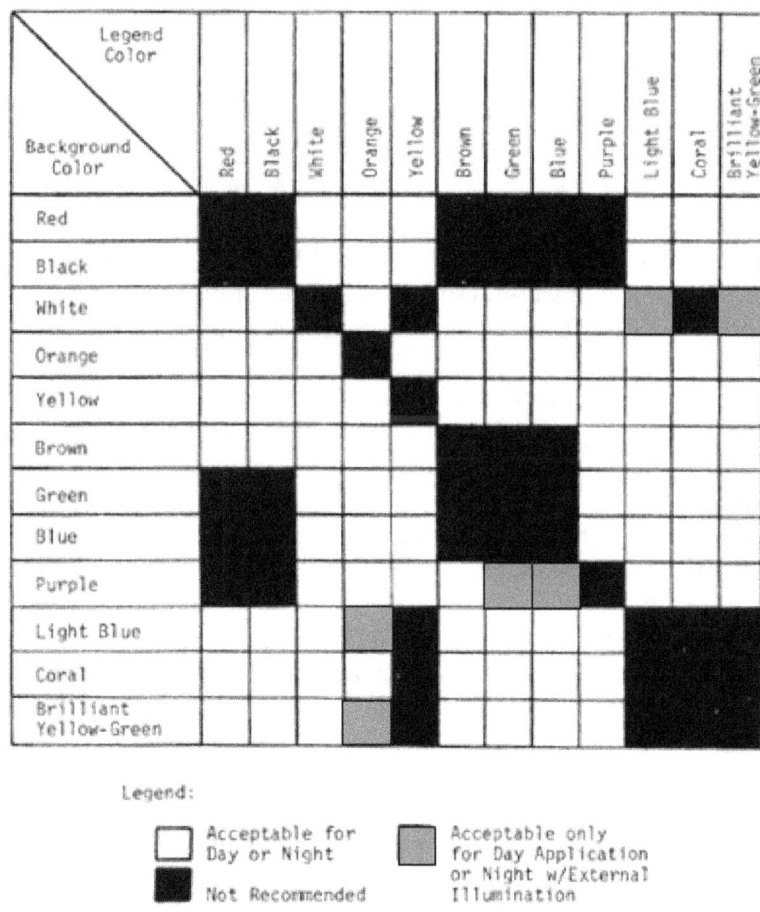

Figure 5: Acceptable color combinations given brightness contrasts.

161

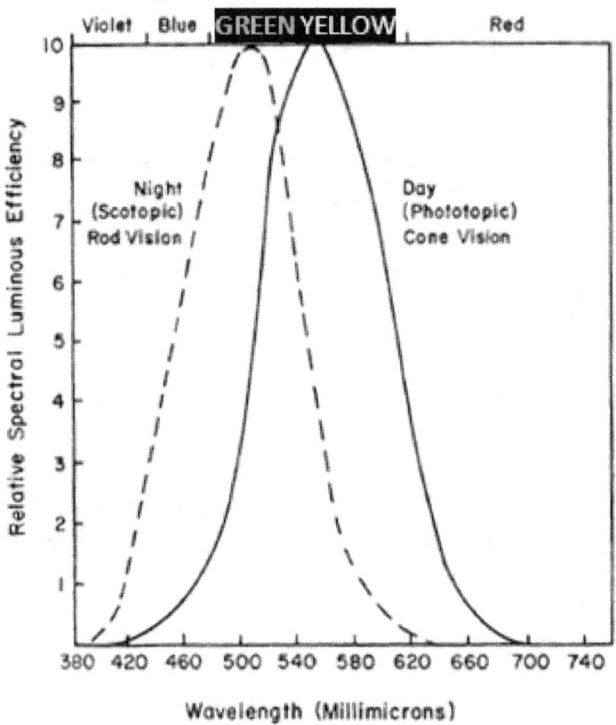

Figure 6: Human sensitivity to color.

In action message statements, the words USE, TAKE, and FOLLOW are synonyms and showed no preference; however USE is used most often because it is the shortest. TAKE has a more active connotation than USE. The verb GO is not good for visual signage, but can be used in audio messages.

Research shows that for words of between four to eight characters, the exposure time should be at least one second per word, excluding prepositions. This is recommended for messages of up to eight words. Long words such as street names or places should be considered two words. Field experience has proven that an eight-word message broken into two phrases and alternated on a two line display functions adequately.

To direct people to diverge from the normal path (of action), the message in its simplest form should contain the incident type, the location of the incident and the action that is required to be taken.

The rule of thumb for letter size indicates that for each one inch of letter height the sign is legible another 50 feet farther away. Spacing between lines should be three-quarters the average height of upper-case letters, and the borders above and below the text should equal the average letter height. The width of the side borders should equal the height of the largest letter.

The color red means stop, especially in traffic. Green means go, or that movement is allowed. Blue indicates general driving information, and yellow signifies a general warning. Black and white are used for regulations. Orange is used for construction and maintenance warnings. Brown indicates that there is a public recreation or scenic area ahead. Purple, yellow-green, light blue and coral are unassigned. The advantage of color coding is that the message can be seen further away.

There are three main language styles when orally giving a message: conversational, short form, and staccato. Short form and staccato are easier to recall. The first statement in an alert message should be an attention statement to alert the audience that a message is about to be given. The second statement should describe the problem. The final statement should present the action that is required to react to the problem. Conversational messages are presented in a style akin to radio commercials, using correct grammar. Short form does not use "carrier" phrases such as adjectives or modifiers, and is similar to the language used for secondary headlines in newspapers. Staccato is terse and telegraphic, like the langauge that is written on highway signs. For maximum recall, people must hear the message at least twice.

To reduce the message content and shorten the message, the description of the impact of the event can be shortened. This can be done by describing one effect rather than several. When repeating the message for the second time, it is only necessary to repeat the action statement. Limiting the alert signal to three seconds will also reduce the message length.

The language of an audio message should be concise rather than conversational. If the format of the message is standardized, people can anticipate what piece of information will come next. The vocabulary should be limited so that people can anticipate which words will be used and can process the information more quickly.

The voice in an audio message should be official, applicable to the audience at the time it is received, and perceived as relevant for the situation. The person delivering the message should have clear enunciation wtihout an obvious dialect. More time should be spent on syllables and less on pauses. Greater speech intensity and greater pitch variability are desirable. The speaker should speak loudly at a moderately fast rate without speaking in a monotone. The message should be read in a "calm, matter of fact, and dignified manner," at a speed of about 175 words per minute; a speed under 110 words per minute sounds too slow, and the message may be perceived as less important. No differences in intelligibility have been found between high and low pitched voices, although low-pitched female voices are commonly used in audible warning systems with good results. Both men and women speak in a frequency range of 200 Hz to 3000 Hz.

Outside noise can easily mask the sound of the warning. Standardized language should be used in order to deliver the maximum amount of information even if only part of the message is heard.

The most common attention-demanding signal uses a frequency of modulation between 1 Hz and 3 Hz. Pitch modulation works better than changes in loudness, with frequency varying by at least 100 Hz. The signal should be no longer than five seconds in duration, followed by 0.5 seconds of silence before the message begins. The alarm signal should sound at 10 dB above the sound level of the message.

Key recommendations:
- [41_Ps_1] Advisory sign messages should contain a problem statement, an effect statement, an attention statement (the audience of the message) and an action statement.
- [41_Co_1] People can remember an eight-word message better when it is broken up rather than displayed in full.
- [41_Co_2] For a DMS with more than one frame, the key word should be repeated in both frames to assist viewers in comprehending the sign.
- [41_Ps_2] Presenting a keyword or phrase in a different color makes it more likely to be

read first and increases its urgency.

- [41_Co_3] Only three units of information should be displayed at once.
- [41_Pc_1] For a light-reflecting sign, the contrast ratio between the text and the background should be at least 30%, although recommended values are 40% in daylight and 50% at night. The required contrast also depends on the size of the letters and the viewing distance.
- [41_Co_4] For words of between four and eight characters, the exposure time should be at least one second per word, excluding prepositions. This is true for messages of up to eight words.
- [41_Ac_1] To direct people to diverge from the normal path, the message in its simplest form should contain the incident type, the location of the incident and the action that is required to be taken.
- [41_Pc_2] If a sign is meant to be visible at night or with low lighting, colors in the middle of the spectrum should be used (light greens and yellows).
- [41_Pc_3] The rule of thumb for letter size indicates that for each one inch of letter height the sign is legible another 50 feet farther away. Spacing between lines should be three-fourths the average height of upper-case letters, and the borders above and below the text should equal the average letter height. The width of the side borders should equal the height of the largest letter.
- [41_At_1] [41_Ps_3] [41_Ac_2] Short form and staccato messages are easier to recall. The first statement in an alert message should be an attention statement to alert the population that a message is about to be given. The second statement should describe the problem. The final statement should present the action that is required to react to the problem. For maximum recall, people must hear the message at least twice.
- [41_Co_5] The language of an audio message should be concise rather than conversational. If the format of the message is standardized, then people can anticipate what piece of information will come next.
- [41_Co_6] The vocabulary of a message should be limited so that people can anticipate which words will be used and can process the information more quickly.
- [41_Co_7] The person delivering the message should have clear enunciation wtihout an obvious dialect.
- [41_Co_8] A speaker should spent more time on syllables and less on pauses.
- [41_At_2] Greater speech intensity and greater pitch variability are desirable.
- [41_Ps_4] The speaker should speak loudly at a moderately fast rate without speaking in a monotone. The message should be read in a "calm, matter of fact, and dignified manner," at a speed of about 175 words per minute
- [41_At_3] [41_Ps_5] A speed under 110 words per minute sounds too slow, and the message may be perceived as less important.
- [41_Co_9] No differences in intelligibility have been found between high and low pitched voices, although low-pitched female voices are commonly used in audible warning systems with good results.
- [41_At_4] The most common attention-demanding signal uses a frequency of modulation between 1 Hz and 3 Hz. Pitch modulation works better than changes in loudness, with the frequency varying by at least 100 Hertz.
- [41_At_5] The signal should be no longer than five seconds in duration, followed by 0.5 seconds of silence before the message begins.
- [41_At_6] The alarm signal should sound at 10 dB above the sound level of the message.

 # 42. Dudek CL, Ullman BR, Trout ND, Finley MD, Ullman GL. (2006, March). *Effective message design for dynamic message signs* (FHWA/TX-06/0-4023-5). Texas Transportation Institute, College Station, TX.

T		Pc	At	Co	Cr	Ps	Ac

This document reported on a project whose goals were to develop the logic needed to automate the decision support of the dynamic message sign (DMS) message design process and to prototype a DMS where the message was automated. The use of DMS in the State ofTexas was reviewed.

To be effective, a DMS must communicate a message in a short time. The design factors to be considered are content, length, load, and format.

From their experience with DMS, drivers have developed an expectation of message content. Information on the problem or reason for the sign is expected first, followed by the location of the problem. The understanding of DMS messages can be improved by the use of standardized ordering of words and message lines. In many cases, the message is too long to be displayed on one frame. Longer messages can be displayed on two frames, but using three or more frames is not considered acceptable. Abbreviations may be used when needed, but these should be widely understood abbreviations. More complicated messages are often harder to abbreviate as well.

Key findings:
- [42_Co_1] Drivers expect DMS to provide information on the problem or reason for the sign first, followed by the location of the problem. Using standardized ordering of words and messages lines can also enhance understanding.

 # 43. Dudek CL, Ullman GL. (2001, November). *Guidelines for changeable message sign messages: Annotated bibliography* (DTFH61-96-C-00048). Texas Transportation Institute, College Station, TX.

| T |

| Pc | At | Co | Cr | Ps | Ac |

Dudek and Ullman reviewed 53 reports and papers relating to messages and signs, specifically addressing transportation issues. These were derived from U.S. material during the period 1970 to 2001. A summary of their key findings are presented.

Source	Key Findings
Heathington, Worrall and Hoff (1970)	Examined the opinion of Chicago drivers regarding information presented to them on electronic signs in heavy, moderate and no congestion. They found that information on the cause of the congestion and the movement speed was rated highly during congested periods.
Dudek, Messer and Jones (1971)	Questioned 505 Texan motorists on the design of real-time information systems, including the need for real-time information, preferences on what information should be presented and where and how, and the potential driver response. Motorists favored real-time information, preferred changeable message signs and radio to phone and television, preferred to receive the information prior to experiencing the conditions (e.g., information on congestion before entering the affected road), and desired information on the nature and impact of the congestion. They expected to return to their normal routes as soon as practicable after responding to the traffic information.
Stockton, Dudek, Fambro and Messer (1976)	Evaluated three lamp matrix message signs in an attempt to establish whether the information provided was understandable and useful. This took place in Houston, 1973. They found that changeable messages were an effective tool as drivers acted to avoid congestion, as long as options were available. Of those that understood the information, most motorists acted upon the information.
Weaver, Dudek, Hatcher, and Stockton	Deliberately attempted to divert a section of traffic on its way to an event. They did this by introducing matrix signs. They found that 52.4 % of the drivers diverted their route specifically in response to the sign. This percentage had been corrected to take into account the normal routes used.
Dudek, Weaver, Hatcher and Richards (1978)	Diverted traffic from previously organized events. This was tested on events with specific and variable starting times. Lamp matrix changeable message signs (CMS) were used. For the event with variable starting times, 63 % to 83 % of the motorists used the diversion route, although when corrected, 30 % to 67 % of the motorists used this route specifically because of the sign. For the first event with a fixed starting time, 79 % to 89 % used the diversion, with 34 % to 44 % doing so because of the sign. For the second event with a fixed starting time, 39 % to 86 % used the diversion route, with 10 % to 58 % responding directly to the sign.
Dudek, Huchingson, Koppa, and	Performed laboratory research in several key topic areas in Texax, Minnesota, and California. These studies related to message design criteria, including the following topics: (1) minimum traffic information; (2) traffic state indicators;

Edwards (1978)	(3) congestion conditions; (4) lane blockage and availability; (5) incident types; and (6) temporal information. They found that:
	(a) drivers unfamiliar with the route requested information
	(b) the severity of the problem influenced whether information was required
	(c) the location of the problem (congestion) was necessary
	(d) the need for temporal information
	(e) the need for information on delay duration and alternative actions (e.g., diversions)
Stockton and Dudek (1978)	Looked at the legibility of lamp matrix signs in controlled studies. The character height was 18 inches, and the sign was legible at a mean distance of 840 feet (47 feet per inch of letter height).
Upchurch, Armstrong, Baaj and Thomas (1992)	Evaluated flip disk, LED and fiber optic VMS systems. Hired observers evaluated twelve signs from a vehicle according to target value and legibility distance. The measurements were conducted under midday, night, washout and backlight lighting conditions. Fiber-optic signs performed better than LED signs under most conditions for target value, legibility distance and viewing comfort.
Emmerink, Nijkamp and Van Ommeren (1996)	Examined the effects of radio messages and VMS on motorist route choice behavior based on a survey of road users conducted in the Netherlands. They found that women were less likely to be influenced by traffic information than men. Motorists were willing to pay for in-car information, with this factor positively linked to male drivers and people on business trips.
Benson (1996)	Surveyed 500 motorists in Washington DC regarding the message content of VMS systems. Approximately half of motorists stated that they regularly relied on the information provided on VMS. Demographics appeared to have only a modest impact, if any. Motorists favored messages that were simple, reliable and useful. This included the location of incidents, traffic information, and guidance (e.g., to avoid rubbernecking). Message contents that met with a less favorable response included delay estimates, safety messages, and the posting of alternative routes.
Wardman, Bonsall and Shires (1997)	Examined the impact of VMS systems on route choice selection. They found that the route choice of motorists can be influenced by the provision of information, assuming that viable, similar alternatives are available.
Proffitt and Wade (1998)	Examined a number of variables related to the comprehension of VMS signs. They found that 25 % of the target population (adults in Virginia) were weak readers who encountered problems reading textual messages. They also found that reading derives much of its speed from implicit knowledge produced through familiarity with the material. They deduced that VMS messages should use standardized content where possible to promote familiarity. They went on to examine warning signs, finding that warning signs should be concise in order to be effective.
Hustad and Dudek (1999)	Examined the extent to which abbreviations could be understood when used in CMS systems. Laboratory studies were conducted in New Jersey and Texas using driver volunteers. Eighteen abbreviations were presented and understood by 85 % of the motorists (considered ready for use), while others were understood by 75 % to 84 % of the motorists (deemed to require an

	educational effort), while the others were considered unacceptable. In general, the proper noun abbreviations were understood by drivers who lived in the same area as the facility or structure to which the sign referred.
Antila, Luoma and Rama (2000)	Conducted a field study to establish the visual demand of viewing three different VMS messages: a sign alternately displaying in Finnish and Swedish (two seconds each), a sign displaying the same messages simultaneously, and a sign displaying air and road surface temperatures in Finnish. The eye movements of 38 drivers were recorded. The alternating bilingual messages were no more demanding that the simultaneous displays.
Luoma, Rama, Penttinen and Anttila (2000)	Interviewed 225 drivers to examine their response to two VMS systems reporting road conditions. A message reporting slippery road conditions reduced the mean speed by 1 km/h to 2 km/h, and some reduction was reported for a sign that recommended minimum headway. However, the signs also had some negative effects, including refocusing attention, encouraging the drivers to look for evidence of the conditions, testing the conditions, and more conservative passing behavior.
Peeta, Ramos and Pasupathy (2000)	Studied drivers' diversion decisions and found that the level of detail of the relevant message content provided to drivers influenced their willingness to divert. This was also influenced by socioeconomic characteristics, familiarity with the road network, and the driver's confidence in the information being provided.
Dudek, Trout, Dunlop, Booth and Ullman (2000)	Conducted a series of laboratory studies to examine the impact of alternative words/phrases, interpretation of phrases, operation practices and use of dynamic message signs with lane control signals. They found that flashing words increased reading times.
Parentela and Eskander (no date)	Surveyed motorists in CA to examine their response to CMS systems. The results indicated that while most motorists paid attention to the CMS messages provided, 28 % considered them a distraction.
Dudek and Ullman (2001)	Examined the impact of frame rates upon information retention. Typically, CMS systems are displayed at a rate of one second per word or two seconds per line of information. Participating in the trials were 104 New Jersey motorists and 260 Texas motorists. The results showed that a display rate of four seconds per frame and one of two seconds per frame with one repetition did not make a difference in the comprehension of a two frame message by motorists.

Key findings:
- [43_Cr_1] [43_Co_1] [43_Ac_1] Information on the cause of an incident was appreciated by those affected.
- [43_Co_2] Real-time information is appreciated.
- [43_Co_3] Motorists prefer receiving information by changeable message sign (CMS) and radio rather than telephone and television.
- [43_Co_4] [43_Ac_2] Drivers prefer to receive information before experiencing the conditions. Information on the nature and impact of the situation is appreciated.
- [43_Ac_3] When relevant information was provided prior to an incident, motorists acted upon it.

- [43_Ac_4] Approximately 50 % modified behavior accordingly when provided with new information via a sign.
- [43_Ac_5] The effectiveness of a sign on influencing response depended on whether the motorists had to be at a place at a particular time or had some flexibility.
- [43_Ac_6] Drivers unfamiliar with the route requested information.
- [43_Ac_7] The severity of an incident influences the need and desire for relevant information.
- [43_Co_5] Information on the location of an incident is desired.
- [43_Ac_8] Suggestions for alternative responses are required.
- [43_Co_6] Under most conditions, fiber-optic signs performed better than LED signs for target value, legibility distance and viewing comfort.
- [43_Ac_9] Women appeared to be less likely to be influenced by information than men.
- [43_Ac_10] Motorists were willing to pay for in-car information (men more than women).
- [43_Co_7] Fifty percent of motorists regularly rely on the information provided by CMS. They favor messages that are simple, reliable and useful.
- [43_Ac_11] Route choice can be influenced by information, assuming alternatives exist.
- [43_Ac_12] Abbreviations can be used (with the understanding of 75 %), with proper nouns and familiarity aiding understanding.
- [43_Co_8] Alternating bilingual messages were no more physically demanding on the viewer than simultaneous displays.
- [43_Co_9] Reports on upcoming conditions had a direct impact on behavior (reduction in speed). There were also some negative effects, including refocusing attention, encouraging the drivers to look for evidence of the conditions, testing the conditions, and more conservative passing behavior.
- [43_Co_10] [43_Ac_13] The level of detail provided in a message influenced willingness to modify behavior.
- [43_Co_11] [43_Ac_14] Familiarity with the available routes influenced willingness to modify behavior.
- [43_Ac_15] [43_Cr_2] Confidence in the information influenced willingness to modify behavior.
- [43_At_1] Some motorists (28 %) considered the dynamic provision of information by CMS a distraction.
- [43_Co_12] A display rate of four seconds per frame and one of two seconds per frame with one repetition did not make a difference in the comprehension of a two-frame message by motorists.
- [43_Co_13] Flashing words increase reading times.

44. Dutta A, Fisher DL, Noyce DA. (2004, July-September). Use of a driving simulator to evaluate and optimize factors affecting understandability of variable message signs. *Transportation Research Part F: Traffic Psychology and Behaviour, 7*(4-5), 209-227.

The purpose of this study was to determine factors that affect the readability and comprehension by drivers of multiple phase messages displayed on variable message signs (VMS). The scenario for the investigation was traffic manipulation during incident management. Concerns include the limitations on sign size and the number of display characters, the order in which the multiple phases are read by the driver, and obstructions due to road curvature in horizontal and vertical directions and to large vehicles in traffic. The optimal duration of message phase displays under the cognitive burden of driving is uncertain, as is the best way to phrase a message over multiple phases.

The participants in this study were 25 male and 23 female university students with a mean age of 23 years old. Driving performance was evaluated for each participant during 24 trips through virtual tunnels in a driving simulator. These trials tested the effects of obstruction of the sign by traffic or road geometry, the sequencing of a two-phase message, message content, and the number of lane changes required by the message. Two sets of message phase timing were studied, with calculations based on either 0.5 seconds or one second per word. Distracting integers in the field of vision tested the attention level of participants and added to the mental load. Obstructions were represented by cutting the usual viewing time for the messages in half. Destinations were assigned for each trip, and road signs provided information about which lane to be in. Message efficacy was measured by the percentage of drivers responding appropriately by changing lanes and by the time (distance from the virtual VMS) at which the action was started.

Drivers were three times more likely to not take the appropriate action (changing lanes) when the message was displayed once than for a single repetition. In every case, drivers took longer to respond for the No Repetition compared to the Single Repetition condition. The smallest differences occurred when messages were longer. Drivers responded faster to a requirement for multiple lane changes than for a single lane change. The best driver performance was seen when message phases were displayed for a relatively short duration and viewed twice while the message was visible.

Key findings:
- [44_At_1] [44_Co_1] [44_Ac_1] Better driver response (faster, correct action) was seen when each of multiple message phases was displayed for a relatively short duration and repeated.

 # 45. Dziekan K, Kottenhoff K. (2007). Dynamic at-stop real-time information displays for public transport: Effects on customers. *Transportation Research Part A 41*(6): 489-501.

Pc	At	Co	Cr	Ps	Ac

Real-time information systems are becoming more and more common in public transportation, where they show the next departure time of a train of bus. This paper examined passenger needs and how passengers can be provided with useful data. An analysis of the international literature was perfomed regarding the effects of real-time information systems. According to the Science Applications International Corporation (SAIC, 2003), passengers looked at real-time displays more often than printed at-stop information. When real-time displays were available, travelers perceived their wait time as being shorter. The displays reduced uncertainty by informing the passengers of when the train or bus would arrive. Travelers prefer the real-time information signs, and there is higher customer satisfaction.

Two studies were evaluated in this paper. The first study looked at perceived wait time reduction. Questionnaires were given to travelers in the Netherlands one month before, three months after and sixteen months after real-time information systems were implemented. No other changes were made to the subway line during this period. The results showed that the perceived wait time was reduced. The second study looked at how people react to real-time displays outside of subway stations. The displays showed the final destination of the upcoming trains and the length of time before each train would reach the station. Two independent observers sat at the entrance to the station and counted the number of passengers running and walking into the subway. The results showed that more people ran into the station when the displays were turned on than when the displays were turned off. The effect on behavior was strongest when the number of minutes until the next train arrived was small.

The real-time information systems must work properly in order to demonstrate the positive effects discussed in this paper.

Key findings:
- [45_Ps_1] Real-time information systems in transportation systems cause travelers to perceive that their wait time is reduced and reduces their anxiety.
- [45_Ac_1] Real-time information systems influence behavior by encouraging passengers to run when the time before the arrival of the next train is small.

46. Easterbrook JA. (1959). The effect of emotion on cue utilization and the organization of behavior. *Psychological Review, 66*(3), 183-201.

P		Pc	At	Co	Cr	Ps	Ac

The traditional approach for an auditory warning is a very loud signal, on the assumption that sufficiently annoying alarms attract and maintain attention. Some recent alarms are designed to inform rather than alarm. In some places, special care baby units in hospitals, for example, it is best to inform the staff only. In factories and other noisy workplaces, the alarm needs to compensate for the use of hearing protectors. An auditory warning needs to not only sound at a suitable volume for the location; it also needs to be effective and appropriate for the particular population being warned.

This paper recommends a user-centered approach to warning design. In this approach, knowledge of the relevant user population enables design of an appropriate warning. Two important considerations for an alarm are that it can immediately be recognized as a warning and that the action in response to the alarm is clear. There are six steps in alarm design: Observe, Accept, Analyze, Investigate, Correct, and Monitor. Auditory icons can be divided into two types: abstract and representational. Abstract icons are similar to traditional alarm types, for which an arbitrarily chosen sequence is designed to have a particular meaning that is understood through training. Representational alarms incorporate either a sound with a known association, the actual sound itself, or a sound with similar qualities (e.g., heartbeats). In the absence of a general emergency, only specific people need be alerted to the situation.

Sometimes warning messages are confused because they are very similar. Auditory messages are more likely to be confused if they consist of only single continuous tones, have the same temporal pattern, or share a similar on/off pattern. Warning systems should consist of a wide range of different sound types to reduce confusion.

Key recommendations:
- [46_At_1] An auditory warning needs to not only sound at a suitable volume for the location; it also needs to be effective and appropriate for the particular population being warned. For instance, in special care baby units in hospitals, only the staff should be informed.
- [46_At_2] In factories and other noisy workplaces, the alarm needs to compensate for the use of hearing protectors.
- [46_Ps_1] A user-centered approach to warning design should be used.
- [46_Co_1] Auditory messages are more likely to be confused if they consist of only single continuous tones, have the same temporal pattern, or share a similar on/off pattern. Warning systems should consist of a wide range of different sound types to reduce confusion.

47. Edworthy J. (1998). What makes a good alarm? *IEE Colloquium Digest on 'Medical Equipment Alarms: The Need, the Standards, the Evidence'* (pp. 5-8). Ref. No 1998/432, The Institution of Electrical Engineers.

B

Pc	At	Co	Cr	Ps	Ac

Alarms are used in a variety of environments to warn individuals of danger or the potential for danger. Unlike visual alarms, it is possible for audible warnings to attract the attention of the individual independent of what they are doing and where they are looking. The author presents information on the characteristics of a successful alarm. Although the emphasis is on medical care settings, many of these characteristics apply to all types of buildings and settings.

A good alarm gets people's attention without startling them. People have been known to turn off alarms that are startling and forget to reactivate them. The alarm should be resistant to masking by any other sounds and should be easily to localize. For medical situations, the staff and the patients should all hear the alarm. The meaning of the alarm should be clear to the medical staff, and its characteristics should convey the level of urgency. A good alarm sounds only in the presence of danger, with no false alarms. Alarms should be 15 dBA to 25 dBA above the masked threshold level to be easily heard. In a normal office setting, 70 dBA to 75 dBA is a suitable level. Alarms should not be much louder than this level because there is nothing to gain except startling the people in the building. In a location where there is an unpredictable fluctuating noise level, adding harmonics within the 500 Hz to 4000 Hz band should minimize the chances of not being heard. Alarms with a lot of harmonics in this frequency range can be played at a lower signal-to-noise ratio than more simple tones, allowing the noise levels to be minimized. These alarms are also easier to localize. Once an alarm is detected, it is important that the meaning is understood and that communication can still take place. Also, it is important that alarms can be heard within all environments of a building, such as in operating rooms. Standardizing the meaning of alarms within similar settings (e.g., hospitals and other health care facilities) would reduce confusion and increase the understanding of medical alarms. Traditional continuous alarms are inappropriate because they can easily be confused, are hard to localize, and stay on until they are manually deactivated. They also distract from the task at hand and interrupt communication. A key factor that causes confusion in alarms is sharing similar temporal patterns like the same rhythm or repetition rate. Temporal patterning can help to communicate the level of urgency. Research has not determined definitively which sounds make the best alarms. In addition to traditional types such as horns, bells, and buzzers, more natural sounds that signal a specific action may be advantageous, especially if they reduce the need to learn a set of alarms. In environments where there are many natural sounds, however, an emergency alarm or alert that differs significantly from the ambient sounds may be best. As people become familiar with an alarm's reliability, they will match their response rates to the reliability of the alarm; if the alarm is 90 % reliable, they will respond 90 % of the time.

Key recommendations:
- [47_At_1] Alarms should attract attention but not startle.
- [47_Co_1] [47_Ps_1] The meaning of the alarm should be clear to the staff, and its characteristics should convey the level of urgency.
- [47_Cr_1] A good alarm sounds only in the presence of danger, with no false alarms.
- [47_Pc_1] Alarms are easily heard when they sound at 15 dBA to 25 dBA above the masked

threshold level. In normal office settings, this is achieved by setting the level of alarm between 70 dBA and 75 dBA.

- [47_Ps_2] In environments with unpredictable and fluctuating noise, the masking of alarms can be minimized by including several harmonics within the 500 Hz to 4000 Hz band. Alarms with rich harmonic content within this frequency range can be played at a lower signal-to-noise ratio than more simple tones, allowing the noise levels to be minimized. These alarms are also easier to localize.
- [47_Ps_3] Alarms need to be audible within all environments of a building, such as in operating rooms. The sound level should allow people to communicate.
- [47_Co_2] Standardizing the meaning of alarms within similar settings (e.g., hospitals and other health care facilities) would reduce confusion and increase the understanding of medical alarms.
- [47_Co_3] Traditional continuous alarms are inappropriate because they can easily confuse, are hard to localize, and stay on until they are manually deactivated. They also distract from the task at hand and interrupt communication
- [47_Co_4] A key factor that causes confusion in alarms is sharing similar temporal patterns like the same rhythm or repetition rate. Alarms for different purposes should have different temporal patterns.
- [47_Ps_3] Temporal patterning can help to communicate the level of urgency.
- [47_Co_5] [47_Ps_4] In addition to traditional alarms such as horns, bells, and buzzers, more natural sounds that signal a specific action may be advantageous, especially if they reduce the need to learn a set of alarms.
- [47_At_2] An emergency alarm or alert should differ significantly from the ambient sounds. In environments where there are many natural sounds, a more traditional alarm may be a better choice.
- [47_Cr_2] As people become familiar with an alarm's reliability, they will match their response rates to the reliability of the alarm; if the alarm is 90 % reliable, they will respond 90 % of the time.

 # 48. Edworthy J, Clift-Matthews W, Crowther M. (1998). Listener's understanding of warning signal words. In MA Hanson (Ed.), *Contemporary Ergonomics 1998: Proceedings of the Annual Conference of the Ergonomics Society* (pp. 316-320). London, UK: Taylor & Francis.

E		Pc	At	Co	Cr	Ps	Ac

This set of two studies builds on previous work on the acoustic characteristics of effective nonverbal auditory warnings to determine whether the same characteristics denote urgency for spoken warnings. The work explored the interaction between the semantic and the acoustic content of a message (its meaning and the way it is delivered) in determining how the message is perceived.

Two studies were conducted in which 43 participants rated spoken words for urgency, appropriateness, and believability. In the first study, a male and a female speaker read eight signal words (LETHAL, DEADLY, POISON, DANGER, BEWARE, WARNING, ATTENTION, and DON'T) in person to each participant. The speakers were instructed to read the words once in an appropriate voice and once in an inappropriate voice, although they were not instructed on how to do this. Each word was read twice, in random order. Words spoken "appropriately" were rated as being significantly more urgent, appropriate, and believable, as were words with more negative connotations. The word DEADLY was rated at the high end of the urgency scale and DON'T was rated at the bottom.

In the second study, synthetic male and female voices were used to pronounce the same eight words as had been used in the first study. The inappropriate voice tone was achieved manually through manipulation of the speed and pitch of the words spoken in an appropriate manner. The results for urgency were the same as in the first study. The ratings of appropriateness and believability, however, did not show a significant difference between appropriate and inappropriate voice tones for the synthetically manipulated voices.

Key findings:
- [48_Ps_1] A message delivered in a higher pitched voice and at a faster speech rate is perceived as being more urgent.
- [48_Ps_2] Although an "appropriate" delivery of warning words by human speakers gave them the perception of being more believable and appropriate than words spoken "inappropriately," this was not found to be related to pitch or speech rate
- [48_Ps_3] Words with a highly negative connotation are considered more urgent, appropriate, and believable.
- [48_Ps_4] Synthetic voices can be manipulated to increase their level of urgency, but it is more difficult to change the perceived appropriateness and believability.

 # 49. Edworthy J, Hellier E, Rivers J. (2003). The use of male or female voices in warnings systems: A question of acoustics. *Noise & Health, 6*(21), 39-50.

Pc	At	Co	Cr	Ps	Ac

Speech warnings must be alerting and intelligible, even in distracting environments with high noise levels and high workloads. Noisy environments pose both acoustics challenges (signal masking and loss of intelligibility) and cognitive challenges (arousal, task performance, attention). Studies have investigated whether the male or female voice is better suited for a communication system. Male and female voices differ by both acoustic quality and by association with social roles. Although gender roles are blurring, female voices are generally associated with nurturing and security, while male voices are associated with authority and strength. In certain environments, either male or female voices would be unusual, and therefore may attract attention. Research has indicated that female listeners are more sensitive to the sex of the speaker, and that female voices are preferred in some settings, although the results are not definitive. In noise, the intelligibility of a message depends on the overlap of the voice spectrum with the noise spectrum. Multiple voices with different characteristics (e.g., alternating male and female) may be advantageous. Digital enhancement and filtering can improve intelligibility in noise, as can technology (such as neural networks) that promotes the signal-to-noise ratio by adapting to the current noise level. Communication of emotional state can be very important in conveying urgency. The acoustic properties that communicate urgency have been determined and demonstrated in speech manipulation. Female voices have been found to communicate urgency better than male voices. Rare and specific messages need to be more intelligible than common messages; attention needs to be given to the content of these messages. Nonverbal acoustic signals and verbal signals may be necessary in some environments.

This study looks at whether the male-female voice range is a continuum, which would suggest that the selection of voice is a question of acoustics rather than sex. Identification of the sex of the speaker has been found to depend on acoustic (pitch, pitch range, level, speed, spectral structure) and phonetic (vowels, aspiration or breathiness) cues. Thirty undergraduate students (24 females and six males, ranging in age from 18 to 48 with mean age of 22 years) at the University of Plymouth listened to 40 prerecorded words and ranked the urgency from one to 10, in a quiet environment. The words consisted of 10 signal words spoken urgently (approximately 80 dBA) and nonurgently (approximately 65 dBA to 70 dBA) by two female speakers. One of the female speakers had an unambiguously female voice (fundamental frequency approximately 600 Hz) while the other was ambiguous in terms of the speaker's sex (fundamental frequency approximately 400 Hz). The subjects were evenly divided into three experimental groups that were either told that both speakers were female, told that one voice was female and the other voice was male, or given no information as to the sex of the speakers. Words spoken urgently and the unambiguous female voice were both rated as being statistically more urgent. There was no effect between the three instruction groups, so the difference in perceived urgency was based on the acoustic characteristics of the voices rather than on the perceived sex of the speakers. The choice of speaker should therefore depend on the acoustic qualities of the voice relative to the noise spectrum, not the sex of the speaker.

Key findings:
- [49_Ps_1] Urgency is better perceived in messages that are delivered urgently and in a higher pitched voice.
- [49_At_1] Voice acoustics matter more than the sex of the speaker. Voices for warning messages should be selected based on minimal overlap with the ambient noise spectrum.
- [49_Ps_2] Female voices may retain more emotional content (urgency) after acoustic manipulation to improve intelligibility in noise.
- [49_Ps_3] The higher perception of urgency in unambiguous female voices was based on acoustic characteristics rather than on the perceived sex of the speaker.

Key recommendations:
- [49_Co_1] Intelligibility should be tested under the ambient noise conditions.
- [49_Co_2] Rare and specific messages need to be more intelligible than common messages to be understood.
- [49_At_2] In a complex noise environment, two voices with different characteristics (e.g., male and female) can be alternated to attract attention.
- [49_Cr_1] [49_Ps_4] The message content (including the choice of signal words to convey importance, an explicit description of the risks, and the use of personal pronouns) is more influential than the sex of the speaker in promoting compliance.
- [49_At_3] [49_Co_3] The choice of speaker should depend on the acoustic qualities of the voice relative to the noise spectrum, not the speaker's sex.

 # 50. Emergency Lighting Section of the National Electrical Manufacturers Association. (2001). *Exit Sign Brightness For Visibility and Safety* (LSD 13-2001). A NEMA Lighting Systems Division Document.

I		Pc	At	Co	Cr	Ps	Ac

NEMA conducted a review of exit sign technologies that specifically addressed issues of brightness. The minimum analytical luminance associated with a visibility from 100 feet has been removed from the NFPA 101 standard. This standard has shifted from analytical measurements for establishing compliance with minimum visibility requirements to a subjective observation test; i.e., moving from a prescriptive-based to a performance-based test.

The Emergency Lighting Section evaluated the brightness of exit signs employing different light sources in clear air. The exit sign brightness test demonstrates that there is a significant difference in the brightness levels provided by different technologies. The LED exit sign clearly provided far higher levels of brightness than the other technologies examined. The luminance levels produced by LED signs (measured over 30 minute, zero minute, and 90 minute periods) were typically two orders of magnitude greater than those produced by photoluminescent signs (either charged with incandescent or fluorescent sources) or tritium signs. LED luminances ranged from 29.3 cd/m^2 to 36.5 cd/m^2.

Collins (NIST) and Boyce (Lighting Research Center) conducted testing into exit sign visibility. Collins concluded that a minimum level of 10 cd/m^2 is required for reasonable visibility in both clear and smoky conditions, while Boyce concluded that the minimum should be 8.6 cd/m^2, with an average requirement of 15 cd/m^2.

Key findings
- [50_Pc_1] The luminance of LED lights decreased over time from 36.49 cd/m^2 after 30 minutes to 29.27 cd/m^2 after 90 minutes.
- [50_Pc_2] The luminance levels of LED signs were typically two orders of magnitude greater than those of photoluminescent signs or tritium signs.
- [50_Pc_3] The minimum luminance of exit signs should be between 8.6 cd/m^2 and 15 cd/m^2 to be effective.

 # 51. Friedman K. (1988). The effect of adding symbols to written warning labels on user behavior and recall. *Human Factors, 30*(4), 507-515.

E		Pc	At	Co	Cr	Ps	Ac

This study specifically addresses four main issues:
1) Whether consumer product warning labels that contain symbols are more effective than warning labels that do not;
2) Whether warning symbols affect the user's behavior when using the product;
3) Whether familiarity with a product affects behavior; and
4) Whether the type of precautionary action a user must take before using a product affects their behavior.

For the experiment, labels were varied to see the influence of adding symbols to written warning information. Symbols were either proactive or reactive, where proactive symbols showed the appropriate action to avoid the danger associated with the product and reactive symbols showed the potential consequence if the precautionary action was not taken. The 144 undergraduate students participating in the study selected a familiar and an unfamiliar product, each bearing a warning message. The warning was noticed by 88 % of the subjects, but only 50 % of those who noticed actually read the entire warning. The more hazardous the product, the more likely the warning was read. A warning symbol on the label significantly increased the perceived hazard of the product.

Key findings:
- [51_At_1] Although a warning label on a product was noticed by 88 % of the subjects, only 50 % of those who noticed actually read the entire warning.
- [51_Ps_1] The more hazardous the product, the more likely the label warning was read.
- [51_Ps_2] The use of warning symbols on consumer products increases the perception of hazard.

 # 52. Gat IB, Keith RW. (1978). An effect of linguistic experience: Auditory word discrimination by native and non-native speakers of English. *International Journal of Audiology, 17*(4), 339 – 345.

	Pc	At	Co	Cr	Ps	Ac
A						

The ability of listeners to discriminate between words is affected by their linguistic experience. The frequency of use in their native languages of sounds, phonemes and syllables that overlap with English is related to the ability of non-native speakers to discriminate sounds used in English speech. Noise signals that are louder than speech by 8 dB or more have been found to interfere with intelligibility, although the ability to discriminate speech in the presence of noise varies considerably from one individual to another. This study looks at the difference between native and non-native English speakers in their ability to understand speech in noise.

The subjects in this study were 18 male graduate students: six participants (23 to 26 years old) who had been raised in the United States, six participants (23 to 34 years old, from Korea, Brazil and Israel) who had been in the United States for three to four years, and six participants (23 to 28 years old, from Japan, Korea, Brazil, Egypt, and Ethiopia) who had been in the United States for one year. Lists of single syllable words were prerecorded and played at 76 dB SPL through a headphone with four noise levels: silence and white noise at signal-to-noise ratios (SNR) of 12, 6, and 0 dB. The subjects were instructed to listen to and write down the words as they heard them. All groups scored high in the quiet condition. Non-native speakers showed a statistically more significant decline during the background noise conditions; the scores of both non-native speaking groups declined about twice as fast with noise as the native speaking group. Individual ability to discriminate speech in noise showed a wide range for the native speakers (standard deviation of 16.7 out of 100 words) but a much narrower range for non-native speakers. At the highest noise level (SNR of 0 dB), native speakers were able to correctly write down about 73 % of the words on average, while the more and less experienced groups of non-native speakers were able to identify about 46 % and 36 % of the words respectively. The number of errors also depended on the commonality of word usage in the English language.

Key findings:
- [52_Co_1] The ability to discriminate words in noise deteriorates about twice as fast for non-native speakers as for native speakers.
- [52_Co_2] Individual native speakers differ significantly in their ability to discriminate speech in high noise levels.
- [52_Co_3] The ability to discriminate words in speech depends on the commonality of their use.

Key recommendations:
- [52_Co_4] Masking sounds need to be reduced as much as possible to improve the comprehension of auditory warnings by both non-native and native speakers.

 # 53. Gilbert T. (1957). Overlearning and the retention of meaningful prose. *Journal of General Psychology, 56*(2), 281-289.

P		Pc	At	Co	Cr	Ps	Ac

This paper was directed at improving the definition of overlearning, in order to broaden its applicability and therefore its usefulness. Overlearning was originally defined as the ratio of the number of training trials after mastery to the number of trials required for mastery. However, this did not explain the apparent higher retention of slow learners, who spend more time with all material and thus greatly overlearn some parts while mastering the rest. The author argues that defining overlearning as the ratio of the number of correct responses to items in a training task (the responses after mastery) to the total number of items increases the utility of the concept. The intent of the experiment is to examine the retention of verbal material as a function of the degree of initial training and the lapse of time.

Twenty-seven enlisted male soldiers with 12 to 13 years of education and ranging in age from 20 to 36 were read a passage that contained 22 facts about a fictitious country. The passage was read at a rate of about two words per second. After hearing the passage, they were asked questions about the different facts in the story. Once a fact had been mastered according to the criteria for a given trial, that fact was omitted from the passage and the new passage was read to the subject. This continued until all facts had been adequately mastered. One third of the subjects had facts removed after answering the question correctly one time (0 % overlearning). The other two equally sized groups had the facts removed after two and three sequentially correct responses to the question (100 % and 200 % overlearning respectively). All subjects were asked to come back later to take part in an unrelated study (to avoid further thinking about the material) taking place either 15 minutes, 24 hours, or 48 hours later, with the same number of subjects for each time delay. Overlearning resulted in significantly greater recall for all time intervals. The 200 % group recalled all 22 items after 15 minutes, compared with 19.3 items for 100 % overlearning and 15.0 items for no overlearning. A substantial decrease in recall occurred between 15 minutes and 24 hours, with no significant change afterwards. The curves showing the number of items recalled as a function of time are parallel for the three overlearning conditions, indicating that the method of retention is independent of overlearning. This task did not demonstrate diminishing returns with greater overlearning.

Key findings:
- [53_At_1] The amount of information recalled, both immediately and after one or two days, increases with the number of repetitions.

54. Glover BL, Magurno AB, Murray LA, Wogalter MS. (1996). Pictorial negation: Preferences for different circle-slash variations. *Proceedings of the Human Factors and Ergonomics Society 40th Annual Meeting*, 910-914.

Pictorials can be used to communicate both positive and negative messages. Pictorials with a negative message usually show a symbol portraying an action in a red circle with a slash over the picture. This experiment was designed to compare variations in the characteristics of the slash. The following negation signs were examined in the study: a red ring with a slash over a symbol, a red ring with a slash under the symbol, a red ring with a partial slash, a red ring with no slash, and a translucent slash. Pictorial orientation was also a factor in this experiment. Sixty volunteers participated in the study. Sixteen pictorial concepts were used. The pictorials were printed on laminated cards. The area of red took up 35 % of the total area inside the outer rim of the circle, leaving 65 % of the area for the symbol, as recommended by ISO 3864. Participants were given cards in sets of eight versions of the same pictorial concept and were asked to lay them on the table in order of how effectively they believed the pictorial would convey the message. The circle with the slash over the symbol and the slash under the symbol were the most preferred. The translucent slash was next, and the least preferred was the partial slash. Familiarity could have influenced the preferences of participants. The placement of the slash within the pictorial should be selected so as not to block the symbol. The problem of concealment can sometimes be solved by changing the orientation of the symbol; however, some designs must be modified.

Key findings:
- [54_Co_1] Circles with a solid slash covering the symbol and a solid slash under the symbol were preferred to translucent or partial slashes for negative pictorials.

55. Haas EC, Casali JG. (1995). Perceived urgency of and response time to multi-tone and frequency-modulated warning signals in broadband noise. *Ergonomics, 38*(11), 2313-2326.

E		Pc	At	Co	Cr	Ps	Ac

The design of auditory warning signals can affect operator performance. Certain signal characteristics can be modified to match the urgency perceived by the listener with the urgency required by the situation. Warnings must also be designed to be audible within the environment in which they are used, without being so loud and distracting that people endanger themselves by shutting them off. Many industrial and military environments contain continuously-operating machinery, which generate noise that can mask certain warning signals and make them difficult to hear. One type of warning signal that can be tailored to specific situations uses a short pulse of sound as a building block that can be combined with intervals of silence to generate a sound burst. Many acoustical and timing parameters can be adjusted to improve the audibility in a noisy environment as well as elicit a desired response from the listener. This set of experiments studied the effect of pulse parameters in these auditory warning signals on perceived urgency and response time in the presence of background noise.

In this study, the pulse parameters studied were pulse format (a sequential pulse, a simultaneous pulse, and a sawtooth frequency-modulated pulse format), pulse level (65 dBC and 79 dBC) and inter-pulse interval (0, 150, and 300) milliseconds. The simultaneous pulse consisted of four pure tones at (500, 1000, 2000 and 3000) Hz, sounding simultaneously during a single pulse duration. The sequential pulse consisted of four pure tones at (500, 1000, 2000, and 3000) Hz, sounding sequentially during a single pulse duration. The sawtooth frequency-modulated pulse format consisted of a pure tone carrier that rose and fell in a sawtooth pattern between 500 Hz and 3000 Hz during a single pulse duration. The inter-pulse interval is defined as the time duration between the end of the offset of one pulse to the onset of the next. The pulse level is defined as the root mean square (RMS) sound pressure level (SPL) of the pulse in dBD as measured at the center of the participant's head.

The environments of interest for this study were military or industrial settings that contain machinery generating steady-state noise, in which operators experience a high level of demand for their attention. The noise in these environments can be represented by pink noise, which contains equal amounts of energy in octave bands. In the experiment, 36 participants (18 men and 18 women), ranging in age from 18 to 22, were exposed to pink noise at 68 dBC while seated in a semi-reverberant acoustic chamber in the U.S. Army Research Laboratory at Aberdeen Proving Ground in Maryland. They were given a standard workload task in order to impose demands on their attention. Participants were presented with auditory signals containing combinations of the three independent variables tested in this study (pulse format, pulse level, and inter-pulse interval), which resulted in 18 different auditory signals (i.e., a 3 x 3 x 2 full factorial study), each of which was presented twice in random order. Each signal consisted of a train of eight pulses, each pulse lasting 350 milliseconds. Participants first rated the urgency of each signal. After a 10-minute break, their response time to each signal was measured by how long it took to press a button on a one-button-keypad with their dominant hand while performing the workload task. After a 30-minute break, the participants were presented with 153 signal pair combinations (all unique combinations of 18 signals taken two at a time) in random order and asked to indicate which signal from each pair was more urgent. The workload task was assessed to be sufficiently demanding on the attention of the participants for the purposes of this experiment.

The results showed that perceived urgency increased strongly with increasing pulse level. Signals with shorter inter-pulse intervals were perceived as significantly more urgent, with signals with no inter-pulse interval (zero milliseconds) considered the most urgent. For each signal type, urgency was rated highest when the pulse level was the highest and inter-pulse interval was the shortest. Finally, sequential signals were rated as significantly less urgent than simultaneous and frequency-modulated signals.

Response time was significantly shorter for the highest pulse level than for the lowest, with a difference of 60 milliseconds. Although this is a very short delay, it can make a difference in a job requiring immediate response, such as piloting a fighter jet. The mean response time for sequential signals was significantly greater (up to 40 milliseconds) than for simultaneous and frequency-modulated signals. Response time was independent of inter-pulse interval. The correlation between results for perceived urgency and response time indicates that response time is fastest when perceived urgency is greatest.

Through variation of parameters, proper design of pulse-based signals can assist listeners in deciding how urgently they need to respond.

Key findings:
- [55_Ps_1] For pulse-based signals in background noise consistent with continuously-operating machinery, perceived urgency increases with increasing pulse level and with decreasing inter-pulse interval, and also depends on pulse format.
- [55_Ps_2] Certain signal characteristics can be modified to match the urgency perceived by the listener with the urgency required by the situation.
- [55_Ps_3] Through variation of parameters, proper design of pulse-based signals can assist listeners in deciding how urgently they need to respond.
- [55_Ac_1] For operators engaged in a task in a noisy environment, response time was shorter by 60 milliseconds for a higher pulse sound level compared to that for a lower pulse level. This is enough to improve the performance of a fighter jet pilot.

Key recommendations:
- [55_Pc_1] Warnings must be designed to be audible within the environment in which they are used, without being so loud and distracting that people endanger themselves by shutting them off.

 # 56. Haas EC, Edworthy J. (1996, August). Designing urgency into auditory warnings using pitch, speed, and loudness. *Computing & Control Engineering Journal, 7*(4), 193-198.

A			Pc	At	Co	Cr	Ps	Ac

This article explores the effects of pitch, speed, and loudness of auditory signals on a listener's perception of urgency and response time. The goal is to develop alarms whose perceived urgency matches the situation, whose meaning is easily understood, and that are not so annoying that the operator shuts them off, potentially leaving the operator unwarned the next time the problem occurs. The findings show that auditory signals with high frequency, fast speed, and high sound levels led to the highest ratings of perceived urgency and the fastest response times. The article introduces a pulse-oriented methodology for the design of auditory warning signals for aircraft and other environments where fast response time and attention to the level of urgency are critical. Pulse design guidelines can be found in several international standards on warning signal design, including ANSI S3.41-R2008 (Audible Emergency Evacuation Signal). An advantage of the signal-pulse method is that different levels of urgency can be built into the warning itself, through selection of parameters including the fundamental frequency, pulse level, and inter-pulse interval.
*

Thirty college students (15 males and 15 females) between the ages of 18 and 40 participated in this study. Participants listened to 27 auditory signals, each consisting of a train of four pulses that varied in pulse fundamental frequency, inter-pulse interval, and pulse level. The inter-pulse interval was the time between the end of the offset (decay) of one pulse and the beginning of the onset of the next pulse, and the pulse level was the RMS sound pressure level (SPL) of the pulse measured in dB LIN at the center of the participant's head. The three fundamental frequencies tested were 200, 500 and 800 Hz, with each signal pulse containing the first four harmonics in addition to the fundamental. The inter-pulse intervals tested were zero, 250 milliseconds and 500 milliseconds, and the pulse levels were set to five, 25, and 40 dB LIN SPL above the ambient noise level of the chamber. Each participant performed 54 tasks, with the 27 signals repeated once. The two dependent measures were the participant's verbal estimation rating of signal urgency and his/her response time in milliseconds. The response consisted of pressing a pushbutton with his/her dominant hand as soon as the signal sounded.

All three parameters were found to affect the perception of urgency. Signals with a fundamental frequency of 200 Hz sounded significantly less urgent than 500 Hz or 800 Hz. Participants rated signals with an inter-pulse interval of zero milliseconds as significantly more urgent than signals containing an interval of 250 milliseconds, which was rated as significantly more urgent than signals containing a 500 milliseconds interval. Finally, participants rated the signals 40 dB LIN above ambient as significantly more urgent than signals at the two lower sound levels. The three variables interacted together in the following way: signals with the highest fundamental frequency (800 Hz), the highest sound pressure (40 dB LIN) and the shortest inter-pulse interval (zero milliseconds) were rated as significantly more urgent than any of the other signals.

Participant response time was affected by the pulse sound level and fundamental frequency only. The response time was significantly shorter for pulse fundamental frequencies of either 500 Hz or 800 Hz than for 200 Hz, and response times for signals 40 dB LIN above ambient were the shortest.

The difference in response time between signals at 5 dB LIN and 40 dB LIN above ambient was 60 milliseconds, which may be important for a fighter pilot.

In the absence of limitations on sound pressure levels above ambient, signal-pulse alarms that need to communicate the highest level of urgency and promote the shortest response time should employ signals with high fundamental frequency (800 Hz), no time between pulses, and high sound pressure level above ambient (40 dB LIN). For intermediate and low levels of urgency, the information within this paper can be used to design a series of signal-pulse alarms. To follow signal design recommendations of 15 dB SPL to 30 dB SPL above ambient (described within ISO 1986), urgent alarms can employ pulse signals with a fundamental frequency above 500 Hz and zero milliseconds inter-pulse interval.

* An in-depth analysis of pulse parameters (i.e., fundamental frequency, the degree of inharmonicity of the pulse, and pulse amplitude envelope) and burst parameters (i.e., burst rhythm, speed, melodic pattern, and pitch contour and length) can be found in an article by Edworthy et al. [Edworthy J, Loxley S, Dennis I. (1991). Improving auditory warning design: Relationship between warning sound parameters and perceived urgency. *Human Factors, 33,* 205-231]. These parameters have been shown to have clear and consistent effects on perceived urgency of auditory warnings.

Key findings:
- [56_Ps_1] Auditory warnings consisting of signal pulses can built different levels of urgency into the warning itself, through selection of parameters including the fundamental frequency, the pulse sound level, and the inter-pulse interval.
- [56_Ps_2] Auditory signals with high frequency, fast speed, and high sound levels result in the highest ratings of perceived urgency.
- [56_Ac_1] Auditory signals with high frequency, fast speed, and high sound levels result in the fastest response times.

 # 57. Hancock HE, Rogers WA, Fisk AD. (1999). Understanding age-related differences in the perception and comprehension of symbolic warning information. *Proceedings of the Human Factors and Ergonomics Society 43rd Annual Meeting, 43*, 617-621.

E		Pc	At	Co	Cr	Ps	Ac
				▓			

Visual acuity, visual searching, working memory capacity, and language comprehension have been shown to decline with increased age. This study looks at age-related differences in the comprehension of warning symbols and whether they are perceived as helpful.

A questionnaire about warning symbols was sent to 5000 people in the greater Atlanta area. The participants were asked how helpful they believed different types of warning symbols to be. One symbol identified a hazard, one prohibited a course of action, and one specified an action to perform. The participants could identify the symbols as very helpful, somewhat helpful, or not very helpful. The second part of the study assessed the recognition and comprehension of 12 specific warning signs. The participants were asked whether they had ever seen the symbol before and to write down in their own words what they believed each symbol meant. Overall, the majority of the respondents believed that each symbol was helpful in comprehending the meaning.

Significant age-related differences were found in the response for some symbols, but not all. In all cases where there was an age-related difference, younger adults had a better understanding of the symbols than the older adults. Adults of all ages believed symbols to be very helpful in general. The older group exhibited significantly poorer comprehension for half of the symbols compared to the younger group.

Key findings:
- [57_Co_1] Older adults report that warning symbols helped them to understand the hazard, but when asked to identify the meaning of the symbols, older adults showed poorer comprehension than younger adults for a number of commonly used symbols.

 # 58. Hancock HE, Rogers WA, Schroeder D, Fisk AD. (2004). Safety symbol comprehension: Effects of symbol type, familiarity, and age. *Human Factors: The Journal of the Human Factors and Ergonomics Society, 46*(2), 183-195.

E

Pc	At	Co	Cr	Ps	Ac

Age-related differences in symbol comprehension were studied using people divided into groups younger and older than 52 years old. Safety symbols were presented on a page in a booklet, with the instruction for participants to write down as many phrases as came to mind when viewing each individual symbol. The first phrase that the participants wrote down was considered an indication of their assessment of the concept associated with that symbol. A total of 40 safety symbols of four types (hazard alerting, mandatory action, prohibition, and information) were evaluated. Hazard symbols warned against specific hazards. Mandatory action symbols prescribed a specific action to be taken. Prohibition symbols showed a specific action to avoid because of the hazard. Information symbols showed information related to general safety, such as equipment location. The symbols were presented in paper test booklets, with each symbol presented on a separate page.

The younger adults generated more phrases than older adults. The first phrase generated for a symbol by younger participants was significantly more accurate in identifying the meaning than the first phrase generated by older participants. First phrases generated for prohibition symbols were the most accurate, followed by those for information symbols and then mandatory action symbols. The younger group also had better overall comprehension of the ANSI safety symbols tested in the study.

Key findings:
- [58_Co_1] People older than 52 years old generated fewer phrases associated with safety symbols than younger participants.
- [58_Co_2] The first phrase that was written by younger participants was a more accurate description of the meaning of safety symbols than those written by older participants.

 # 59. Hellier E, Edworthy J, Weedon B, Walters K, Adams A. (2002, Spring). The perceived urgency of speech warnings: Semantics versus acoustics. *Human Factors: The Journal of the Human Factors and Ergonomics Society 44*(1), 1-17.

E		Pc	At	Co	Cr	Ps	Ac

This study looks at the ways in which the meaning of a speech warning and the tone in which it is delivered interact to convey urgency. This paper addresses the following questions:
- Can speakers imbue urgency in their utterances when asked to do so?
- Can listeners understand the level of urgency the speaker is trying to convey?
- To what extent do acoustic variables underpin the relationship between speaker and listener?
- To what extent do semantic effects occur independently of acoustic effects?

In Experiment 1, a male and a female actor were asked to speak a series of ten warning signal words (DEADLY, DANGER, WARNING, CAUTION, RISKY, NO, HAZARD, ATTENTION, BEWARE, and NOTE) using three different styles: urgent, nonurgent (normal speaking voice), and monotone. When presented in visual warnings, the arousal strengths of these ten words are known to vary in a consistent manner. The 31 participants in the experiment were 15 males and 16 females between the ages of 18 and 36. They were asked to listen to a recording of the 60 signal words (10 words in three speaking styles by two speakers), presented in one of four random arrangements, and judge the level of hazard implied by each word. The style in which the signal words were spoken was found to have a significant effect on urgency ratings, with words spoken in the urgent style rated as being significantly more urgent than those spoken in the nonurgent style. This suggests that speakers are reliably able to convey different levels of urgency to listeners by intent. The meaning of the signal words also had a significant effect on the urgency ratings. DEADLY was perceived as being more urgent than DANGER, which was more urgent than WARNING, with NOTE at the bottom of the list. The ranking order agrees with the ordering of the same words in visual warnings. The female speaker's words were generally rated as more urgent than the male's, although this was not true for all words (e.g., NOTE and ATTENTION). The range of urgency ratings across the three presentation styles was much larger for some words than for others, which may reflect a wider range of common usage for those words (NO, for example). Finally, the female voice was found to produce a greater range of urgency across the three styles than the male voice.

Acoustic analysis was performed on the signal words for each presentation style. Three measures of acoustic properties, the mean fundamental frequency (Hz), the pitch range (Hz), and the amplitude (dBA), were found to vary consistently as a function of presentation style, while the speed of the word (seconds per syllable) did not. The urgent versions of the words were higher in pitch, louder, and had a wider pitch range than the nonurgent versions. This indicates that listener response may be partially based on changes in acoustic patterns. The female speaker produced a higher fundamental frequency and greater pitch range than the male speaker.

Participating in Experiment 2 were 31 individuals, 20 female and 11 males, ranging in age from 18 to 41 years old. The participants listened to and rated urgency for five signal words (DEADLY, DANGER, WARNING, CAUTION, and NOTE) that were produced using male and female synthesized voices. The amplitude, fundamental frequency, and pitch range of synthesized signal words were altered individually from a nonurgent to an urgent value. For example, one set of stimuli consisted of a male voice speaking the five signal words with an urgent mean amplitude, nonurgent mean frequency, and nonurgent mean pitch range. No effect due to the sex of the speaker was found in the results. Overall, increases in the amplitude, fundamental frequency, and pitch range were associated with increases in perceived urgency, although the effect was not statistically significant for pitch range. Urgency ratings again depend on the meaning of the signal word, with DEADLY perceived as most urgent and DANGER more urgent than WARNING.

Changes in all three parameters were found to produce larger changes in perceived urgency than changes in only one or two parameters. This is not unexpected, since in real speech increases in urgency are a result of systematic changes in all three parameters. This experiment also demonstrated that it is possible to imbue synthesized speech with urgency through the use of three acoustic parameters: amplitude, fundamental frequency, and pitch range. Intelligibility was a problem for a few of the words in Experiment 2, and the authors acknowledge the importance of messages that are realistic, intelligible, and believable. Word meaning was again found to be important, with the words DEADLY and DANGER rated as more urgent that other words. Finally, although a difference in urgency rating was found between real male and female voices in Experiment 1, no difference was found between male and female synthesized voices in Experiment 2.

Key findings:
- [59_Ps_1] Speakers are reliably able to convey different levels of urgency to listeners.
- [59_Ps_2] Urgency is transmitted through both the acoustic characteristics of the message and its meaning.
- [59_Ps_3] A female voice transmits a greater range of urgency than a male voice over styles from monotone to nonurgent to urgent.
- [59_Ps_4] Urgency can be imbued into synthesized speech through the use of three acoustic parameters: amplitude, fundamental frequency, and pitch range.

Key recommendations:
- [59_Ps_5] Synthesized speech messages need to be realistic, intelligible, and believable.

 # 60. Hellier E, Weedon B, Adams A, Edworthy J, Walters K. (1999). Hazard perceptions of spoken signal words. In MA Hanson, EJ Lovesey, SA Robertson (Eds.), *Contemporary Ergonomics 1999* (pp. 158-162). London, UK: Taylor & Francis.

E		Pc	At	Co	Cr	Ps	Ac
						▓	

This is a continuation of work to understand how the acoustic characteristics of a message affect the impact on the listener. Previous work with signal words spoken in three styles (urgent, non-urgent, and monotone) demonstrated that style, word meaning, and the sex of the speaker have a significant effect on the perception of hazard by the listener.

Fifteen males and 16 females between the ages of 18 and 36 years old listened to ten signal words recorded by a male and female actor in urgent, non-urgent, and monotone voices. The order of word presentation was random, and the subjects heard each word in each condition twice for a total of 120 words. The subjects were asked to score the urgency of the words, with higher values being assigned to more urgent words. Significant effects were found for word choice, speaking style (with words said urgently being scored as the most urgent and those said in monotone being scored as the least urgent). A dependence on the sex of the speaker was also found to be significant, with the female voice perceived as more urgent.

Key findings:
- [60_Ps_1] Words that have a stronger negative connotation and are spoken in an urgent, female voice are more effective in conveying urgency.

61. Hemphälä H, Kihlstedt A, Eklund J. (2010). Vision ergonomics at recycling centres. *Applied Ergonomics, 41*(3), 368-375.

E			Pc	At	Co	Cr	Ps	Ac

Swedish citizens are frequently asked to bring recyclables to recycling centers. They can drop off the recyclables at a variety of times, including at night. This requires facilities to be lit. When citizens enter the recycling center they are required to sort their recyclables. Some people sort at home prior to arriving, while others sort at the center. The purpose of this paper is to present research-based recommendations on lighting systems and sign design for these recycling centers.

Good outdoor lighting has proven to reduce crime, increase safety, and improve the efficiency of carrying out tasks. There are no specific recommendations for lighting in recycling centers; however, lighting for roads and outdoor workplaces is covered by Commision Internationale de l'Eclairage (CIE) standards, including CIE Publication 15, *Lighting of Outdoor Workplaces* (2005). These standards recommend an illuminance level that is as even as possible to prevent dark areas. Lighting should not produce glare, because this results in a measurable change in visibility due to a reduction in luminance contrasts in the human retinal image. Both white metal halide lamps and yellow high-pressure sodium lights are used for high intensity outdoor lighting. Previous research showed that the response time for peripheral vision is shorter under the white light from metal halide lamps, and white light is also advantageous for tasks that require separating objects by color. The visibility of signs depends on their position, shape, color, and the complexity of the surroundings. They should be well-lit, and lighting for adjacent signs should not affect the visibility or legibility of the opposite sign. The shape of the sign can simplify a message; for example, instead of having a rectangular-shaped sign with an arrow pointing in the right direction, the sign itself can be shaped like an arrow. The best visibility for signs during the day has been found for signs using negative polarity with white characters on a dark green background, and the detection distance of green signs is shorter if the sign is highly reflective. Positive polarity (dark characters on a white background) is recommended for signs during the night.

Data on lighting, signage, facility usage, and attitudes were collected from 16 recycling centers in Sweden through measurements of illuminance and luminance, observation of the activities of 163 users, questionnaires from 317 users, and interviews with 77 users. Questionnaires were also distributed to 122 employees at 42 recycling centers. Detailed lighting measurements at night were collected throughout two facilities. Users and employees saw a need for brighter lighting to reduce the risk of accidents and for larger, more brightly lit signs to improve efficiency on tasks. Illuminance and luminance values varied considerabl y, and in many areas were found to be inconsistent with European lighting standards for similar tasks in workplaces and on roads. For traffic moving at less than 10 km/h, for example, standards require street lights to supply ground luminance of at least 1.0 cd/m^2 with uniformity better than 0.4. For large container areas where sorting takes place, illuminance and luminance should be above 100 lx and 1 cd/m^2 respectively, with uniformity exceeding 0.4. Few signs were in compliance with standards that recommend characters on signs to be 15 cm to 25 cm high in order to be visible at a distance of 40 meters to 60 meters. Final recommendations from the study include that signs and information boards where work is performed should be illuminated at a minimum level of 100 lx, with a minimum mean luminance of 10 cd/m². A frame should be placed around the signs. Negative polarity (light characters with dark background) and a font without serifs should be chosen for daytime signs.

Key findings:
- [61_Pc_1] Glare results in a measurable change in visibility due to a reduction in luminance contrasts in the human retinal image.
- [61_Pc_2] Illuminance on signs and information boards should be at least 100 lx.
- [61_Pc_3] Mean luminance on signs where work is performed should be at least 10 cd/m².
- [61_Pc_4] The best visibility for outdoor signs during the day has been found for negative polarity with white characters on a dark green background. At night, positive polarity is preferred. The detection distance of green signs is shorter if the sign is highly reflective.
- [61_Pc_5] Characters on signs should be 15 cm to 25 cm high in order to be visible at a distance of 40 meters to 60 meters.

Key recommendations:
- [61_Pc_6] Glare on or near outdoor signs should be eliminated.
- [61_Pc_7] Outdoor signs should be well-lit, and lighting for adjacent signs should not affect the visibility or legibility of the opposite sign.
- [61_Co_1] The shape of a sign can simplify a message; for example, instead of having a rectangular-shaped sign with an arrow pointing in the right direction, the sign itself can be shaped like an arrow.

62. Heskestad AW. (1999). Performance in smoke of wayguidance systems. *Fire and Materials, 23*(6), 375-381.

Heskestad AW, Pedersen KS. (1998). Escape through smoke: Assessment of human behaviour and performance of wayguidance systems. *Proceedings of the First International Symposium on Human Behavior in Fire*, Belfast, UK, 631–638.

F		Pc	At	Co	Cr	Ps	Ac

A set of five experiments on fire egress carried out during the 1990's in Norway was re-analyzed to evaluate performance parameters for fire safety engineering. The purpose of wayfinding systems is to move people to safety as quickly as possible. Their performance can be evaluated not only through measurements of visibility in smoke and in clear air, as has been studied elsewhere, but also by the speed of evacuees and by how effectively the system encourages the use of the appropriate egress path. Directional information and the exit sign must not only be sensed under conditions of reduced visibility, but must also be capable of guiding evacuee behavior. This paper quantifies two parameters: movement speed and the probability of correct decisions during egress.

Four key parameters were identified to assess wayguidance performance: movement speed, time taken to change direction, time taken to make a decision, and probability of making the correct decision. In these five experiments, movement speed was estimated as between 0.2 m/s and 0.5 m/s. The Bleivik experiments measured the speed of test subjects in white smoke under three levels of luminance (0.2, 10, and 30 cd/m^2) provided by electrically powered low location lighting. A slight increase was found in movement speed as the luminance increased from 0.2 cd/m^2 to 10 cd/m^2, but speed decreased slightly at the highest luminance. This may be due to the reflection of light from the white smoke (smoke veiling), or it may simply reflect the large degree of variation among the small number of test subjects.

The probability of making correct route decisions was studied using data from the Hobøl experiments, which studied the effect of a variety of wayguidance systems on egress through an escape path containing 11 decision elements. The probability of making the correct decision was estimated to be between 69 % and 79 % for electrically powered and photoluminescent low location lighting and greater than 90 % for an electrically powered cold cathode system emitting green light and tactile systems (with notches indicating the direction towards the nearest exit). The best performance of 98 % was found for a combination of tactile system and cold cathode system.

Key findings
- [62_Ac_1] The movement speed through a smoky escape path is relatively insensitive to luminance.
- [62_At_1] [62_Ac_2] The highest probability of making correct route decisions (98 %) was found for a wayguidance system consisting of a combination of tactile and green light cold cathode lighting systems.
- [62_At_2] [62_Ac_3] Electrically powered and photoluminescent low location lighting wayguidance systems resulted in the lowest percentage of evacuees (69 % to 79 %) making correct escape route decisions.

63. Honeywill P. (1999). Designing icons for the graphical user interface. *Digital Creativity, 10*(2), 67-78.

M		Pc	At	Co	Cr	Ps	Ac
		▓	▓				

Principles from the field of graphic design can provide insights into the design of icons for a graphical user interface. Influences outside of an icon can negatively impact its effectiveness; the surrounding context/contrast is important. If the icon is yellow, for example, placing it in a yellow background would negatively affect its noticeability. Since computing allows the user to manually change the background, it is important to use edge definition for the icon in order to decrease the effect of changing the background color. The icons should be in an asymmetric arrangement. This forces the user to focus on the design, rather than becoming habituated to the expected location of icons. A serif typeface contributes to the asymmetry of the design.

Key findings:
- [63_Pc_1] The effectiveness of the design of a sign is influenced by the context of the surroundings and its contrast with the sign.
- [63_Pc_2] Color contrast between an icon and its background aids in the icon being noticed. In the absence of control over the background, edge definition is necessary to decrease the effect of color changes.
- [63_At_1] An asymmetric arrangement of icons (symbols) attracts attention and forces the user to focus on the design.

64. Howes DH. (1957). On the relation between the intelligibility and frequency of occurrence of English words. *Journal of the Acoustical Society of America, 29*, 296-305.

A		Pc	At	Co	Cr	Ps	Ac

The intelligibility of spoken words was studied as a function of commonality of use and word length. In this study, 282 words were selected based on their frequency in four popular magazines and on word length. Up to four words were randomly selected for each pairing of sixteen frequency categories and eleven word length categories. A recording was made of a male speaker reading each word. Each word was preceded by a pure tone for 2.5 seconds. The words were divided into four approximately equal sets. The tapes were played for the subjects through earphones with background noise at 80 dBA and the words (depending on the trial) at (-9, -6, 0, +6, +9, +12, and +20) dBA from ambient. The subjects consisted of five college students. The required threshold for intelligibility decreased by 4.5 dBA per logarithmic unit of word frequency. For word length, at a given word frequency, the threshold for intelligibility was found to decrease nearly linearly by approximately 1 dBA per letter.

Key findings:
- [64_Co_1] Common words are intelligible at a lower sound threshold in noise. Under test conditions, the required threshold for intelligibility was found to decrease by about 4.5 dBA per logarithmic unit of word frequency.
- [64_Co_2] Shorter words are intelligible at a lower sound threshold in noise. Under test conditions, the intelligibility threshold was found to decrease by approximately 1 dBA per letter.

 # 65. Huchingson RD, Koppa RJ, Dudek CL. (1978). *Human factors requirements for real-time motorist information displays, Vol. 13: Human factors evaluation of audio and mixed model variables* (Report No. FHWA-RD-78-17). Washington, DC: U.S. Department of Transportation.

T		Pc	At	Co	Cr	Ps	Ac

The research described in this document deals with informational displays used to prompt route diversion and to guide traffic when accidents occur. While other guides in the series (Volumes 8, 10, and 11) are concerned with visual displays, this volume discusses audio and mixed model displays. Of primary interest are the human factors aspects of the display message, including content, format and element ordering, redundancy, placement, and message quantity, load and length.

Identical studies were performed in three locations: College Station, Texas, St. Paul, Minnesota, and Los Angeles, California. Twelve distinct messages were prepared and arranged in random order in a taped message for presention to each subject. Four messages followed a conversational language style with correct English and appropriate syntactic elements, in a style similar to a spoken message that a motorist might receive at a tourist information center to direct him or her to some attraction. Four messages followed a short-form style, omitting adjectives, modifiers, and conventional 'carrier' phrases that have low informational content. The final four messages were spoken in a staccato style, using the terse and telegraphic written language of highway signs. The subjects at each site were divided into three groups, with each message presented once, twice, or three times, respectively. The dependent variable in this study was the number of correct recalls, as a function of language style and the number of repetitions. At the end of each study, the subjects were asked their preference for each of the three language styles.

The first study was carried out with 85 subjects in a laboratory at Texas A&M University in College Station, Texas. Each message was presented once to 30 subjects, twice to 32 subjects, and three times to 23 subjects. Message retention was found to increase from one to two presentations of the message, but not from two to three. Listening only once to each message consistently resulted in poorer recall. A few more subjects recalled messages they heard three times than those who heard messages twice, but the difference in recall between two and three presentations was not significant. The language style was not found to affect message recall. The part of the message that was recalled best was the information on what is wrong and the information on what action one should take next. The least well-recalled was the emergency 'condition'; for example, the traffic situation that resulted from the problem. The staccato language style was preferred by 52 % of the subjects, followed by the short-form style, which was preferred by 31 %, and by the conversational style, which was preferred by only 17 %.

The second study, in St. Paul, Minnesota, had 72 participants. The results in this case were somewhat different; subjects in this study performed better with each additional repetition for both conversational and short-form styles. For the staccato style, recall increased from one presentation to two but did not change significantly from two presentations to three. Message recall in this study depended on language style. The conversational style resulted in worse recall than short form or staccato when messages were presented once and twice. For three presentations, the short-form style resulted in better recall than conversational or staccato styles. Another difference from the first study was found in the recall of specific parts of the message; in this study the 'condition' statement was recalled about as well as the other components. These participants preferred the staccato and short-form equally well (39 % and 37% respectively), while only 24% preferred the conversational style.

In the third study, 65 subjects participated in Los Angeles, California. Some odd results were found in the case of the short-form message, with two presentations resulting in poorer performance than one or three presentations. For the other two language styles, recall increased with each presentation. Messages given in the conversational style were recalled the worst after a single presentation, while the short form was recalled the best. After three presentations, the short form and staccato styles were also superior to the conversational style. As in the second study, the 'condition' information was least well-recalled compared with the other message components. The most preferred language style in the third study, with 49 % of the subjects in agreement, was the staccato style, while 28 % preferred the short form and 22 % preferred the conversational style.

In summary, the conversation language style was the least preferred and was also associated with the poorest recall. Subjects in all three studies preferred the staccato style, although the participants in St. Paul equally preferred the short-form style. Overall, retention was found to depend on the language style, with the short form language style found to be superior for one presentation of the message and the short form and staccato styles equally effective for three presentations. The improvement of recall from one to two presentations was found in two studies but not the third. The "condition" statement was recalled less frequently than other information, including the nature of the problem and the action to be taken.

Based on these results, design recommendations for auditory message content include:
- Avoid conversational style when presenting a complex message over the radio.
- Repeat the message at least once.

A separate set of studies examined the effects on recall of internal and successive redundancy of audio information. Internal redundancy implies that information within the message is repeated while the entire message is presented once, and successive redundancy implies repetition of the entire message with each piece of information within the message mentioned only once. In this study, the taped messages advised motorists of an obstruction along the route and described a diversion route via a succession of turns at designated intersections. Subjects were presented with one of two conditions: a six-unit problem consisting of three turns and three street names or an eight-unit problem consisting of four turns and four street names. No maps or schematics were provided. Participating were 17 females and 23 males between the ages of 21 and 55. To test recall of the directions, participants were given a book in which each page displayed a sign. Each participant was asked to indicate whether he/ she would continue straight, turn left, or turn right, and was referred to the next page based upon this decision.

The results of this study showed that internally redundant audio messages resulted in significantly more failures to complete the diversion than the successively redundant message, with 31% of subjects successfully completing the former task and 57 % successfully completing the latter task. The eight-unit problem was completed successfully by 36 % of the subjects and the six-unit problem by 53 %. The majority of errors (by 54 %) occurred at the first decision point. (It should be noted that additional studies in the same series that are not included in this annotated bibliography acknowledge the potential benefits of presenting a message twice in succession using internally redundant messages.)

Additional experiments indicate that eight units represent the retention limit of auditory messages for design purposes. For routes that require more than four turns and recall of more than four street names, a trailblazer system is recommended.

Key findings:

- [65_At_1] For auditory messages about driving route changes, staccato (terse and telegraphic) and short-form (brief and concise but grammatically complete) styles were preferred to a conversational style, which was additionally associated with the poorest recall.
- [65_At_2] Recall of an auditory message generally improved from one to two presentations. The effect on recall of increasing the number of presentations from two to three is inconclusive.
- [65_At_3] Recall of a statement about the traffic consequences of an incident was worse than recall of other information, including the nature of the incident and the action to be taken by drivers.
- [65_At_4] [65_Ac_1] Internally redundant audio messages, with repetition only within the message, resulted in significantly less success in completing a set of instructions (31% of participants) than successively redundant messages, with repetition only of the entire message (57 % of participants).
- [65_At_5] [65_Ac_2] An eight-unit set of instructions (four turns and four street names) was completed successfully by 36 % of participants, while a six-unit set (three turns and three street names) was completed successfully by 53 %.
- [65_At_6] Eight units represent the retention limit of auditory messages for design purposes.

Key recommendations:

- [65_At_7] Avoid a conversational style when presenting a complex message over the radio.
- [65_At_8] Repeat an auditory message at least once for improved recall.
- [65_At_9] [65_Ac_3] For routes that require more than four turns and recall of more than four street names, a trailblazer system is recommended.

 # 66. Idzikowski C, Baddeley AD. (1983). Fear and dangerous environments. In: Hockey GRJ, (Ed.). *Stress and Fatigue in Human Performance* (pp. 123–144). New York: John Wiley and Sons Ltd.

P		Pc	At	Co	Cr	Ps	Ac

This article summarizes findings on behavior under the threat of physical danger. Measures of anxiety include subjective reports, physiological measurements (such as heart rate and galvanic skin response), and assessments of individual predisposition. Low to moderate degrees of stress have been found to improve performance, while high levels impair it. Fear during combat is heightened by lack of communications and loneliness. Many soldiers fail to fire their weapons in battle, and navigator errors increase as bombers approach their targets.

Social interactions can be positive or negative. Under threatening conditions, people may behave in a more cooperative manner than under normal circumstances. During civilian bombings in WW II, morale stayed high, and mental illness and suicides actually decreased. Different groups of people may respond differently to the same level of fear. The authors speculated that how the group responds to the situation is due to its structure and to the quality of the leadership, but what defines beneficial leadership is unknown. However, threatening conditions can inhibit correct responses to assigned tasks.

Measured physiological and biochemical changes indicate that the degree of arousal for experienced and competent parachutists is lower than for novices and occurs well before the jump. An increase in the perception of danger moves the physiological response of experienced parachutists closer to that for novices. Performance of military parachutists on a task requiring sensory and motor coordination depended on the degree of training. The performance of divers on manual dexterity tasks was impaired with both increasing depth and worse diving conditions. On the day of the jump, novice parachutists had a significantly higher auditory threshold and higher reactivity to anxiety-provoking words. Visual recognition of numbers composed of dots against a distracting background deteriorated as parachutists approached the jump point. Swedish airline pilots who were slow to perceive threats in pictures presented for very short times were more likely to have accidents. Performance in filling out a complex form deteriorated for subjects in a military plane who were told they would have to ditch in the ocean. Performance on a secondary task deteriorated for divers in a pressure chamber who thought they were experiencing a 60-foot dive.

Performance measures do not always track subjective and physiological measures of arousal. One explanation could be that performance is discontinuous and reaches a breakpoint in anxiety at which it deteriorates rapidly. Another is that arousal narrows attention, so that peripheral stimuli are ignored and peripheral tasks are performed poorly. This could be due to a division of attention between the external demands of the task and internal monitoring (worry, self-blame). The effects of anxiety on manual dexterity may be due to the difference in physiological state from that during training, from the physiological effects themselves, or from regression to more automatic responses.

Key findings:
- [66_Ac_1] The response of a group to a dangerous situation varies, and may be related to

the structure of the group and to the quality of its leadership.

- [66_Ac_2] The response of individuals to fear and anxiety varies considerably. Swedish airline pilots who were slow to perceive threats in pictures presented for very short times were more likely to have accidents.
- [66_Ac_3] On the day of the jump, novice parachutists had a significantly higher auditory threshold and higher reactivity to anxiety-provoking words.
- [66_Ac_4] The performance of divers on manual dexterity tasks was impaired with both increasing depth and worse diving conditions.
- [66_Ac_5] Visual recognition of numbers composed of dots against a distracting background deteriorated as novice parachutists approached the jump point.
- [66_Ac_6] Physiological effects of anxiety are reduced by experience.
- [66_Ac_7] Low to moderate degrees of stress tend to improve performance, while high levels impair it.
- [66_Ac_8] Fear during dangerous situations is heightened by lack of communications and isolation.
- [66_Ac_9] Deterioration in performance, including manual dexterity, sensory-motor tasks, and secondary tasks that require divided attention, can be expected when physical danger is perceived.
- [66_Ac_10] Threatening conditions have been shown to inhibit correct responses to assigned tasks for subjects in military planes and dive chambers who were told they were in danger. Reducing the perception of danger may improve performance.
- [66_At_1] Arousal may narrow attention, so that peripheral stimuli are ignored and peripheral tasks are performed poorly. Attention may be divided between the external demands of the task and internal monitoring (worry, self-blame).

 # 67. Jamson SL, Tate FN, Jamson AH. (2005). Evaluating the effects of bilingual traffic signs on driver performance and safety. *Ergonomics, 48*(15), 1734-1748.

E		Pc	At	Co	Cr	Ps	Ac

In bilingual countries, signage often needs to be written in both languages. Because road signs are only visible to drivers for a short time as they pass by, and because drivers are occupied in a task that requires their attention, it is important to understand the impact of bilingual signs on driving performance and safety. The process of perceiving and comprehending a road sign can be modeled by three stages: 1) the perceptual processes that encode the text, 2) the parsing state that transforms the text into a mental representation of its meaning, and 3) the utilization stage, in which the driver converts this mental representation into action. The process can be made more efficient by limiting the length of the message, presenting it in an uncluttered environment, and providing context, enabling the message to be understood using less information from the sign itself. Processes that are more practiced and more automatic require less attention and allow the driver to concentrate more on driving safely.

Variable Message Signs (VMS) are electronic traffic signs that can give drivers information about special events like accidents or road closings. For bilingual VMS signs, the driver needs to read and comprehend only half the information presented. When the two languages use the same letters, a monolingual reader may need to process the entire sign to determine the relevant text. This may cause distraction. Some ideas to reduce the overall reading time include improving the sequencing of the languages or differentiating them by colors or fonts. This study looked at the impact of different types of mono- and bilingual VMS on driving performance (speed and headway) in a driving simulator while the driver was engaged in low and high workload tasks. An equal number (12) of monolingual and bilingual participants was recruited to see whether their use of the road signs differed. Since the study was carried out in England, English and Welsh languages were used on the signs.

Neither mean travel speed nor minimum headway was affected by one-line or two-line monolingual signs (in English) or by two-line bilingual signs. However, both four-line monolingual and four-line bilingual signs affected driver performance by slowing travel speed (by about 11 kph) while approaching and passing the sign. Recovery of the original speed was slower for the four-line bilingual signs, possibly reflecting a longer time to process the information. Under the low workload condition, four-line signs resulted in increased minimum headway between their car and the one in front, consistent with decreasing speed. Under the high workload condition, however, minimum headway decreased by an average of one second while reading four-line signs. This creates a more hazardous condition, increasing the possibility and severity of collisions. The response time to carry out instructions from the road sign increased with the number of lines of text.

Key findings:
- [67_Ac_1] One or two-line monolingual and two-line bilingual variable message signs (VMS) did not affect driving speed or distance to the car in front (headway).
- [67_Ac_2] Both four-line monolingual and four-line bilingual signs affected driver performance by slowing travel speed (by about 11 kph) while approaching and passing the

sign.
- [67_Ac_3] Recovery of the original speed was slower for four-line bilingual VMS road signs, possibly reflecting a longer time to process the information.
- [67_Ac_4] Under the low workload condition, four-line signs resulted in increased minimum headway between the driver's car and the car in front, consistent with decreasing speed. Under the high workload condition, however, minimum headway decreased by an average of one second while reading four-line signs.
- [67_Ac_5] The response time to carry out instructions from the road sign increased with the number of lines of text.

Key recommendations:
- [67_Co_1] The process of perceiving and comprehending road signs can be made more efficient by limiting the length of the message, presenting it in an uncluttered environment, and providing context.
- [67_Co_2] The overall reading time for bilingual signs may be improved by better sequencing of the two languages or by differentiating them by colors or fonts.

68. Jang PS. (2007, March). Designing acoustic and non-acoustic parameters of synthesized speech warnings to control perceived urgency. *International Journal of Industrial Ergonomics, 37*(3), 213-223.

E

Pc	At	Co	Cr	Ps	Ac

An understanding of the acoustic and non-acoustic characteristics of warning messages that convey urgency can improve their effectiveness in stimulating appropriate action. Synthesized voice warnings have been shown to be inherently alerting and can therefore both alert and inform simultaneously, unlike non-verbal warnings that require training to decode. In addition, their content can be well-controlled, which makes them useful tools for exploring the effect of various parameters on perceived urgency. In this set of four experiments, the effectiveness of synthesized voice warnings was studied by varying the parameters of speech rate, pitch, message content, message format, sex of the speaker, and message interval. Note that sound level is not included, since urgency is not the most important factor in its determination; the warning must be loud enough to attract attention and be heard over ambient noise, but not so loud that it is painful and distracting. These considerations may leave a narrow range for manipulation to affect perceived urgency. Quantitative analysis of speech rate, pitch, and message interval using Steven's power law, which relates objective measures to subjective sensations, results in a model that quantifies the effects of these variables on perceived urgency.

In the first experiment, 25 participants (14 male and 11 female) between the ages of 19 and 40 (mean 27) were asked to listen and respond to 128 messages spoken by two artificial voice generating systems. The six variables in this study were: message type (cabin pressure dropping, collision warning), message format (semantic context, keyword context), sex, average fundamental frequency (120 Hz, 190 Hz), speech rate (0.201 seconds per syllable, 0.276 seconds per syllable), and interval between messages (0.1 seconds, 0.8 seconds). Participants listening in a quiet room rated the urgency of each message using both a line scale and a number to estimate the magnitude. All six variables were found to be statistically significant, and there were no interaction effects between the different variables. The collision message, keyword format, male voice, 190 Hz (higher) mean frequency, 0.201 seconds per syllable (faster) speech rate, and 0.1 seconds (shorter) message interval were all scored as being more urgent than the other value of each variable. Other research has found that female voices are generally rated as more urgent than male voices. Since synthesized male and female voices in this study were set to the same fundamental frequency and loudness, the differences in urgency ratings must reflect uncontrolled factors such as smoothness and timbre rather than pitch. Indeed, some listeners reported that the male voice sounded harsher and more unnatural than the female voice.

The second, third, and fourth experiments analyzed the effects of the three quantifiable variables on perceived urgency to develop a quantitative model. The second experiment involved 28 participants consisting of 18 male and 10 female subjects between the ages of 19 and 39 years old. The procedures were similar to the first experiment except that only four base messages, consisting of the most and least urgently rated messages and voices, were used. The fundamental frequency was varied in 10 % increments from 70 % to 130 % of the original level, for a total of 28 messages, presented in random order. Using linear regression in log-log coordinates, all four voice and message combinations were found to be statistically similar and could be explained by an exponent of 0.701 in Stevens' power law. Experiment 3 used 27 participants (18 males and nine females) between the ages of 19 and 38. The experimental design was the same as for Experiment 2 except that the fundamental frequency was fixed and the speech rate was varied from 70 % to 130 % of that in the original experiment. As was the case with the previous experiment, all four voice and message combinations could be reduced to a single log-log space regression equation with a coefficient of -1.189. Experiment 4 used 27 participants (18 males and nine females) ranging in age from 19 to 39 years old. In this experiment, fundamental frequency and speech rate were held fixed while the spacing between messages was varied from 0.01 seconds to 1.8 seconds in increments of 0.3 seconds. In log-log coordinates, across the four voice and message combinations, the relationship between perceived urgency and message spacing could be represented by a single regression coefficient of -0.127. For Stevens' power law, the higher the absolute value of the exponent, the greater the change in perception for a specified change in the objective measure. The most economical way to change the perceived urgency is therefore to modify the speech rate, followed by the fundamental frequency and finally the message interval. Quantification of the relative effects of multiple variables enables the transmission of equivalent levels of urgency through different parameters, as well as providing a technique to appropriately scale suggested prioritization of tasks.

Key findings:
- [68_Ps_1] Message type, message format, sex of the speaker, average fundamental frequency, speech rate, and interval between messages all have a significant effect on perceived urgency for synthesized voice messages. Perceived urgency increases for keyword context (compared to syntax context), higher average frequency, faster speech rate, and shorter message interval.
- [68_Ps_2] Synthesized male voices were perceived as more urgent than synthesized female voices with the same fundamental frequency, possibly because the male voice sounded more unnatural.
- [68_Ps_3] The relative effects of multiple variables in audible warnings on the perception of urgency can be quantified using Stephens' power law.
- [68_Ps_4] Quantification of the relative effects of multiple variables in audible warnings enables the transmission of equivalent levels of urgency through different parameters, as well as providing a technique to appropriately scale suggested prioritization of tasks.
- [68_Ps_5] For synthesized voice messages, the most economical way to change the perceived urgency using quantitative variables is to modify the speech rate, followed by the fundamental frequency and finally the message interval.

Key recommendations:
- [68_At_1] [68_Ac_1] Synthesized voice messages can both alert and inform simultaneously.
- [68_Pc_1] [68_At_2] [68_Ps_6] An audible warning must be loud enough to attract attention and be heard over ambient noise, but not so loud that it is painful and distracting. This may leave a narrow range for manipulation to affect perceived urgency.

69. Jaynes LS, Boles DB. (1990). The effect of symbols on warning compliance. *Proceedings of the Human Factors Society 34th Annual Meeting*, Santa Monica, CA, 984-987.

E		Pc	At	Co	Cr	Ps	Ac

This study looked at the effect of various written warning designs on compliance rates for donning personal protective equipment to perform a chemistry laboratory experiment.

Eighty students were given a printed set of instructions for a chemistry task, with warnings placed in the middle of the instruction sheet. The four warning designs were text, a pictograph with a circular enclosure, a pictograph with a triangular enclosure, and text plus a pictograph with a triangular enclosure. The text warning was "Warning: wear goggles, mask and gloves while performing the task to avoid irritating fumes and possible irritation of skin." All pictographs consisted of an image of goggles, mask and gloves. Compliance was quantified as zero, one, two, or three, depending on the number of types of protective gear donned by the subject. A questionnaire after the task determined whether subjects could recall the warnings.

Among subjects as a whole, 89 % noticed the warning, 64 % recalled the precautionary steps to take, and 43 % donned all three types of protective gear. Compliance was highest for the warning design that included both text and pictograph, followed by the text warning alone, and finally by the pictograph warnings without text. No difference in compliance rate was found between circular and triangular enclosures for the pictorial.

Key findings:
- [69_At_1] [69_Ac_1] Compliance with safety instructions was higher for a warning message that combined a pictograph with text than for either pictograph or text alone.

 # 70. Jin T. (2002). Visibility and human behavior in fire smoke. *The SFPE Handbook of Fire Protection Engineering (3rd Edition),* Eds: PJ DiNenno et al., National Fire Protection Association, Quincy, MA, pp. (2–42)-(2–54).

Jin T, Yamada T. (1989). Experimental study of human behavior in smoke filled corridors. *Proceedings of the Second International Symposium On Fire Safety Science,* Eds: T Wakamatsu, Y Hasemi, A Seizawa, P Seeger, P Pagni, ISBN 0-891168648.

Jin T, Yamada T. (1985). Irritating effects of fire smoke on visibility. *Fire Science and Technology, 5*(1), 79–90.

Jin T. (1976). *Visibility through fire smoke, Part 5: Allowable smoke densities for escape from fire. (*Report of Fire Institute of Japan No. 42).

Jin T. (1978). Visibility through fire smoke. *Journal of Fire & Flammability, 9,* 135-155.

F		Pc	At	Co	Cr	Ps	Ac

Research on wayfinding in smoke is reviewed in this set of articles. The visibility of signs in black smoke was found to be better than in white smoke at the same smoke density. This is likely due to the difference in interaction of ambient light with the two types of smoke. White smoke produces more of a veiling effect than black smoke of an equivalent extinction coefficient. Relationships were found between the visibility of various types of signs and the extinction coefficient of the smoke. The visibilities of light emitting and reflecting signs are given by the following empirically derived formulations:

$$V = (5-10)/C \text{ (m), light-emitting sign}$$

$$V = (2-4)/C \text{ (m), reflecting sign}$$

where C is the extinction coefficient of the smoke.

The visibility of signs also depends on color. Assuming uniform smoke with an extinction coefficient of 0.5 m^{-1}, the visibility of red and blue signs of each type can be expressed as a ratio of the extinction coefficients, such that

$$\frac{V_{red}}{V_{blue}} \approx \frac{C_{s,blue}}{C_{s,red}}$$

Several sets of ratios were obtained and associated with burning materials and fire conditions. Red signs were found to have a visibility 20 % to 40 % greater than blue signs.

The authors derived safety criteria based on tenability limits, which are the conditions required to allow safe egress. The tenability limits for visibility (neglecting the physiological effects of smoke) are based on the extinction coefficient of the smoke and the familiarity level of the population with the route being traveled. A tenability level of 0.15 m^{-1} (assumed visibility of 13 meters) was suggested for individuals unfamiliar with an environment. A more severe tenability limit of 0.5 m^{-1}

(assumed visibility of four meters) was suggested for individuals familiar with an environment. A compilation of visibility tenability limits suggested by other researchers (Kawagoe, Togawa, Kingman, Rasbash, LAFD, Shern) shows a range of extinction coefficients from 0.1 m^{-1} to 0.4 m^{-1}, corresponding to assumed visibilities between about 20 meters and four meters.

The effectiveness of flashing lights as part of the wayfinding system was studied. The experimental variables included whether a sign was flashing or static, the observation distance, sign luminance, and the size of the sign. (Three sizes of signs were used in the study, but their dimensions were not provided.) Participants were asked to grade the conspicuousness of each sign on a five-point scale. The results indicated that a flashing sign is considered more conspicuous under almost all circumstances. For the small and medium signs, the advantage of flashing signs is clear. For a large sign, especially at a 20-meter distance, the results are equivalent. Flashing signs thus performed equivalently or better under all circumstances and for all sign sizes.

Further trials were performed using traveling flashing signs to aid evacuation from a smoke-filled environment. A corridor (1.4 meters × 6.3 meters × 2.5 meters) was filled with smoke, and ambient lighting at a level of 200 lx was provided by fluorescent lamps. Flashing green lights were located at 0.5 meter, one meter, and two meter intervals along the corridor. Twelve participants traversed the corridor and were asked to comment on the conspicuousness of the signs. The results indicated that the system was effective in clear air and up to an extinction coefficient of 1.0 m^{-1} when the spacing between the lights was less than two meters. At a spacing of two meters between the lights, the effectiveness of the system declined rapidly as the smoke conditions worsened.

Key findings
- [70_Pc_1] Red signs have a visibility 20 % to 40 % greater than blue signs in smoke.
- [70_Pc_2] Flashing lights are more conspicuous than static lighting systems.

 # 71. Jin T, Yamada T, Kawai S, Takahashi S. (1991). Evaluation of the conspicuousness of emergency exit signs. *Fire Safety Science-Proceedings of the Third International Symposium,* 835-841.

F			Pc	At	Co	Cr	Ps	Ac

The first experiment measured the conspicuousness of an ordinary exit sign in an underground shopping mall during business hours. The observation distance, size, and luminance of the exit sign were varied. The exit signs were green with a pictograph of a person leaving a room, and three sizes were tested: small (with a luminance of 325 cd/m²), medium (785 cd/m²) or large (986 cd/m²). The exit signs were located 2.5 meters above the floor and were fixed to the ceiling. The 33 participants were asked to evaluate the noticeability of the exit signs at distances of (60, 50, 30, 20, and 10) meters.

In the second experiment, the same procedures were used to evaluate the conspicuousness of a flashing exit sign. Results were then compared to the results from the ordinary exit sign in the first experiment.

In the first experiment, the large exit sign was found to be more conspicuous than the smaller one. The larger sign can be understood 80 meters away, the medium sign is typically understood 30 meters away, and the small sign can be understood at eight meters. Increasing luminance increases the noticeability of the exit sign. Visibility is independent of size when the luminance is over 300 cd/m². For the large sign, the level of conspicuousness is unchanged when the ratio of exit sign luminance to observation distance is held constant. This relationship does not hold for the small and medium sized signs. A flashing light has a more significant affect on the noticeability of the sign for smaller signs than for larger ones. This could be because the large sign is already relatively conspicuous.

Key findings:
- [71_At_1] Conspicuousness of an exit sign can be improved by adding a flashing light.
- [71_Pc_1] Flashing lights have a more significant affect on the noticeability of a sign for smaller signs than for larger ones.
- [71_Pc_2] The visibility of an exit sign is independent of size when the luminance is over 300 cd/m².

72. Jin T, Yamada T. (1994). Experimental study on effect of escape guidance in fire smoke by travelling flashing of light sources. *Proceedings of the Fourth International Symposium of Fire Safety Science*, 705-714.

F		Pc	At	Co	Cr	Ps	Ac

An effective escape guidance system using light and/or signs is a good way to ensure quick and safe exiting from a building. Flashing lights can be used to direct people to an exit. An original design using flashing green lights 25 cm to 325 cm apart was previously developed and tested. The results of the previous experiments showed that the effectiveness of the escape guidance improved as the spacing between light sources decreased. Effectiveness also improved with increasing luminance, up to a plateau beginning at 2000 cd/m^2, and with increasing size of the light source, up to a plateau beginning at (5 x 5) cm. To improve guidance effectiveness when the light source is (5 × 5) cm with a fixed luminance of 2000 cd/m^2, the traveling speed of the flashing lights should be at least 2 m/s.

In the experiment reported in this paper, a passageway with dimensions 1.4 meters high × 6.3 meters long × 2.5 meters wide was filled with smoke. Four fluorescent 40 W lamps were located on the ceiling, providing approximately 200 lx in the absence of smoke. Sixteen flashing light unit boxes were located on the right side of the floor at a height of 0.5 meters. Twelve subjects participated in the study. Each subject walked along the hallway under varying conditions of smoke concentration, light spacing, and flashing light traveling speed. The subjects were asked to rank the effectiveness of the escape guidance on a scale from one to seven. The lamps for escape guidance were 18 W compact fluorescent lamps installed in wooden grey boxes. The chromaticity was controlled by a milk white acrylic resin plate with a green silk screened surface. The luminance was adjusted to 270 cd/m^2, and the traveling speed of the flashing lights was 2, 4, or 8 m/s. The smoke for the tests was generated by smoldering Japanese cedar.

The results show that the guidance effectiveness iss more sensitive to the spacing of the lights than to the traveling speed of the flashes. Decreasing the spacing improves the effectiveness of the guidance lights. When the traveling speed is fast and the spacing between lights is small, the lights can appear to point in the opposite direction. Smoke decreases the effectiveness of the traveling flashing lights. Smoke with an optical density per meter path length of 0.2 m^{-1} resulted in a higher effectiveness rating than the no smoke condition, possibly because the blocking of some noises by the smoke allowed subjects to focus more on the lights.

Key findings:
- [72_At_1] [72_Ac_1] Traveling flashing lights provide effective wayguidance in both light and thick smoke, although smoke decreases the effectiveness.
- [72_Ac_2] The effectiveness of traveling flashing lights is more sensitive to the spacing of the lights than to the traveling speed.
- [72_Ac_3] Decreasing the spacing improves the effectiveness of traveling flashing guidance lights.
- [72_Co_2] [72_Ac_4] When the traveling speed is fast and the spacing between lights is small, flashing wayguidance lights can appear to point in the opposite direction.

 # 73. Keating JP. (1982, May). The myth of panic. *Fire Journal,* 57-62.

F		Pc	At	Co	Cr	Ps	Ac

In most fires, people do not panic. However, they may become intensely focused and be able to process only major elements in the environment that are perceived as relevant to the situation. Under stress, they tend to fall back on familiar responses that may or may not serve them well in an emergency. The author presents recommendations for notification that takes these factors into account. First, messages must be simple so that they represent a predominant cue that can be processed under stress. Second, international symbols should be used in signage, both to simplify the information and to allow occupants not familiar with the language to know what they are to do. Finally, occupants should be given all needed information. Instructions telling occupants to use the stairs, for example, should also indicate where the stairs are.

The effectiveness of notification is enhanced by education in advance of the emergency situation and by human factor considerations in building design. In this light, tragedies may be seen as opportunities to educate people about lifesaving behavior should they find themselves in a similar situation.

Key findings:
- [73_At_1] Under stress, people may become intensely focused and be able to process only major elements in the environment that are perceived as relevant to the situation. They also tend to fall back on familiar responses that may or may not serve them well in an emergency.

Key recommendations:
- [73_At_2] Messages must be simple in order to represent a predominant cue that can be processed under stress.
- [73_Co_1] Use international symbols in signage to both simplify the information and inform occupants not familiar with the language.
- [73_Ac_1] Give occupants all needed information, such as directions on where to go in addition to what to do.

74. Keating JP, Loftus EF. (1975). *People Care in Fire Emergencies – Psychological Aspects, 1975* (Technology Report 75-4). Society of Fire Protection Engineers, Boston, MA.

F		Pc	At	Co	Cr	Ps	Ac

The authors designed and then tested a voice alarm system for the Seattle Federal Building in 1974. The building evacuation plan called for area evacuation, in which only people on the affected and adjacent areas are evacuated, rather than total building evacuation. In this plan, people immediately above the fire floor are told to evacuate to floors above the fire; movement thus occurs in both the upward and downward directions. The system utilized taped messages played over a public address system rather than live voice instructions. The FCC warning signal (1000 Hz pure sine wave oscillating tone) was used to get the occupants' attention, as a sound that occupants are preconditioned to recognize as one that precedes safety instructions. The initial announcer was a female voice, followed by instructions given in a male voice. The change of voices was believed to draw greater attention. Since the FCC tone followed by a male voice had become associated with "false alarms", it was hypothesized that the female voice would be better at making people believe the alarm as well as pay attention. The use of a male voice for instructions was expected to come across as more authoritative and thus make it more likely that the directions would be followed. A higher range male voice delivers more of the message in frequencies above 1000 Hz, which makes the voice more easily understood.

The authors presented several key concepts to be incorporated into messages. First, tell occupants what is happening, what to do, and why alternatives (e.g., the elevator) should not be used. Second, indicate that an actual person (e.g., the building manager) is in charge and that the individuals are not being singled out. Third, use specifics about the emergency to reduce ambiguity. Fourth, word the message to promote calm behavior, and finally, repeat essential components twice. For this particular test scenario, the word "evacuate" was not used since people were not to leave the building; the authors stressed the importance of using words that will not mislead the occupants into unintended actions.

To test their system, two unannounced drills were conducted during the first week the building was open. One drill involved occupants that had been trained to know what the alerting sound would be and what they were to do, and the second involved occupants that had not received training. The first drill was conducted at 3:15 pm, and the second took place at 3:40 pm. Firefighters in gear were present to make the occupants believe that the drill was authentic. In both cases, all floors were cleared of the building occupants in one minute to 1.5 minutes. Questionnaires were administered after each drill to determine how people interpreted the alarm, of which 205 were returned. No differences were found between the two occupant groups from either the survey results or the observations by researchers present during the drill. Occupants understood what they were to do and stated a preference for voice instruction over the warning signal alone.

Key findings:
- [74_At_1] [74_Ac_1] A voice alarm system consisting of an alarm tone followed by a female voice to attract attention and a male voice to deliver instructions resulted in a successful area evacuation, with personnel vacating to receiving floors either above or below their own.

- [74_Ac_2] A voice alarm system is capable of directing an evacuation in which speed and proper relocation of personnel do not depend on specific training.

Key recommendations:

- [74_Ac_3] An effective voice alarm system includes the nature of the emergency and directions (what is happening, what to do, and why alternatives should not be used), assurance that someone is in charge, specifics, information to promote calmness, and repetition of essential components.
- [74_At_2] Change voices during the emergency broadcast message to attract attention.
- [74_Cr_1] [74_Co_1] A higher pitched male voice is both authoritative and more easily understood.

 # 75. Keselman A, Slaughter L, Patel VL. (2005, August). Toward a framework for understanding lay public's comprehension of disaster and bioterrorism information. *Journal of Biomedical Informatics 38*(4), 331-344.

M		Pc	At	Co	Cr	Ps	Ac

This article reviews research relating to emergency messaging and how it is understood by the public. Insights are obtained from articles on comprehension and decision-making (cognitive science) and on the effects of anxiety on cognitive performance.

Lay people (non-experts) typically receive disaster-related information from a variety of sources, including public health and government organizations, news media, and friends or coworkers. All accumulated information is then processed into a representation of the event that serves as a basis for decision-making and action. Basic principles are needed for effective disaster communication and coverage. News media accounts may emphasize the sensational over the delivery of information about risk, actions that are being taken by authorities, and actions that the public can take to protect themselves. The degree of comprehension of emotionally-laden news by lay people is not obvious.

Public health guidelines state that emergency messages should be brief, clear, and effective, since during a crisis people might have trouble focusing on technical aspects of the message. Recommendations include short, simple sentences and messages aimed at specific populations (e.g., those in imminent danger). Guidelines for risk communication take into account that non-experts perceive lower risk for natural disasters over manmade disasters, voluntary over involuntary risks, and habitual over exotic risks. However, a greater appreciation is needed of lay misconceptions and gaps in understanding and how these influence the representation of the crisis situation as a basis for decision-making.

Even when text complexity is at an appropriate readability level, the difference in background knowledge between experts and lay people often results in miscommunication. A study demonstrated that the health concepts described by physicians are not interpreted in the same way by their patients, who may therefore discount the explanation and devise an alternative, resulting in poor compliance. Before a message is used, there is a need to "walk through" the message to identify every place where misconception by the target audience can occur. The background of individuals can make a large difference in interpretation. Prior knowledge provides context in which meaning is understood and assumptions from which to make inferences. This was demonstrated by a study of indigenous Kenyan mothers, for which comprehension of instructions for administering medication was found to depend on knowledge of basic health concepts. Prior knowledge also affects what details are remembered and how those details are organized for reasoning and decision-making. Comprehension can be studied through analysis of propositions, or units of thought connected by a relationship. When asked to recall the details of a clinical case, expert clinicians recalled and inferred more relevant information than novices. While intermediate level med students were found to recall more relevant facts than experts, they are not yet expert diagnosticians, since the information is not yet organized into the necessary coherent knowledge structure. The ability to sort relevant from irrelevant information is key to separating the novice and intermediate from the expert. Emergency communication needs to help lay people make this distinction. Experts are also able to use heuristics as shortcuts for routine situations or under time pressure. Science concepts taught to lay people are not easily integrated into practical knowledge, but may instead compete with other explanations more deeply rooted in tradition or experience, as demonstrated in studies with Indian mothers and inner-city adolescents. This suggests that communication with the public needs to address possible discrepancies between common lay theories and science.

The way people process information is different when they are in danger. When physical danger is imminent, reactions are quick, with little information processing. Before and after these crisis situations, decisions are more deliberate. From studies of news comprehension, people remember names and places best but have more difficulty recalling events, and specific numbers are not typically remembered. Recall is improved when the information is relevant, repeated, local, sensational or tragic, which is hopeful for public health disaster communications but less so for preparedness. News recall is also higher when there is previous knowledge. Although cognitive performance (e.g., attention, response speed, memory) under emotional stress has been studied, no known studies have looked at comprehension under these conditions. When arousal is high, the focus of attention is very narrow, and some relevant information may be missed. Worry introduces distracting thoughts. Alertness is heightened by arousal and information is more readily encoded into long-term memory although short-term memory access is diminished. Caution is advised in extrapolating the effects of anxiety on cognition from laboratory results to actual disasters. The effects of physical danger on cognition include learned helplessness (impairment in the ability to learn new tasks) and deterioration in memory retrieval, visual recognition of numbers, and performance on secondary tasks. Increased fear can also decrease accuracy, as demonstrated by an increase in navigation calculation errors by bombers as the aircraft approaches the target in wartime. Expertise and experience mitigate these effects.

Key findings:
- [75_Co_1] Message readability alone does not explain comprehension problems.
- [75_Co_2] Emergency messages that can be comprehended by lay people must take background knowledge into account.
- [75_Co_3] [75_Cr_1] Due to differences in background knowledge, the health concepts described by physicians are not interpreted in the same way by their patients, who may therefore discount the explanation and devise an alternative, resulting in poor compliance.
- [75_Co_4] [75_Ac_1] Prior knowledge provides context in which meaning is understood and assumptions from which to make inferences. It also affects what details are remembered and how they are organized for reasoning and decision-making.
- [75_Ps_1] Emergency communication needs to help lay people sort relevant from irrelevant information.
- [75_Cr_2] Science concepts taught to lay people are not easily integrated into practical knowledge, but may instead compete with other explanations more deeply rooted in tradition or experience. Communication with the public needs to address possible discrepancies between common lay theories and science.
- [75_Ac_2] When physical danger is imminent, reactions are quick, with little information processing. Knowledge is critical for medium urgency decisions.
- [75_At_1] [75_Ps_2] Recall of news items is best for names and places, lower for events, and very low for specific numbers. Recall is improved when the information is relevant, repeated, local, sensational or tragic, or matches previous knowledge.
- [75_At_2] When arousal is high, the focus of attention is very narrow, and some relevant information may be missed. Alertness is heightened and information is more readily

encoded into long-term memory, although short-term memory access is diminished.

- [75_Ac_3] The effects of physical danger on cognition include impairment in the ability to learn new tasks and deterioration in memory retrieval, visual recognition of numbers, and performance on secondary tasks.
- [75_Ac_4] Increased fear can decrease accuracy, as demonstrated by an increase in navigation calculation errors by bombers as the aircraft approaches the target in wartime.
- [75_Ac_5] Expertise and experience mitigate the effects of physical danger on performance.

Key recommendations:

- [75_Co_5] Before a message is used, there is a need to "walk through" the message to identify every place where misconception can occur.
- [75_Co_6] Comprehension can be studied through analysis of propositions, units of thought that are connected by a relationship.

 # 76. Kline PB, Braun CC, Peterson N, Silver NC. (1993). The impact of color on warnings research. *Proceedings of the Human Factors and Ergonomics Society 37th Annual Meeting*, 940-944.

E		Pc	At	Co	Cr	Ps	Ac

This study evaluated the appropriateness of grey-scale (achromatic) and multicolored labels in product warnings. Research has shown that the use of color coding in displays provides many advantages, including aesthetic appeal and changes in visual search time and cognitive processing. [Christner and Ray. (1961). An evaluation of the effect of selected combinations of target and background coding on map-reading performance. *Human Factors*, 131-146.] Color has been shown to positively influence visual search and counting tasks and to influence memory.

Thirty-three undergraduate students rated color and achromatic versions of twelve labels that varied by product class and signal word. The labels were from four types of common commercial household products and used one of three different signal words: DANGER, WARNING, and CAUTION. The combinations of product class and signal words resulted in the twelve different label conditions. Labels were sorted into four random presentation orders. Participants evaluated the labels based on six attributes: salience, readability, hazardousness, likelihood of injury, carelessness, and familiarity, rating each attribute on a nine-point scale. Labels presented in color were perceived to indicate more hazard than those presented in black and white.

Key findings:
- [76_Co_1] Color has been shown to positively influence visual search tasks and counting, and to influence memory.
- [76_Ps_1] Colored signs are perceived to indicate more hazard than black and white signs.
- [76_Co_2] Colored signs must have high contrast between the background and the text.

77. Koreimann S, Strauß S, Vitouch O. (2009). Inattentional deafness under dynamic musical conditions. *Proceedings of the 7th Triennial Conference of European Society for the Cognitive Sciences of Music (ESCOM 2009).* Jyväskylä, Finland, 246-249.

P		Pc	At	Co	Cr	Ps	Ac

Inattentional blindness, in which a simple attention task causes people to overlook large unexpected changes in a scene, was first demonstrated in 1979. Inattentional deafness is the analogous phenomenon for listening, in which a distracting task causes listeners to overlook an unexpected sound that is obvious to others. This is different from, and in some sense opposite to, the "cocktail party effect," in which distracting voices pull the attention of listeners away from what they are focused on.

This study demonstrates inattentional deafness to music under controlled laboratory conditions with 58 non-musicians and 57 amateur musicians. Participants listened to the first 1'50" of Thus Spake Zarathustra, a piece of classical music with which all were familiar. The experimental group was asked to count the number of tympani beats, and the control group just listened to the music. An unexpected electric guitar improvisation was added to the last 20 seconds of music to test whether the subjects would notice the change. Only one non-musician out of 29 in the experimental group heard the e-guitar, while 15 of 29 (52 %) in the control group noticed it. For musicians, 11 out of 29 (38 %) in the experimental group and 19 out of 28 (68 %) in the control group heard the e-guitar. A second study, with only non-musicians participating, investigated the effects of task difficulty by mixing the same musical piece with a more obvious e-guitar take (less embedded and incorporating the slide or bottleneck technique). With this easier task, 12 out of 24 in the study group and 19 out of 23 in the control group identified the e-guitar solo.

Key findings:
- [77_At_1] A distracting task can cause listeners to overlook an unexpected sound that is obvious to others. This inattentional deafness has been demonstrated using a well-known musical piece.
- [77_At_2] Expertise (musical training) helped but was not fully successful in drawing the attention of those occupied in a listening task to a striking change (e-guitar improvisation) in a well-known musical piece.
- [77_At_3] A striking musical modification drew the attention of 50 % of non-musicians occupied in a listening task, compared to over 80 % for those not otherwise occupied and to one out of 29 for a less obvious modification.

Key recommendations:
- [77_At_4] In order to attract everyone's attention for an auditory message, other activities that are occupying them should be halted.
- [77_At_5] When people are engaged in a task, an auditory message must be an obvious change from ambient noise to draw attention.

 # 78. Koster EHW, Crombez G, Verschuere B, Van Damme S, Wiersema JR. (2006, December). Components of attentional bias to threat in high trait anxiety: Facilitated engagement, impaired disengagement, and attentional avoidance. *Behaviour Research and Therapy, 44*(12), 1757-1771.

P		Pc	At	Co	Cr	Ps	Ac

Previous research has proposed that individual differences in attention to threatening information can be sorted into two categories: high trait anxious (HTA) and low trait anxious (LTA). Attention to threat is seen as a normal and adaptive process that depends on both task demands and stimulus input. Stimuli are judged to be considerably more threatening by HTA individuals than by LTA individuals. This attentional bias may be related to anxiety disorders. A model of attending to a new stimulus suggests three steps: 1) initial shift of attention, 2) engagement with the stimulus, and 3) disengagement from the stimulus. The purpose of this study was to determine whether the attentional bias of HTA individuals relates more to ease of engagement with the threat or difficulties with disengaging from the threat. The methodology was the exogenous cueing task, in which participants indicate detection of a visual target at the left or right side of a fixation cross, preceded by a peripheral cue at the same or opposite location from the target. A cue at the same location shortens the response time, while a cue at the opposite location lengthens it. Varying the emotional content of the cue (neutral or threat) also changes the response time, with the difference in same side cues providing a measure of attentional engagement and the difference in opposite side cues providing a measure of attentional disengagement.

The 47 participants in the first study consisted of 24 low anxiety (LTA) psychology students (seven males, 17 females) with a mean age of 19 years old and 23 high anxiety (HTA) students (two males, 21 females) with a mean age of 20 years old. The subjects were shown a high threat (HT), mild threat, or neutral picture for either 100 milliseconds or 500 milliseconds and then asked to identify the location (right or left) of the target (a small black square) that appeared randomly in the same or opposite location. Each subject responded to 240 trials. HTA individuals were found to have enhanced attentional engagement and impaired disengagement for highly threatening information displayed for a short time (100 ms). Unexpectedly, HTA individuals showed attentional avoidance (longer response times) for HT pictures displayed for a longer time (500 ms). Results for LTA individuals did not show the expected attention to HT pictures.

A second study was carried out to investigate the unexpected results from the first study. An identical approach was used, except that pictures were presented for 200 milliseconds or 500 milliseconds. The 38 participants in this study consisted of 19 LTA students (seven males, 12 females) with a mean age of 20 years old and 19 HTA students (nine males, 10 females) with a mean age of 18.5 years old. Attentional avoidance of HT pictures was observed for both durations, with a stronger effect for HTA individuals. This is in opposition to other research, which suggests impaired attentional disengagement rather than avoidance for HTA individuals. There are procedural differences in methodology among studies; one possible explanation is that the pictures shown as cues here were in color, and therefore more graphic than black-and-white pictures used by others.

Key findings:
- [78_At_1] A model of attending to a new stimulus suggests three steps: 1) initial shift of

attention, 2) engagement with the stimulus, and 3) disengagement from the stimulus. Attentional bias, in which the stimulus is judged to be more threatening than it is in reality, may relate to any of these steps.

- [78_At_2] Individuals differ in their response to threatening information, with some people alerting to lower levels of stimulus and remaining in an anxious state longer than others.

- [78_At_3] Information indicating a high level of threat may cause people to be more engaged or, conversely, may result in attentional avoidance.

 # 79. Kuhn BT, Garvey PM, Pietrucha MT. (1997). Model guidelines for visibility of on-premise advertisement signs. *Transportation Research Record, 1605*, 80-87.

T		Pc	At	Co	Cr	Ps	Ac

Advertising signs on the road play an important role in directing drivers. The purpose of this research was to synthesize existing literature on sign visibility.

The most important factor in sign detection is sign placement. The signs should be placed within 30 degrees of the driver's line of sign. The optimal cone of vision is approximately 10 degrees to 12 degrees along the horizontal axis and five degrees to eight degrees along the vertical axis. Detection of signs located 20 degrees and 30 degrees outside the fovea (the part of the eye that is responsible for sharp central vision) required the viewer to stand at a distance one-third and one-fourth, respectively, ofthe detection distance for a sign within the line of sight. If a sign is not surrounded by many other objects in a driver's cone of vision, it is more likely to be noticed.

As the number of objects in the driver's cone of vision increases, the conspicuity of a sign decreases. External contrast is the difference between the luminance of the sign and the area immediately surrounding the sign. As the external contrast ratio of the sign increases, so does the sign's conspicuity.

For freeway signs, it was found that high-intensity background sheeting with high-intensity stick on copy, opaque sheeting with button copy, and engineering grade background sheeting with button copy all have acceptable freeway guide sign detection distances. Black on orange and white on green signs were detected at greater distances than black on white signs. Previous research concluded that black on white signs were being confused with white light sources and that it was necessary to got closer before recognizing the object as a sign. Green signs with high internal contrast improve sign detection. Edge definition or the use of borders enhances sign conspicuity. The internal contrast ratio between legend and background should be 12:1, where luminances for both are measured in cd/m².

It is suggested that sign legend luminance should be 75 cd/m² and must be at least 2.4 cd/m² for black on light signs at night. With light on dark signs, the luminance should be 30 cd/m² to provide the maximum nighttime legibility distance. Signs at angles greater than 20 degrees should be manipulated to appear normal on the road. Text is more legible if it is not all uppercase but mixed case. The stroke width to height ratio to both positive and negative contrast letters is most legible at a ratio of 1:5. However, light letters on a darker background should have a thinner stroke (a ratio of 1:7), and dark letters on a lighter background should have a bolder stroke. The first syllable can generally be used as an abbreviation for words having nine letters or more, unless the first syllable itself is a word. For five to seven letter words, abbreviations consisting of the key consonants are suggested, such as 'frwy' for freeway. Abbreviations should only be used as a last resort because they may lead to errors in interpretation. Legibility distance increases as letter height increases; however, there is not a linear relationship between increasing height and legibility distance. Positive contrast signs (light text on dark background) have longer legibility distances than negative contrast signs (dark text on a light background).

Key findings:
- [79_Pc_1] Black on orange and white on green signs were detected at greater distances than black on white signs.
- [79_Pc_2] Borders increase sign conspicuity.
- [79_Pc_3] The internal contrast ratio between legend and background should be 12:1, where luminances for both are measured in cd/m^2.
- [79_Pc_4] The most legible stroke width to height ratio for both positive and negative contrast letters is 1:5.

 # 80. Lablale G. (1990). In-car road information: Comparisons of auditory and visual presentations. *Proceedings of the Human Factors Society 34th Annual Meeting.* Santa Monica, CA: Human Factors Society.

E		Pc	At	Co	Cr	Ps	Ac

In-car message systems are designed to provide navigation and road condition data in auditory and visual formats or a combination of both. The main objective of this research is to identify ways in which to reduce driver mental workload and stress and to make driving more comfortable and safer.

Sixty-two drivers participated in the experiments highlighted here. Thirty participated in the first experiment regarding road information messages. The route used in these experiments was two kilometers in length, with four turns and very little traffic. Speed was restricted to 60 km/h.

The purpose of the first experiment was to test three types of sensory modality (visual, auditory, and repeated auditory) with four categories of message length. The visual messages appeared on the visual in-car display. The auditory messages were transmitted by loudspeakers, and each message was repeated three seconds after initial transmission for the repeated auditory modality. The length of road information messages depended on the number of information units in each message. Road information units contained several different categories of information, including geography (region, city), type of road, position or direction, event-case and event-consequence, time and distance, and proposed action. The four types of message length were: Level one (four information units), Level two (seven to nine information units), Level three (10 to 12 units), and Level four (14 to 18 units). Each driver was instructed to memorize each road information message presented to him/her.

The results of the study showed an inversely proportional relationship between message length and memorization performance. The recall of a level one message was 100 %, whereas the score for a Level four message dropped to 48.4 %. No statistically significant differences in recall were found among modalities (visual, auditory, repeated auditory). For longer (Level four) messages, however, visual presentations were found to result in better recall than single auditory messages. Recall of longer (Levels three and four) auditory messages was found to increase when the message was repeated. For these longer messages, 63 % of drivers preferred the repeated auditory messages, 33 % preferred visual presentations, and 3 % preferred single auditory messages. For the shorter (Levels one and two) messages, drivers divided about evenly in preference among the three modalities.

Key findings:
- [80_At_1] An inversely proportional relationship was found between the length of in-car road information messages and recall. The recall of a message with four units of information was 100 %, whereas the score for a message with 14 to 18 units dropped to 48.4 %.
- [80_At_2] For long in-car road information messages (containing 14 to 18 units of information), visual presentations result in better recall than single auditory messages.
- [80_At_3] Recall of longer auditory in-car road information messages (containing 10 to 18

units of information) increases when the message is repeated.

- [80_At_4] Repeated auditory messages were preferred by 63 % of drivers for longer auditory in-car road information messages (containing 10 to 18 units of information), and visual presentations were preferred by 33 %.
- [80_At_5] No clear preference was found among visual, auditory, and repeated auditory modalities for short in-car road information messages (four to nine units).

81. Lai CJ. (2010). Effects of color scheme and message lines of variable message signs on driver performance. *Accident Analysis & Prevention 42*(4), 1003-1008.

The study evaluated the effects of color schemes and message lines on Chinese variable message signs (VMS) when viewed by drivers. The color schemes were single color (green, yellow or red), two color (green and yellow, green and red, or yellow and red), and three color (green and yellow and red). Seven types of color schemes were used in the study, with single, double, or triple sets of message lines. Each participant viewed 84 randomized VMS presentations, with each sign shown twice. Thirty university students between the ages of 19 and 28 years old participated. The experiment was carried out in a virtual driving environment, with subjects in the driver's seat of a car with a screen in front of them to simulate driving. The participants were asked to identify the message on the VMS by turning the steering wheel. After the steering wheel was turned, a new sign appeared and a computer recorded the response time.

Results showed that the two-color scheme resulted in the shortest response time, and surveys indicated that the participants preferred a two-color scheme sign. The response time for the double message line was shorter than for single or triple message lines, and the participants preferred the double message line. The double message line VMS displayed the same message on two lines as the single message line VMS displayed on one line. The shorter response time for the single line message may therefore be explained by recognizing that shorter line lengths are easier to scan than longer ones.

Key findings:
- [81_Co_1] [81_Ac_1] A two-color scheme for VMS road signs resulted in the shortest response time and was preferred by drivers.
- [81_Co_2] [81_Ac_2] Variable message signs (VMS) with a double message line were better understood and resulted in shorter response times than those with single or triple lines.

82. Lesch MF. (2008). A comparison of two training methods for improving warning symbol comprehension. *Applied Ergonomics 39*(2), 135-143.

E		Pc	At	Co	Cr	Ps	Ac

This study compared the effects of two different types of training on the understanding of warning symbols. The participants were people in two age groups, from 20 to 35 years and from 50 to 70 years. Participants completed a pre-training test to determine their prior knowledge of symbols. Then twenty participants in each age group were given two types of training: verbal label training or accident scenario training. For verbal label training, a computer screen displayed a symbol with a label describing the symbol. For accident scenario training, the symbol with a label describing it was displayed along with the possible consequences if the warning is not followed. In each case, participants were allowed to view the symbol as long as they wished. When they were finished with one symbol, they pressed the number three on the keyboard and a new symbol appeared. Ninety symbols were used in the study, with a text label and accident scenario training written for each. Each participant was presented with 36 symbols for each training type and was tested on 30 of them. In the test, participants were shown a symbol and a statement and were asked to decide as quickly and accurately as possible whether the statement matched the meaning of the symbol. If they answered that the symbol did match the statement, they were asked to rank the answer from six to 10, where 10 indicated that they were very confident and six that they were not confident. If they answered that the symbol and statement don't match, they ranked the answer from one to five, where one indicated that they were very confident and five that they were not confident.

The pre-training test showed that many warning symbols are not well understood before training. Participants got an average of 45 % correct with a confidence score of 65 %. The younger group was found to benefit from the training more than the older group. Accident scenario training improved processing of warning symbols better than verbal label training, with both quicker and more confident responses after this type of training.

Key findings:
- [82_Co_1] Training is beneficial to the understanding of new symbols.
- [82_Co_2] Younger populations benefit more from training than older groups.

83. Lesch MF, Rau P-LP, Zhao Z, Liu C. (2009). A cross-cultural comparison of perceived hazard in response to warning components and configurations: US vs. China. *Applied Ergonomics, 40*(5), 953-961.

E

Pc	At	Co	Cr	Ps	Ac

This study compared the reaction of students from the United States and from China to warning words and warning colors in order to assess the applicability of current standards across different cultures. Twenty undergraduate students in China and twenty students in the United States were used in this study. Each participant ranked colors (Blue, Green, Yellow, Orange, Red, and Black) and hazard words (NOTICE, BEWARE, CAUTION, DANGER and DEADLY) from one (not at all hazardous) to nine (extremely hazardous). These words and colors appear in the ANSI standards. Each Chinese student had previously passed a standard English test and, for the experiment, rated the hazard words in both English and in Chinese. The Chinese students generally perceived that the words suggested a lower level of hazard than the students from the United States. For United States students, the perceived hazard ranking of the colors from low to high was Blue, Green, Yellow, Black, Orange, and Red; and for China it was Yellow, Blue, Green, Red, Black and Orange. The Chinese students rated the signal words lower (both in English and in Chinese) than the students from the United States. This indicates that Chinese people require stronger signal words and colors to communicate the same hazard level communicated to the Americans. Neither the signal words nor colors communicated the exact meaning suggested in the ANSI standards. The most notable discrepancy is the color yellow. The Chinese viewed it as the lowest level of danger while ANSI expects it to be viewed as a middle level of hazard.

When testing more complex configurations, (containing combinations of colour and wording), it was found that the relative perceived hazard was similar for both US and Chinese students. The response to these configurations was also consistent to ANSI standards.

Key findings:
- [83_Ps_1] Colors suggest different levels of hazard.
- [83_Ps_2] Signal words suggest different levels of hazard.
- [83_Ps_3] The interpretation of hazard words and colors is not consistent across cultures. United States students ranked colors from low to high hazard in the order Blue, Green, Yellow, Black, Orange, and Red; while Chinese students ranked them as Yellow, Blue, Green, Red, Black and Orange. The Chinese students rated signal words lower in hazard than the United States students.

 # 84. Leung AKP, Hellier E. (1998). Perceived hazard and understandability of signal words and warning pictorials by Chinese community in Britain. In MA Hanson (Ed.), *Contemporary Ergonomics 1998: Proceedings of the Annual Conference of the Ergonomics Society* (pp. 321-327). London, UK: Taylor & Francis.

Pc	At	Co	Cr	Ps	Ac

This study compared the ability of people in the Chinese and English populations in London to understand visual warnings and perceive hazard. This tests the comprehensibility of signal words that appear in standards and guidelines, which usually recommend DANGER, WARNING, and CAUTION to denote the highest to lowest hazards respectively.

Forty eight Chinese participants that could read and speak English and 48 native English speakers rated 43 signal words for understandability and the level of hazard being communicated. The subjects were also presented with twelve pictorial warnings, for which they had to write down the meaning and then rate the understandability. No significant differences between Chinese and English groups were found in hazardousness or understandability ratings for eight common signal words (NOTE, ATTENTION, NOTICE, CAREFUL, DANGER, CAUTION, WARNING, and DEADLY). Across all 43 words, the Chinese population displayed significantly lower levels of understandability (especially with some signal words: HALT, LETHAL, PROHIBIT, and FORBIDDEN). Based on higher levels of understandability, small standard deviations, and similar results across subject groups, the authors suggested using these 12 signal words: CAREFUL, CAUTION, DANGER, DANGEROUS, EXPLOSIVE, FATAL, HARMFUL, HAZARD, HAZARDOUS, POISON, SERIOUS, and TOXIC. Regarding the perception of hazard for specific words, DEADLY was rated significantly higher on hazardousness than DANGER, which was rated significantly higher on hazardousness than either WARNING or CAUTION. WARNING was not rated significantly higher on hazardousness than CAUTION. This agrees with other studies that did not find a distinction between the perceived meanings of WARNING and CAUTION.

Key findings:
- [84_Co_1] Common warning signal words should be used when non-native speakers may be present.
- [84_Co_2] Populations of non-native English speakers should be tested on their comprehension of words used in warning messages.
- [84_Co_3] The signal words WARNING and CAUTION are not perceived as denoting significantly different levels of hazard.

85. Levine TR, Anders LN, Banas J, Baum KL, Endo K, Hu ADS, Wong NCH. (2000, June). Norms, expectations, and deception: A norm violation model of veracity judgments. *Communication Monographs, 67*(2), 123 – 137.

M		Pc	At	Co	Cr	Ps	Ac

There is a question about whether people judge veracity by a finite set of nonverbal behaviors or whether any abnormal behavior will raise a question of honesty. This study examines the extent to which people who exhibit unexpected strange behavior (violations of expectations) are more likely to be judged as honest or deceptive (veracity judgments). There is a distinction between norms, which are socially appropriate behaviors, and expectations, which vary with the individual. Four models of veracity judgments were tested by the experiment: a normative expectation model, an expectancy violation sufficient model, Expectancy Violation Theory, and a norm violation model.

The subjects were 128 undergraduate students at the University of Hawaii, 62 female, 58 male, and eight unknown, ranging in age from 18 to 40 years (mean age 21 years old), and ethnically diverse. Each subject interviewed a confederate, with crosses between norms, expectations, truthfulness, and confederate identity. The abnormal behavior that violated social norms consisted of strange eye movements, excessive stretching, several volume shifts in voice, and teeth picking. After the interview, the subjects completed a questionnaire that contained an assessment of honesty, filler having to do with communication anxiety, and manipulation check questions. Although all violations of norms were assessed as abnormal behavior, the rating of abnormality varied significantly. The data was found to be most consistent with the norm violation model, with aberrant behavior causing subjects to be more likely to distrust the answers given by the confederate. In evaluating trustworthiness in face-to-face communication, behavior was found to be more important than prior expectations.

Key findings:
- [85_Cr_1] In evaluating trustworthiness in face-to-face communication, behavior was found to be more important than prior expectations.

Key recommendations:
- [85_Cr_2] Officials delivering messages should behave in accordance with social norms to be more credible.

 # 86. Lindell MK, Prater CS, Peacock WG. (2005, June). Organizational communication and decision making in hurricane emergencies. *Prepared for the Hurricane Forecast Socioeconomic Workshop*, 16-18 February 2005, Pomona, CA.

Lindell MK, Prater CS, Peacock WG. (2007, August). Organizational communication and decision making for hurricane emergencies. *Natural Hazards Review 8*(3), 50-60.

C		Pc	At	Co	Cr	Ps	Ac

A review of social science research on the response of households, businesses, and special facilities to communication from emergency organizations suggests that local officials dealing with an approaching hurricane need information on evacuation times, costs, and potential loss of life if evacuation is delayed. They also need improved decision support systems. Although hurricane prediction has improved over time, the population at risk has increased. An emergency response system has four functions: emergency assessment, hazard operations, population protection, and incident management. Information reaches the public through official (broadcast) channels, such as emergency management agencies, internal web sites, and electronic and print news media. A study in 2000 found that 58 % of respondents reported maintaining constant access to news sources within two to three days of a hurricane strike, with 89 % monitoring the Weather Channel and 30 % the internet. Information may flow in the opposite direction from the public to news sources. Information also reaches the public through informal (contagion) sources from friends, neighbors, and coworkers.

During an emergency, people seek confirmation of warnings from multiple sources. They develop their own independent emergency assessments (risk perception) and pass information on to others. Population segments differ in their access to warning messages and in their perception of the credibility of information sources. Emergency managers should use multiple pathways for disseminating information, including hotlines to provide additional information. They should attempt to obtain feedback from the public. They should monitor news media to assess the situation and to make sure that messages being given to the public are accurate. The most effective warnings (highest level of compliance) are from a credible source and sufficiently specific that individuals can determine whether the threat pertains to them personally. Traffic management, transportation alternatives (for about 8 % without personal vehicles), and shelters must be arranged and also communicated to the public. The public can help by acting to protect structures from flooding, controlling debris, and warning others. A significant time delay after an evacuation warning almost always occurs, due to psychological preparation (including information seeking) and logistical preparation. Compliance is highest for mandatory evacuations, and for people who consider themselves at risk for flooding, and for residents of mobile homes. Early evacuation and spontaneous evacuation from regions considered safe can be high.

Research is needed on the speed of information flow and on the ability of recipients to utilize emergency information. Evacuation decisions are difficult because of the uncertainty of hurricane forecasts and the significant costs involved, although specific data and methodology for estimating costs are lacking. A major uncertainty in evacuation models is the rate of traffic flow when the demand on evacuation routes exceeds capacity. An accurate forecast of the delay time before evacuation is needed. Research to improve communication of hurricane risk is needed, including a model of formal and informal warning networks and better ways to display information on personal consequences and location.

Key findings:
- [86_Cr_1] During an emergency, people seek confirmation of warnings from multiple sources. These can include emergency management agencies, internal web sites, electronic and print news media, web sites, and friends and neighbors.
- [86_Cr_2] During an emergency, people develop their own independent emergency assessments (risk perception) and pass information on to others.
- [86_Pc_1] [86_Cr_3] Population segments differ in their access to warning messages and in their perception of the credibility of information sources.
- [86_Cr_4] [86_Ps_1] The most effective warnings (highest level of compliance) are from a credible source and sufficiently specific that individuals can determine whether the threat pertains to them personally.
- [86_Ac_1] A significant time delay after an evacuation warning almost always occurs, due to psychological preparation (including information seeking) and logistical preparation.

Key recommendations:
- [86_At_1] Emergency managers should use multiple pathways for disseminating information, including hotlines to provide additional information.
- [86_Co_1] Emergency managers should attempt to obtain feedback from the public.
- [86_Ac_2] Emergency managers should monitor news media to assess the situation and to make sure that messages being given to the public are accurate.
- [86_Ac_3] Traffic management, transportation alternatives, and shelters must be arranged and communicated to the public.

 # 87. Mathews A, Mackintosh B. (1998). A cognitive model of selective processing in anxiety. *Cognitive Therapy and Research, 22*(6), 539–560.

P		Pc	At	Co	Cr	Ps	Ac

This paper develops a new model for selective processing during anxiety states. It has been found that at higher levels of anxiety, people both direct more attention to a possible threat and are more likely to assign threatening meaning to ambiguous information. For example, a task for which threatening cues must be ignored is slowed in an anxious state, while a task that uses threats to find a target is speeded up. Reaction speed and preferences are affected by even brief (15 milliseconds to 20 milliseconds) exposures to threatening and non-threatening pictures, indicating that the emotional meaning of cues is processed automatically rather than at a conscious level.

Evidence from these experiments as well as animal brain studies indicates that the brain uses one of two ways to evaluate the threat level: either higher-level conscious processes, which are slow, or via a short-cut that goes directly from the thalamus to the amygdala. The new model hypothesizes that the dangerous threats identified consciously when first met may then be stored and processed automatically in similar later events. This builds a threat assessment system that can evaluate threat cues automatically to decide whether attention should be drawn from the task at hand. The threshold value for response is different for each individual and increasing fear or anxiety temporarily sets the threshold lower. Sensitive, or high trait-anxious individuals, have stored a wider range of representations of various levels of threat, making an automatic reaction to cues in the environment more likely. A cue at a high level of danger will therefore attract everybody's attention, but a cue at a lower level will attract the attention only of people that are either at a high state of anxiety or anxious by nature. Low trait-anxious people that are concentrating on a task show a tendency to avoid paying attention to threat cues so that they can complete the task. High trait-anxious people pay high attention to threat cues until a point at which the threat level is so high that it overwhelms cognitive resources, at which point these people may show a paradoxical tendency to ignore new threats. As the number of threatening cues in the environment increases, a person can start to focus on a small number of cues, to the detriment of processing new cues. This effect is more pronounced with higher anxiety levels. With higher anxiety levels, therefore, people are more likely to assign threatening meaning to ambiguous information.

Key findings:
- [87_At_1] The threshold value for response to threat cues is different for each individual, and increasing fear or anxiety temporarily sets the threshold lower.
- [87_At_2] In order to capture the attention of both high and low trait-anxious people from a normal low-anxiety state, an alarm needs to communicate a high level of danger.
- [87_At_3] Low trait-anxious people show a tendency to avoid paying attention to threat cues so that they can complete the target task. Warnings need to break through this tendency.
- [87_At_4] As the number of threatening cues in the environment increases, a person may start to focus on a small number of cues, to the detriment of processing new cues.
- [87_Ps_1] With higher anxiety levels, people are more likely to assign threatening meaning to ambiguous information. Messages therefore need to be communicate the level of risk clearly and precisely, encouraging as little inference as possible.

Key recommendations:
- [87_Ps_2] Messages need to be communicate the level of risk clearly and precisely, encouraging as little inference as possible.

 # 88. Mayer DL, Laux LF. (1989). Recognizability and effectiveness of warning symbols and pictorials. *Proceedings of the Human Factors Society 33rd Annual Meeting*, 984-988.

E		Pc	At	Co	Cr	Ps	Ac

The use of pictorials and symbols to communicate hazards is increasingly necessary due to the increasing number of people that are not literate in English. This study had two parts: (1) participants were shown 16 symbols and asked to describe the hazard that the symbol represented and what it would be used to warn about; and (2) participants were shown three symbols: *Electric shock* (wire shocking hand), *Poison* (skull and cross bones), and *Flammability* (Flames), and asked to list all of the ways that they could be hurt, injured or killed while using a product with one of those warnings.

The results for the first part showed that the most recognizable pictograms were *Flammability* and *Corrosive*. For the second part, participants under 40 were more likely to indicate that they would look for warnings on the package or product than the subjects over 40. The researchers found that overly detailed human figures may detract from the effectiveness of the pictogram, but parts of the human body like feet or hands provide a frame of reference. Pictograms depicting protective equipment or those with simple messages provided good recognizability.

Key findings:
- [88_At_1] People younger than 40 are more likely to look for warnings on a package than those over 40.
- [88_Co_1] A pictogram showing specific body parts is more effective than one that shows the entire body. Specific information is better than general information.
- [88_Co_2] Pictograms depicting protective equipment and those with simple messages are easily recognized.

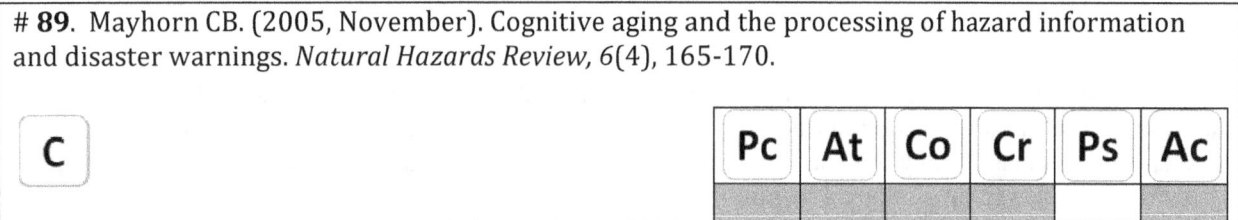

 # 89. Mayhorn CB. (2005, November). Cognitive aging and the processing of hazard information and disaster warnings. *Natural Hazards Review, 6*(4), 165-170.

This paper reviewed findings regarding the cognitive effects of aging in the context of the Protective Action Decision Model (PADM), which defines the relationships among the source (message originator), channel (delivery method), and receiver (decision-maker). Recommendations are made for modifications to public risk communication messages and delivery that will be more effective for older adults. Effective communication of hazards and protective action recommendations can reduce the incidence of trauma and fatalities within this vulnerable population. The cognitive processing of the message by the receiver is affected by many factors, including perception, attention, memory, text comprehension, and decision-making, and may be influenced by the design of the message and its delivery.

Perception: Decreasing visual abilities can lead to difficulties in seeing smaller size fonts, differences between objects of similar colors (especially shades of green and blue), and any image in the presence of glare. Large sans serif fonts should be used in warning texts. For auditory messages, higher frequencies (above about four kHz) become harder to hear due to both noise-induced and age-related hearing losses, and background noise is harder to filter. Information from multiple speakers in a noisy environment (e.g., from the radio, telephone, PA system, or face-to-face) is often missed. Auditory warnings should be transmitted at frequencies below four kHz that are not affected by hearing loss. Multiple modes of warning message transmission are advised, both to reinforce the message and to make sure that at least one message is perceived and understood.

Attention: Older adults have a harder time using selective attention to filter out irrelevant auditory information, and may fail to attend to the important information in a warning. This is especially problematic in noisy or chaotic environments that may be encountered in a disaster. The message could include an announcement that the listener should remove themselves from distractions. Intrusive alarms should be used to disrupt other tasks (including sleeping) and gain attention.

Comprehension: Older adults are less able to draw inferences from novel warnings. Warnings that are specific (hazard, location, time, and guidance), clear, and accurate reduce the need to make inferences and thus improve comprehension. The text of the messages should only include simple words (no technical words or jargon) and sentence structure to avoid confusion. Testing should be done before deployment of the warning to make sure that it is comprehensible by the intended audience.

Memory: Different types of memory are used to process warnings. Semantic memory is the store of information built by experience, which is not affected by age, and can be improved with training. Age does affect working memory, which can be taxed by complex tasks such as safety procedures. Older adults may choose less protective procedures over more complex ones (e.g., with more steps) because they are aware of their limitations. To avoid overtaxing working memory, warning messages should use simple sentence structure and break protective actions to be taken into a few, simple steps. Presenting information in a familiar way both counters memory limitations and improves credibility. Feedback from the community is required to assess what messages are most effective.

Decision-making: More research is needed on decision-making processes for older adults. Studies indicate that older adults consider fewer pieces of information than younger adults and are susceptible to biases. They may be more cautious than younger adults, but the speed and quality of decisions appears to be similar. The number of feasible options available to older adults may be reduced due to physical disabilities, low financial means, and inadequate social networks. Protective action recommendations should include options that allow everyone to be protected regardless of age, ability, or financial means. New technology, such as Reverse 911 to reach people by phone, can be helpful but should be tested before implementation.

Improving risk communications for older adults is likely to improve them for everyone.

Key findings:
- [89_Pc_1] Decreasing visual abilities in older adults can lead to difficulties in seeing smaller size fonts, differences between objects of similar colors (especially shades of green and blue), and any image in the presence of glare.
- [89_Pc_2] For auditory messages, higher frequencies (above about four kHz) become harder to hear for older adults due to both noise-induced and age-related hearing losses, and background noise is harder to filter.
- [89_Pc_3] Information from multiple speakers in a noisy environment (e.g., from the radio, telephone, PA system, or face-to-face) can be missed by older adults.
- [89_At_1] Older adults have a harder time using selective attention to filter out irrelevant auditory information, and may fail to attend to the important information in a warning.
- [89_Co_1] Older adults are less able to draw inferences from novel warnings. Warnings that are specific, clear, and accurate reduce the need to make inferences and thus improve comprehension.
- [89_Ac_1] Semantic memory, the store of information built by experience, is not affected by age and can be improved with training
- [89_Ac_2] Age affects working memory, which can be taxed by complex tasks such as safety procedures. Older adults may choose less protective procedures over more complex ones because they are aware of their limitations.
- [89_Ac_3] When making decisions, older adults consider fewer pieces of information than younger adults and are susceptible to biases.
- [89_Ac_4] Older adults may be more cautious in decision-making than younger adults, but the speed and quality of the decisions appears to be similar.
- [89_Ac_5] The number of feasible options available to older adults may be reduced due to physical disabilities, low financial means, and inadequate social networks.

Key recommendations:
- [89_Pc_4] Use large sans serif fonts in warning texts, especially for elderly populations.
- [89_Pc_5] For elderly populations, do not use similar colors, especially blue and green shades, to distinguish between two different things
- [89_Pc_6] For elderly populations, transmit auditory warnings at frequencies below four kHz that are not affected by hearing loss.
- [89_Pc_7] Eliminate background noise, especially for elderly populations.
- [89_Pc_8] Multiple modes of warning message transmission are advised, both to reinforce the message and to make sure that at least one message is perceived and understood.
- [89_At_2] An announcement that the listener should remove themselves from distractions

can be helpful to older adults.

- [89_At_3] Intrusive alarms should be used to disrupt other tasks (including sleeping) and gain attention.
- [89_Co_2] Messages should be simple and straightforward. The text of the messages should only include simple words (no jargon) and sentence structure to avoid confusion.
- [89_Co_3] Testing should be done before deployment of the warning to make sure that it is comprehensible by the intended audience.
- [89_Ac_6] To avoid overtaxing working memory in older adults, warning messages should use simple sentence structure and break protective actions to be taken into a few, simple steps.
- [89_Cr_1] Presenting information in a familiar way to older adults both counters memory limitations and improves credibility.
- [89_Ac_7] Feedback from the community is required to assess what messages are most effective.
- [89_Ac_8] Protective action recommendations should include options that allow everyone to be protected regardless of age, ability, or financial means.
- [89_Ac_9] New technology may be very useful, but should be tested before deployment.

 # 90. Melton AW. (1970, October). The situation with respect to the spacing of repetitions and memory. *Journal of Verbal Learning and Verbal Behavior, 9*(5), 596-606.

P		Pc	At	Co	Cr	Ps	Ac
			▓				

The effect of repetition of auditory material on recall has been studied by many researchers. This article summarizes research up to the year 1970 on the relative value for learning of practice that is either consecutive (massed) or distributed over time.

In the early 1950's, the test methods changed from tasks that engaged perceptual-motor skills to verbal tasks in order to provide more accurate measurements to test theories. Substantial differences in retention between distributed practice (DP) and massed practice (MP) were seen for lists of paired words with pairs repeated at various intervals within the list. The probability of recall of one word of a pair after being prompted with the other was improved by hearing the pair a second time after a lag, with seven to eight intervening pairs as the optimal lag. For lists of individual items, the benefits of repetition were generally found to be greater as the probability of recalling the word at that point in the list decreased. The free-recall method, in which after hearing the list the subject is asked to list as many words as can be recalled, has shown the largest and most reproducible differences between MP and DP. Researchers have generally found distributed practice to result in higher word recall than massed practice, although there is some disagreement in experiments with lists that are all DP or all MP. The advantage of distributed practice over massed practice is especially true when the number of repetitions is large. Better recall has been observed with visual than with auditory presentation, and with lists consisting of all nouns than with mixed lists. The recall of words that appear twice improves with increasing lag (number of words between presentations), although the effect is smaller for auditory lists of mixed word types. Slower rates of speech always result in higher recall.

Key findings:
- [90_At_1] Recall is higher when items are repeated at intervals (distributed) than when the items are repeated consecutively (massed).
- [90_At_2] The advantage of distributed practice over massed practice is especially true when the number of repetitions is large.
- [90_At_3] Items are recalled better when presented visually rather than verbally.
- [90_At_4] Words in lists consisting of all nouns are recalled better than words in mixed lists.
- [90_At_5] The recall of words that appear twice improves with increasing lag (number of words between presentations), although the effect is smaller for auditory lists of mixed word types.
- [90_At_6] Slower rates of speech always result in higher recall.

Key recommendations:
- [90_At_7] Messages should be repeated at intervals rather than consecutively for maximum recall.
- [90_At_8] Auditory messages should be delivered slowly for maximum recall.

 # 91. Mileti DS, Bandy R, Bourque LB, Johnson A, Kano M, Peek L, Sutton J, Wood M. (2006, September). *Annotated bibliography for public risk communication on warnings for public protective actions response and public education (Revision 4).* Boulder, CO: Natural Hazards Center, University of Colorado at Boulder.
http://www.colorado.edu/hazards/publications/informer/infrmr2/pubhazbibann.pdf

C		Pc	At	Co	Cr	Ps	Ac

This annotated bibliography contains brief summaries and causal findings from approximately 346 published documents on communication of hazard information to the public and the public response during disasters. References include selections from the literature on disaster preparedness, evacuation, human behavior, risk communication, and warning design, including information from natural and man-made disasters. The sources in this document overlap and complement the sources presented here, and many of the same points are echoed in the findings.

Key recommendations:
- [91_Pc_1] [91_At_1] [91_Co_1] [91_Cr_1] [91_Ps_1] [91_Ac_1] An extensive literature is available on the communication of hazard information to the public and the public response during disasters.

 # 92. Mileti DS, Sorensen JH. (1990). *Communication of emergency public warnings: A social science perspective and state-of-the-art assessment.* Oak Ridge, TN: Oak Ridge National Laboratory, U.S. Department of Energy.

C		Pc	At	Co	Cr	Ps	Ac

Based on over 200 studies of warning systems and warning response, this document outlines the state-of-the-art in communication of emergency public warnings from a social science perspective. This document provides important information on:

- Current warning systems in the United States (as of 1990);
- The definition of a warning system, including components and processes for detection, management, and response;
- Ways to build and evaluate a warning system, including guidance on the decision to warn, writing the warning message, disseminating the message, and monitoring human response to that message;
- Organizational aspects of warning systems, including potential organizational problems when warning the public;
- Public response, including processes that people go through before responding to a warning message and factors that influence (positively or negatively) human response to warnings;
- Hazard-specific aspects of warning, including characteristics specific to certain types of hazards and the design of warning systems to address these specific needs;
- Problems and limitations associated with technology, organizational structures, and society, along with suggestions on how to improve the warning systems.

More than 50 years of disaster-based social science research on message content was collected, and the findings were synthesized to determine best practices for the content and dissemination of warning messages during an emergency. Research shows that message content is one of the most important factors in determining the effectiveness of a warning system. An effective message includes the following:

- What? Guidance on what people should do about it,
- When? An idea of when people need to act,
- Where? A description of the location of the risk or hazard,
- Why? Information on the danger (i.e., why is protective action necessary, including the consequences if action is not taken),
- Who? The name or title of the source that is providing the information.

The style of the warning message and its dissemination are also crucial. An effective message is specific, consistent, certain, clear, and accurate. By disseminating messages frequently and through the correct channels, they are more likely to achieve an appropriate public response. Additionally, this document discusses the importance of monitoring human response after the message(s) is disseminated. The main purpose for doing this is to determine whether the warning system is guiding behavior in a manner consistent with current knowledge of the hazard and disaster risks. If the warning is not effective, adjustments should be made in the warning process, including the dissemination of follow-up messages to correct public response. This is especially important if the hazard and disaster risks change – by monitoring human response, follow-up messages can be disseminated to ensure that people are kept safe in a changing environment. There are several means available to monitor real-time public response, including social networks and video or in-person observation of response.

Even when warnings meet all of these standards, they may not be able to provide convincing emergency information for all members of the public. People have inherently different perceptions of the world around them – which they bring to any emergency – predisposing them to different responses, especially when situational information provided by the environment is conflicting. Four main factors are at work: 1) the ability to process risk information (due to education, cognitive abilities, pre-emergency knowledge, and experience); 2) access to social networks and to events that facilitate desirable warning response outcomes (social networks influence whether people receive informal warnings, whether they engage in confirmation of warnings; and whether there are additional options such as staying with a friend); 3) incentives to take the warning seriously (such as responsibility for children or a predisposition toward heightened risk perception); and 4) constraints (such as limited resources, disabilities, illness, and age). There are ways to maximize warning response and minimize the negative impacts of individual characteristics. For example, messages can be provided in multiple languages and via multiple sources (especially those that represent vulnerable groups), and the dissemination of warnings through informal channels can be encouraged.

This document categorizes hazards based upon their prediction times (long and short), impacts (known or unknown), and our current detection capabilities (easy to detect or difficult to detect). Rapid-onset (or short prediction time) events include the following hazards: flash floods, fast volcanos, fast fixed-site hazardous materials, tornados, avalanches, local tsunamis, landslides, hazardous materials, nuclear attacks, terrorist attacks, and sabotage. (Since this document focuses mainly on natural hazards, building fires are not discussed.) According to Mileti and Sorensen, rapid-onset events materialize in less than 15 to 30 minutes from the time of first prediction and detection. It is not impossible to send out warnings within this short time frame, but it may be more expensive and require a higher level of planning.

> Key findings/recommendations:
> - [92_Ps_1] [92_Ac_1] An effective message includes information on what (people need to do), when (they need to do it), where (the danger is), why (the action must be taken) and who (the warning is from).
> - [92_Ac_2] An effective message is specific, consistent, certain, clear, and accurate.
> - [92_At_1] [92_Cr_1] Messages that are disseminated frequently and through the correct channels are more likely to achieve an appropriate public response.
> - [92_Ac_3] Monitoring the public response after message dissemination, such as through social networks and video or in-person observation, provides information on whether the warning system is guiding behavior in a manner consistent with current knowledge of the hazard and disaster risks. If the warning is not effective, adjustments should be made.
> - [92_Ac_4] Follow-up messages should be provided if the hazard and disaster risks change
> - [92_Co_1] People have inherently different perceptions of the world around them – which they bring to any emergency – predisposing them to different responses, especially when situational information provided by the environment is conflicting.
> - [92_Co_2] [92_Ac_5] Four factors affect individual perception of and response to an emergency: 1) ability to process risk information, 2) access to social networks, 3) incentives to take the warning seriously, and 4) constraints on action.
> - [92_At_2] To maximize warning response, provide messages in multiple languages and via multiple sources (especially those that represent vulnerable groups) and encourage

dissemination of warnings through informal channels.

- [92_Ac_6] Warnings for rapid-onset events, which materialize less than 15 to 30 minutes from detection, may be more expensive and require a higher level of planning.

 # 93. Mollerup P. (2001). The way in to the way out: Signage design at Copenhagen Airports. *Information Design Journal, 10*(1), 73-81.

M		Pc	At	Co	Cr	Ps	Ac

During the 1980s, the management of Copenhagen Airport decided to change the design of airport signage to improve the customer experience. Key attributes were identified: sign plate color must contrast with the background and surroundings, the graphics must contrast with the sign plate, the color of the sign plates must be used exclusively for official signs, and the color of a sign must indicate that it is an official sign.

The Copenhagen airport sign designers decided on a dark blue sign that provided a good contrast against potential backgrounds. Against the blue background, white letters were used for Danish text and yellow for English text. Numbers and pictorials were shown in yellow. Ideally, pictograms would be selected from internationally recognized standards, but there were no international standards for many of the desired airport applications. Many of the pictorials in use at the airport were unclear without the presence of text, so the designers reduced the use of icons from 200 to 40 during the renovation.

Many important rules for sign design are discussed in this paper. Signs in a public space should be limited to only those necessary, placed in appropriate locations. Signs should be located where passengers must make choices on where to go, at a height above eye level to maximize the number of people seeing them. A sign's design should indicate its importance; important signs should be large, lighted, and positioned perpendicular to the walking direction. Text on normal signs can be read at a distance 400 to 500 times the height of its lower case letters. Larger letters should be used to stress the important points of a sign. Signs should be surrounded by empty space.

Key recommendations:
- [93_Pc_1] Sign plates must provide good color contrast against the background and surroundings.
- [93_Pc_2] Sign graphics must provide sufficient contrast against the sign plate. .
- [93_At_1] Placement is a key factor; signs in an airport should be placed where passengers must make choices on where to go, at a height above eye level to maximize the number of people seeing them.
- [93_Pc_3] People with normal sight can read text at a distance 400-500 times the height of its lower case letters.
- [93_Co_1] A sign's design should indicate its importance; important signs should be large, lighted, and positioned perpendicular to the walking direction.

#**94**. Monroe LK. (2007, July). Signage advice for zoos. *Buildings, 101*(7), 62-64.						
B	**Pc**	**At**	**Co**	**Cr**	**Ps**	**Ac**
			▨			▨

This article provides information on how to improve signage in public spaces, including commercial buildings. The author designs signs for zoos.   In the United States, there is a right hand bias, such that without other information people tend to turn to the right. This bias should be considered when designing wayfinding signage. If the same sign is used in multiple locations, all signs should be oriented in the same way to avoid confusion. Signs should be placed out of people's reach if possible in order to avoid vandalism.

Key findings:
- [94_Ac_1] In the United States, there is a right hand bias, such that without other information people tend to turn to the right. This bias should be considered when designing wayfinding signage.
- [94_Co_1] If the same sign is used in multiple locations, all signs should be oriented in the same way to avoid confusion.
- [94_Ac_2] In public spaces, signs should be placed out of people's reach if possible in order to avoid vandalism.

95. Morley FJJ, Cobbett AM. (1997). *An evaluation of the comprehensibility of graphical exit signs for passenger aircraft: Phases 1 & 2* (COA Report No. 9706). Human Factors Technology Group, Cranfield University, Bedford, England.

| E | | Pc | At | Co | Cr | Ps | Ac |

This article examined the comprehensibility of graphical exit signs. The first phase of the survey was given to 695 passengers ranging in age from 10 to 75 years (average 40 years old, 74 % male and 26 % female). Their backgrounds represented a range of countries (Western and Eastern Europe, North America, Asia, Africa, Latin America, Middle East and Oceania), with 61 different first languages. The participants were presented with a selection of seven signs indicating exit location (exit here, exit to left/right, or exit further along the aisle) and asked what each sign meant and whether they had seen the sign before. Results are summarized in Table 18 below and in the following bullet points:

Table 18: Exit sign comprehension – Phase I.

Sign		Meaning	Comprehension	Result Acceptability (>66 %)
1a		Actual exit location	76.1	Yes
1b			73.1	Yes
2a		Exits in both directions	83.6	Yes
2b			81.0	Yes
3a		Located further down the aisle	37.4	No
3b			57.6	No
3c			57.4	No

- Signs 1(a,b): The difference in comprension scores was not significant; however, some responses indicated that meaning was inferred from the direction in which the figure was facing.
- Signs 2(a,b): Sign 2a showed a modest advantage over Sign 2b, where two doors are shown, indicating that showing two exits more clearly communicates the alternatives.
- Signs 3(a,b,c): All designs failed. This result may have been influenced by the lack of contextual cues during the survey. The downward-facing 2D arrows in Sign 3a were widely misinterpreted. The upward-facing arrows in Signs 3b and 3c performed better. The addition of 3D effects (Sign 3c) did not improve the comprehension.

- The impact of region upon comprehension was examined through statistical tests. Western Europe, North America and Oceania were found to score significantly better than Asia, Africa, Latin America, or the Middle East.
- Previous exposure of participants to the sign was independent of comprehension except for a moderate relationship found for one of the signs.
- Frequent and moderately frequent travelers were able to comprehend the signs significantly better than infrequent travelers.

In the second phase of the survey, 670 passengers (average age of 40 years, 68 % male and 32 % female) were questioned on six exit signs. The results from Phase II are summarized in Table 19 and in the following bullet points:

Table 19: Exit sign comprehension – Phase II.

Sign		Meaning	Comprehension	Acceptibility (>66 %)
1a – Graphical		Actual exit location	82.2	Yes
1e – Language	EXIT		95.4	Yes
2a – Graphical		Exits in both directions	83.0	Yes
2e – Language	‹EXIT›		74.1	Yes
3d – Graphical		Located further down the aisle	44.4	No
3e - Language	EXIT		41.3	No

- Two of the three sign meanings (1 and 2) were correctly identified to an acceptable degree.
- The current text-based sign for exit location (Sign 1e) outperformed the best graphical sign when indicating location, although graphical variants performed sufficiently well to be used.
- Graphical exit signs indicating an exit to the left/right performed sufficiently well to be used and outperformed the language equivalent.
- The graphical symbol indicating an exit further along the aisle (incorporating 3D sideways arrows) was not comprehended sufficiently well to be used; nor was the equivalent text-based sign.
- Comprehension levels were influenced by region but not related to language.
- Previous exposure was moderately related to comprehension for one sign.
- Frequency of travel did not have a significant effect.

Key findings:
- [95_Co_1] Graphical signs that indicate the location of an exit or that exits are located to the left/ right are well understood. Comprehension is about the same for text and graphical signs.
- [95_Co_2] Graphical and text signs identifying that an exit is located further up the aisle were unsuccessful.

 # 96. Nasrallah M, Carmel D, Lavie N. (2009). Murder, she wrote: Enhanced sensitivity to negative word valence. *Emotion, 9*(5), 609–618.

P		Pc	At	Co	Cr	Ps	Ac

Research has been inconclusive on whether information that is negative in emotional valence (pleasing/non-pleasing) results in more rapid and accurate detection. Enhanced sensitivity to such information would be expected to be advantageous from an evolutionary sense, since it would expedite an appropriate response to a potentially dangerous situation. This study exposes participants to emotional and neutral words to determine whether identification is faster and more accurate for negative words than for positive.

The authors conducted three different experiments. The first used 26 subjects (20 females, 6 males) in the age range 18 to 44 with a mean age of 26 years old, the second used 23 subjects (17 females, seven males) with a mean age of 21 years old, and the third used eight subjects (five females, three males) with a mean age of 27 years old. Subjects in all three experiments were asked to look at a fixation cross on a screen, followed by a mask of eight hash marks for 67 milliseconds, a target word for 22 milliseconds or 33 milliseconds, and another mask for 67 milliseconds. A mixture of emotional and neutral words was presented in blocks of 44 words each (88 words for the third experiment), with no word repeated. Emotional words were either all negative or all positive for each block in the first two experiments, and were mixed within each block for the third. The subjects were asked to respond after each word as to whether the word was emotional or not and how sure they were of their answer on a scale of one to five. Positive or negative words that the subject reported as being neutral were not included in the analysis. After the entire test, the subjects were asked to rate the emotional level of the set of words. In the first two experiments, each block contained half neutral and half either positive or negative words.

In the first experiment, the time length of word presentation varied between either 22 milliseconds or 33 milliseconds. Detection of emotional content and sensitivity were better for negative than for positive valence, and confidence ratings were found to be significantly lower for 22 milliseconds than for 33 milliseconds presentations. To look at subliminal effects, the second experiment kept the presentation of words to a duration of 22 milliseconds and reduced the luminance. The third experiment controlled for the effects of arousal by mixing an equal number of positive and negative words matched for arousal level within each block. In all three experiments, the subjects were able to more accurately identify the emotional content of negative words than the positive words. This supports the conclusion that negative stimuli have preferential access to processing resources, with the brain either processing the information faster or requiring less information to judge the contents.

Key findings:
- [96_At_1] Negative stimuli have better access than positive stimuli to processing resources, with the brain either processing the information faster or requiring less information to judge the contents.

 # 97. National Association of the Deaf. (n.d.). *Emergency warnings: Notification of deaf or hard of hearing people.* Silver Spring, MD: NAD(Author).
http://tap.gallaudet.edu/emergency/nov05conference/EmergencyReports/NADEmergency.doc

D		Pc	At	Co	Cr	Ps	Ac

The National Association of the Deaf provides this document to inform broadcasters and public emergency management about their legal responsibilities to modify information procedures to meet the needs of the deaf and hard of hearing community. Listed here is some of the guidance provided by this document that is specifically related to emergency messages and dissemination practices:

- All television broadcasters, cable operators, and satellite television services that provide local emergency news are required by the Federal Communication Commission (FCC) to relay all essential emergency information via captioning or other visual means. A simple crawl at the bottom or top of the screen is adequate if it matches the information provided to other customers. However, broadcasters must provide critical information in visual form about all emergencies, including circumstances that require interrupting a program with live information and special news updates. A feed to a live real-time captioner or an on-side stenocaptioner is the best way to provide this information. Care must be taken not to block the visual emergency information, including with other captioning.
- Examples of critical information to be communicated during an emergency are: specific details about the affected areas; evacuation orders, including the areas to be evacuated and evacuation routes; approved shelter locations and/or instructions on sheltering in place; instructions on how to protect property; road closures; and instructions on obtaining relief assistance.
- Federal, state, and local governments are legally required to make their emergency notification systems accessible to deaf and hard of hearing people.
- National Oceanographic and Atmospheric Administration (NOAA) Weather Radio broadcasts only an auditory signal. Textual information from NOAA National Weather Service radio broadcasts is now available through modified radio receivers, which provide an alert through a strobe light, an auditory signal, and, if requested, a pillow vibrator or bed shaker to awaken the person. The receiver displays the type of warning (watch, warning, or advisory) that has been issued, although not the full text of the National Weather Service warning. This information is available through more sophisticated units connected to a satellite feed or through a pager system.
- Some communities, especially in areas frequented by tornados, have added a visual cue such as a powerful strobe light to civil defense alarm systems. This can aid in alerting individuals, but is limited to those who are awake and in direct line of sight from the tower.
- Interpreters for the deaf need to be part of emergency training and disaster relief assistance programs for the public.
- A possibility being considered by some communities is the use of Reverse 911 systems to alert deaf or hard of hearing populations. These systems are set up to enable an emergency notification agency to initiate TTY calls to warn individuals in specific areas of an impending emergency.

Key findings:
- [97_Pc_1] All television broadcasters, cable operators, and satellite television services that

provide local emergency news are required by the FCC to relay all essential emergency information via captioning or other visual means in order to reach the deaf and hard of hearing population.

- [97_Pc_2] Federal, state, and local governments are legally required to make their emergency notification systems accessible to deaf and hard of hearing people.
- [97_Pc_3] [97_At_1] Textual information from NOAA National Weather Service radio broadcasts is available through modified radio receivers, which provide an alert through a strobe light, an auditory signal, and, if requested, a pillow vibrator or bed shaker to awaken the person.
- [97_Pc_4] The full text of a National Weather Service warning is available through receivers connected to a satellite feed or through a pager system.
- [97_Pc_5] [97_At_2] A powerful strobe light added to a civil defense siren tower can aid in alerting individuals but is limited to those who are awake and in direct line of sight from the tower.

Key recommendations:

- [97_Pc_6] [97_Co_1] Care must be taken by television broadcasters not to block the visual emergency information, including with other captioning.
- [97_Ac_1] Examples of critical information to be communicated to the public during an emergency are: specific details about the affected areas; evacuation orders, including the areas to be evacuated and evacuation routes; approved shelter locations and/or instructions on sheltering in place; instructions on how to protect property; road closures; and instructions on obtaining relief assistance.
- [97_Ac_2] Interpreters for the deaf need to be part of emergency training and disaster relief assistance programs for the public.
- [97_Pc_7] Reverse 911 systems, which enable an emergency notification agency to initiate TTY calls, can be used to alert deaf or hard of hearing populations.

98. National Council on Disability. (2009, August). *Effective emergency management: Making improvements for communities and people with disabilities.* Washington, DC: NCD (Author). http://www.ncd.gov/publications/2009/Aug122009

D		Pc	At	Co	Cr	Ps	Ac

This document is a review of the current state of policy and practice in the management of disaster situations for people with disabilities within the community, including a recommendation of best practices. Recommendations on warning messages and dissemination are located throughout the document, with specific sections found in Chapter 2 on Preparedness, Chapter 3 on Response, and Chapter 9 on Initiatives in Progress.

All messaging strategies should take into account that there is often significant overlap among different vulnerable groups. A 2005 report by the National Council on Disability (NCD) found that the messages and practices currently used were designed for homogenous populations and did not consider the needs, means, and abilities of vulnerable populations. Innovative technologies and programs, such as adapted weather radios, text messaging, and weather pagers, promote wider dissemination of warning information. Emergency messaging systems for the homogeneous population should remind people to check on people in the vulnerable populations. However, this cannot be the only means of message transmission because other people might not be around or might not properly transmit the message. Social networks, including work units, neighborhood associations, and community organizations, can be arranged in advance to help warn disabled populations. Preparedness plans should include alternative backup networks. Research is needed on rapid onset evacuation of vulnerable populations in the workplace.

The use of words such as 'immediate' and threatens' convey a sense of urgency. Conditions that affect emergency response include mobility disabilities, fatigue, respiratory conditions, emotional and cognitive difficulties, and vision or hearing loss. These conditions may be temporary or permanent. The most accessible exits from a building may be crowded by nondisabled evacuees. Members of the vulnerable communities should be involved in planning and disseminating messages in ways that are amenable to how the vulnerable populations use the technology. Accommodating people with disabilities often improves response for everyone. Messages need to be provided using a wide variety of media sources through both audible and visual means. Multiple repetitions of warning messages may be needed. Messages should include instructions on how to shelter in place for those people that are unable to evacuate. Showing members of the vulnerable populations taking protective actions such as using the shelters encourages people to believe that their needs will be met if they comply with the instructions. The messages should be delivered by people trusted by the vulnerable population community, preferably those that have regular contact with them. Training and drills can be used to help familiarize people with the procedures in the event of an emergency.

This document provides a detailed list of recommendations for research and improvements to practice and policy.

Key recommendations:
- [98_Pc_1] [98_Ps_1] Messages and practices must consider the needs, means, and abilities

of vulnerable populations. There is often significant overlap among different vulnerable groups, and accommodating people with disabilities often improves response for everyone.

- [98_Pc_2] Innovative technologies and programs, such as adapted weather radios, text messaging, and weather pagers, promote wider dissemination of warning information.
- [98_Pc_3] Social networks, including work units, neighborhood associations, and community organizations, can be arranged in advance to help warn disabled populations. Preparedness plans should include alternative backup networks.
- [98_Ac_1] Research is needed on rapid onset evacuation of vulnerable populations in the workplace.
- [98_Ps_2] The use of words such as 'immediate' and threatens' convey a sense of urgency.
- [98_Ac_2] Conditions that affect emergency response include mobility disabilities, fatigue, respiratory conditions, emotional and cognitive difficulties, and vision or hearing loss. These conditions may be temporary or permanent.
- [98_Ac_3] The most accessible exits from a building may be crowded by nondisabled evacuees.
- [98_Pc_4] [98_Cr_1] Members of the vulnerable communities should be involved in planning and disseminating messages in ways that are amenable to how the vulnerable populations use the technology.
- [98_Pc_5] Messages need to be provided using a wide variety of media sources through both audible and visual means.
- [98_At_1] [98_Co_1] Multiple repetitions of warning messages may be needed.
- [98_Ps_3] [98_Ac_4] Messages should include instructions on how to shelter in place for those people that are unable to evacuate.
- [98_Ps_4] [98_Ac_5] Showing members of the vulnerable populations taking protective actions such as using the shelters encourages people to believe that their needs will be met if they comply with the instructions.
- [98_Cr_2] Warning messages should be delivered by people trusted by the vulnerable population community, preferably those that have regular contact with them.
- [98_Ac_6] Training and drills can be used to help familiarize people with emergency procedures.

99. National Organization on Disability. (2009). Communication is the key. *Functional needs of people with disabilities: A guide for emergency managers, planners, and responders* (pp. 31-35). Emergency Preparedness Initiative. Washington, DC: NOD (Author). http://nod.org/assets/downloads/Guide-Emergency-Planners.pdf

D		Pc	At	Co	Cr	Ps	Ac

The National Organization on Disability is a non-profit organization that promotes the full participation of Americans with disabilities in all aspects of life. The Emergency Preparedness Initiative promotes the inclusion of people with disabilities in all levels of emergency management, including mitigation, preparedness, response, and recovery. Surveys have indicated that people with disabilities are less prepared for emergency situations and more anxious about them than those who are not disabled. This document describes the needs of this population that must be considered by emergency managers.

Emergency communication needs to reach individuals with disabilities, understanding that many are unemployed, socially isolated, or in other ways less connected to society. The following considerations should be taken into account by public officials and the media:

1) Television stations must not run text message crawls across any area reserved for closed captioning.
2) Camera operators and editors need to include in the picture any sign language interpreter next to the official spokesperson.
3) Emergency hotlines must include TTY/TDD (text telephone or telecommunication device for the deaf) numbers when available, or the instruction "TTY callers use relay."
   The same information must be provided by officials and on the television and radio.
4) Essential emergency information should be frequently repeated in a simple message to accommodate individuals with cognitive disabilities.
5) Make sure that the emergency website is accessible (details on Web accessibility can be found in this document on page 58).
6) Provide information in alternate formats based on population needs, including Braille, audio recording, large font, text messages, and emails.
7) Begin transmission of information in all formats as early as possible.
8) A well-designed and frequently practiced phone tree is an effective way to reach members of the disability community during an emergency. A reverse calling system allows the community to alert emergency professionals to issues of which they may not be aware.
9) New and improved technology should be used to help those with hearing impairments. Today, alerts can be sent out through relay services, text messaging, blogging, Twitter, email, secure access to websites, videophone calls, and online video conferencing programs such as Skype.
10) Public officials should keep aware of new emergency alert systems, such as Oklahoma's Weather Alert Remote Notification for those who are deaf or hard of hearing, where people can receive emergency alerts via email, pager, or text messaging.

Key recommendations:
- [99_Pc_1] Television stations must not run text message crawls across any area reserved for closed captioning.

- [99_Pc_2] Camera operators and editors need to include in the picture any sign language interpreter next to the official spokesperson.
- [99_Pc_3] Emergency hotlines must include TTY/TDD (text telephone or telecommunication device for the deaf) numbers when available, or the instruction "TTY callers use relay."
- [99_Cr_1] All official and media emergency information sources need to provide the same emergency phone numbers.
- [99_Co_1] Essential emergency information should be frequently repeated in a simple message to accommodate individuals with cognitive disabilities.
- [99_Pc_4] Make sure that the emergency website is accessible to those with disabilities.
- [99_Pc_5] Provide information in alternate formats based on population needs, including Braille, audio recording, large font, text messages, and emails.
- [99_At_1] Begin transmission of information in all formats as early as possible.
- [99_Ac_1] A well-designed and frequently practiced phone tree is an effective way to reach members of the disability community during an emergency. A reverse calling system allows the community to alert emergency professionals to issues of which they may not be aware.
- [99_At_2] New and improved technology should be used to help those with hearing impairments.
- [99_At_3] Public officials should keep aware of new emergency alert systems

 # 100. National Research Council. (2011). *Public response to alerts and warnings on mobile devices: Summary of a workshop on current knowledge and research gaps.* Washington, DC: The National Academies Press.

| C |

| Pc | At | Co | Cr | Ps | Ac |

On April 13-14, 2010, a workshop was held in Washington, DC under the auspices of the National Research Council's Committee on Public Response to Alerts and Warnings on Mobile Devices. The purpose of the workshop was to gather inputs from social science researchers, technologists, emergency management professionals, and other experts on how the public responds to alerts and warnings, with a specific focus on how the public responds to mobile alerting. This document describes the workshop presentations and discussions. The topics most relevant to this project are presented here.

An alert and a warning are two different concepts. An alert indicates that something significant has happened or may happen. A warning usually follows an alert, providing detailed information indicating who is at risk, where the risk resides, who is sending the warning message, and what protective actions should be taken.

The U.S. Department of Homeland Security's Commercial Mobile Alert Service (CMAS) is currently being developed to provide a national capacity to deliver brief (90 character) text alerts to cellular telephone subscribers. The wide usage of cell phones by the U.S. population and the ability to target messages to a cell phone's actual location makes this an attractive opportunity to more precisely target those individuals who would be most at risk in a crisis situation. This would be an opt-out system, in that individuals would receive AMBER and imminent threat alerts unless they opted out of this system. However, it will not be possible to opt out of Presidential alerts.

There is an established process that individuals go through each time they receive information during a crisis or emergency (presented by Joseph Trainor, based on Lindell and Perry 2004; Mileti and Sorensen 1990): Receive the warning, pay attention to the warning, understand the warning, believe that the warning is credible, personalize the threat, determine whether protective action is needed, determine whether protective action is feasible, and decide if you have the resources to take protective action. Along the way, people take steps – actions – to verify that the information they are receiving about the threat is reliable and that a threat could materialize.

The source of an alert or warning influences a person's response. If the source of the information is not identified in the message, the message is unlikely to be deemed trustworthy.

The document cites Mileti and others for guidance on what to include in warning messages (included in this annotated bibliography), but admits that much less is known about effective content for alerts because alert technologies, such as sirens and weather radio alerts, can only convey a small amount of information. More research is needed into effective content for CMAS alerts.

The following are additional topics that were discussed during this workshop:
- CMAS as one of many sources of alerts and warnings

- Social media and its place in emergency communication
- At-risk populations and their current usage of emergency communication technology
- Special considerations for people who are blind or have low vision, including text-to-speech capabilities
- Special consideration for people with hearing impairments, including email, SMS (texts), television text and graphics, National Oceanographic and Atmospheric Administration Weather Radio(NOAA WR) – text displays and flashing lights and vibration functions to awaken a person. The alert and warning need to catch their eyes in order to pull their attention away from what they are doing
- Special consideration for the elderly population
- Special consideration for racial minorities, including that preferred information sources can vary across racial and ethnic lines. It is important to identify which information sources hold credibility and trust for these populations and to use those avenues to provide information. Are the non-English speaking television stations receiving this information, translating it correctly, and providing to their audiences? There are a number of Spanish-language television stations in the U.S. that are watched by a large portion of the U.S. population as their primary source.

Key recommendations:
- [100_At_1] [100_Ps_1] [100_Ac_1] A warning usually follows an alert, providing detailed information indicating who is at risk, where the risk resides, who is sending the warning message, and what protective actions should be taken.
- [100_Pc_1] The wide usage of cell phones by the U.S. population and the ability to target messages to a cell phone's actual location makes the development of the Commercial Mobile Alert Service (CMAS) an attractive opportunity to more precisely target those individuals who would be most at risk in a crisis situation.
- [100_Pc_2] [100_At_2] Some technologies are better suited for delivering alerts (e.g., sirens or CMAS), and others are better suited for warnings (e.g., broadcast radio or television).
- [100_Pc_3] [100_At_3] For time-sensitive disasters, sirens and other immediate alerting systems are required.
- [100_Ps_2] People take steps to verify that the information they are receiving about the threat is reliable and that a threat could materialize
- [100_Cr_1] The source of an alert or warning influences a person's response. If the source of the information is not identified in the message, the message is unlikely to be deemed trustworthy.
- [100_Co_1] More research is needed on effective content for alerts, because alert technologies, such as sirens and weather radio alerts, can only convey a small amount of information.
- [100_Pc_4] Special consideration is needed for at-risk populations, including those who are blind or have low vision, those with hearing impairments, those who are elderly, and racial and ethnic minorities and non-English speakers

 # 101. Nilsson D, Frantzich H. (2004). Evacuation experiments in a smoke-filled tunnel. *Proceedings of the 3rd Human Behavior in Fire Symposium*, Belfast, UK, 229-238.

F		Pc	At	Co	Cr	Ps	Ac

Evacuation experiments were carried out to look at the effectiveness of different wayfinding systems in facilitating egress from a smoke-filled tunnel. The tunnel was 37 meters long, five meters wide and 2.7 meters high, with three exits located along the side of the tunnel. Ambient light was provided by five fluorescent lights with a luminescence of 21 lx. Thirty males and 16 females took part in these trials, with ages ranging from 18 to 29 years. Participants were provided with limited information prior to the experiment.

The effects of flashing orange lights, rows of flashing lights, and floor markings on travel speeds, exit use, and whether or not the signs were noticed by the participants were studied. During the experiments, non-irritant, non-toxic smoke was used to reduce visibility.

Forty two percent of the participants reported seeing the exits associated with flashing lights. Of these, 50 % went on to use the associated exits. None of the participants saw the floor markings. The rows of flashing lights were seen by 38 % of the participants, but none of them used the associated exit.

In a post-experiment survey, participants suggested that an arrow added to the flashing light system would help to direct travel. The potential benefit of a tactile system was also suggested, given that people often guide themselves through smoke using walls. They also proposed that the effectiveness of directional signage would be improved by training to promote familiarity.

Key findings:
- [101_Pc_1] [101_Ac_1] Static flashing lights were able to attract the attention of people unfamiliar with an escape route but were unsuccessful in directing movement in a particular direction. An arrow may have helped.
- [101_Pc_2] Flashing lights were much more noticeable than floor markings in an escape route.
- [101_Ac_2] In smoke-filled conditions people often use walls for guidance, suggesting that a tactile system would be useful in wayfinding.

102. Nilsson D, Frantzich H, Saunders W. (2005). Coloured flashing lights to mark emergency exits – Experiences from evacuation experiments. *Proceedings of the Eighth International Symposium on Fire Safety Science,* 569–579.

F

Pc	At	Co	Cr	Ps	Ac

An evacuation of an IKEA warehouse by individuals who were familiar with the structure illustrated some of the issues with current emergency exit design. This article describes a set of three evacuation experiments that were carried out to study potential design improvements. Of particular interest was whether flashing lights could be used to overcome learned irrelevance, in which familiarity limits the consideration of potential egress routes and may lead away from chosing the nearest exit. Several variables were investigated, including the type of wayguiding system, colors, strobes or flashing lights, and whether daylight can be seen through the exit door. Exit usage and the association of colors with safety were studied.

Three experiments were conducted with 90 adults traveling through a 37 meters long corridor in the absence of smoke. The corridor was lit by four fluorescent lights that generated luminance levels up to 72 lx. Experiment 1 presented a choice of two exits with different wayguidance systems. Each participant was initially located either equidistant from both exits or closer to Exit 2. The results indicated that green flashing lights attracted people to an exit over a standard emergency exit sign, even when the starting point was further away. A green or orange strobe light was less effective. When asked which system they preferred, the participants chose the green flashing light (72 %), followed by the green strobe light (59 %), and finally the orange strobe light (36 %).

In Experiment 2, each participant started the evacuation at one end of the corridor, with an exit located in clear view at the far end and marked by a standard emergency exit sign. A second exit was located half way along the corridor and tangential to the direction of movement. The wayguidance system for this exit was varied. The introduction of a green flashing strobe light was found to markedly increase the use of the second exit. Participants reported that in an emergency they primarily associated the color green with safety and the color red with danger.

In Experiment 3, participants were shown red, blue, green, and orange strobe and flashing lights and were asked to grade each color according to the strength of its association with an emergency exit sign. Both the green strobe and the green flashing light were more often associated with an emergency exit than the blue lights, and the red light also had a significant association, although to a lesser degree.

Key findings:
- [102_At_1] Flashing lights attract attention.
- [102_Ac_1] Green flashing lights attracted people to an exit over a standard emergency exit sign, even when the starting point was further away. A green or orange strobe light was less effective.
- [102_Co_1] [102_Ac_2] Green is more strongly associated with emergency exits than other colors. This association translated into action during evacuation experiments; i.e., exits marked by green flashing lights were used preferentially.

- [102_Co_2] Red is more strongly associated with danger than with safety.
- [102_Co_3] [102_Ac_3] Green flashing lights are more strongly associated with safety than green strobe lights. However, this did not necessarily influence egress behavior.
- [102_At_2] [102_Co_4] [102_Ac_4] Because flashing lights attract attention and green is associated with safety, green flashing lights can be used to overcome familiarity issues and direct people to an emergency exit.
- [102_At_3] [102_Ac_5] Green flashing lights are able to overcome participant proximity; i.e., exit signs with green flashing lights attracted people who were closer to other exits.
- [102_At_4] [102_Ac_6] Green flashing lights are able to overcome a less advantageous angle of observation; i.e., exit signs with green flashing lights on a side exit attracted people who could see an exit at the end of the corridor.

103. Nilsson D, Frantzich H, Saunders WL. (2009). Influencing exit choice in the event of a fire evacuation. *Fire Safety Science- Proceedings of the Ninth International Symposium*, 341-352.

F		Pc	At	Co	Cr	Ps	Ac
		▓	▓				▓

In this experiment, there was an unannounced evacuation of an office building, which was recorded on video. Green flashing lights were placed at exits on some floors to determine whether the lights influenced the choice of exits by evacuees. At these exits, one green strobe light with a diameter of 70 mm was located next to the exit sign. After the occupants exited the building, they were given a questionnaire that asked which floor they were on before the drill and whether something prompted them to use the exit they did. The questionnaire did not mention the strobe lights. An unannounced evacuation was also conducted at a movie theater. Green flashing lights were placed at two of the four exits to see whether they influenced people's choice of exit. When the building occupants exited the theater, they were given a similar questionnaire to that given to the office building occupants.

In the office building evacuation, a larger proportion of people used the exit staircase marked with flashing green lights than used the other staircases, but the difference was not significant. Relatively few people mentioned the flashing lights in their surveys. It was uncertain whether this was because the lights were not perceived as an important feature or because they were not noticed at all. In the movie theater, everyone used the exits with the flashing green lights. This may be because the occupants were less familiar with the movie theater and took the exits that attracted their attention. Many people who filled out the survey at the theater mentioned the flashing lights as a feature that made the exit stand out.

Key findings:
- [103_Pc_1] [103_At_1] [103_Ac_1] Green flashing lights placed next to some exits in an office building were not mentioned by evacuating occupants as a factor influencing exit choice. Although more people evacuated through those exits, the difference was not significant.
- [103_Pc_2] [103_At_2] [103_Ac_2] In a movie theater, all those evacuating used the exits with green flashing lights in preference to those with normal exit signs. Flashing lights may be more effective in influencing the choice of exits in less familiar places.
- [103_Ac_3] Green flashing lights at exits can be used to influence exit choice, but their effectiveness may depend on the setting.

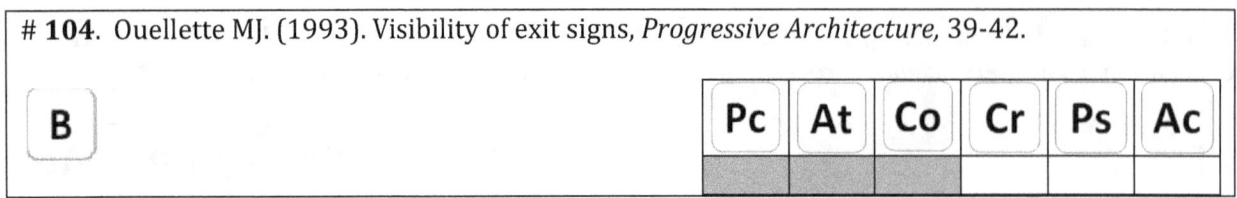

104. Ouellette MJ. (1993). Visibility of exit signs, *Progressive Architecture*, 39-42.

B

Pc	At	Co	Cr	Ps	Ac

The factors that influence signage visibility were examined to develop recommendations for improved building codes. Rea, Clark and Ouellette (1985 – included in this annotated bibliography) studied the visibility of thirteen exit signs in smoke. They found that exit signs that conformed to code requirements in clear air became completely obscured by smoke under experimental conditions. Figure 7 below shows that the exit signs generally became more visible as they grew brighter, where visibility is expressed as the minimum quantity of smoke that obscures the sign. The variability in the data is indicated by the shaded areas. This implies that other factors (e.g., content height, character spacing, contrast, and glare) may have also influenced the results. Sign visibility is consistently better in darkness than with room lights at emergency lighting levels or above.

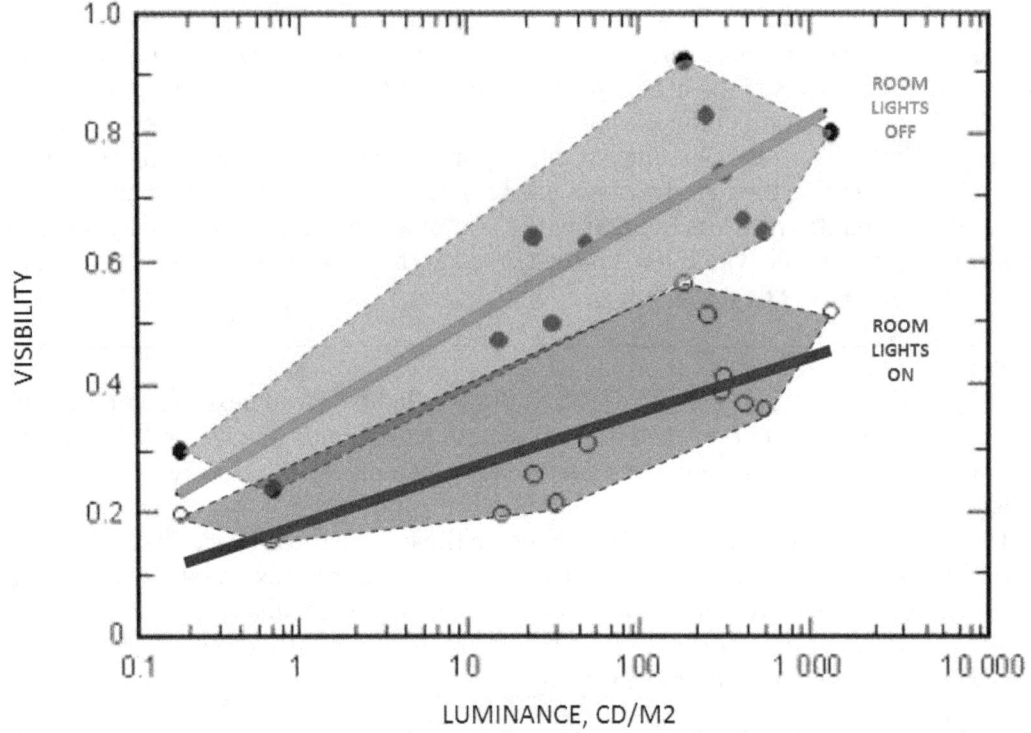

Figure 7: Visibility levels as a function of sign luminance.

Suggestions for overcoming the effects of smoke upon signage visibility include the following:
- *Sign Brightness:*
  - In the presence of smoke, signs should be as bright as possible to overcome smoke obscuration and enhance performance. The visibility threshold for signs with luminances ranging from 0.12 cd/m² to 1300 cd/m² increases with luminance, whether the room lighting is at normal levels, at emergency lighting levels, or off. No upper limit was found; the brighter the signs, the more subjects preferred them.

- In the absence of smoke, very bright signs bothered the participants. An upper limit was suggested under these conditions by 92 % of the participants, with a range from 70 cd/m² to 700 cd/m².
  - Bifunctional signs are available that increase sign brightness during an emergency. These may be able to address issues of both performance and comfort.
- *Ambient illumination:* Light sources at the sign or between the viewer and the sign obscure the sign by scattering the light. Ambient illumination should be reduced.
- *Background Conditions:* Signs with opaque backgrounds appear to perform better in smoke than signs with trans-illuminated backgrounds. In smoke, the large luminous areas on trans-illuminated backgrounds act in a manner similar to ambient lighting – veiling or confusing the sign content.
- *Sign Color:* Color was not a major factor. Humans tend to be more sensitive to the yellow-green region of the spectrum, so less energy is required to operate a green light with the same perceived brightness as a red light. In clear air experiments, the color of the signs (when controlled for brightness) seemed to have a marginal impact on visibility. Color may be more important for the initial attraction of the attention of the viewer, for which it should contrast with the background surface. However, the exact interpretation of the color (i.e., the meaning associated with it) is also important for comprehension under a variety of conditions. The meanings of many colors are culturally dependent.
- *Symbols Employed:* Research by Collins indicates that the chevron style is the most visible for the arrow symbol. The chevron should be at least 60 mm high to be visible from a distance of 30 meters.
- *Sign Placement:* Placement on the floor or low on the wall may be beneficial given the lower smoke densities and lower ambient light levels at this height. However, smoke densities may not always be lower at the floor, and low-level signs may also be more prone to damage. Low level exit signs may therefore be of most benefit as supplemental rather than replacement signage.
- *Performance Reliability:* In a survey involving nearly 1200 participants, Gilbert highlighted the finding that increased maintenance improved the chances of a sign passing an inspection. The likelihood of 90 % of exit signs passing inspection increased from 45 % to 75 % when checks were made monthly rather than as part of a routine plan. Selecting long-life sources also helped with sign reliability.

Key findings:
- [104_Pc_1] In the presence of smoke, signs should be as bright as possible to overcome smoke obscuration and enhance visibility. No upper limit was found; the brighter the signs, the more subjects preferred them.
- [104_At_1] In the absence of smoke, very bright signs bothered the participants. An upper limit was suggested under these conditions by 92 % of the participants, with a range from 70 cd/m² to 700 cd/m².
- [104_Pc_2] Bifunctional signs are available that increase sign brightness during an emergency. These may be able to address issues of both performance and comfort.
- [104_Pc_3] Light sources at an exit sign or between the viewer and the sign obscure the sign by scattering the light. Ambient illumination should be reduced.
- [104_Pc_4] Signs with opaque backgrounds appear to perform better in smoke than signs with trans-illuminated backgrounds. In smoke, the large luminous areas on trans-illuminated backgrounds act in a manner similar to ambient lighting – veiling or confusing the sign content.

- [104_Pc_5] Humans tend to be more sensitive to the yellow-green region of the spectrum, so less energy is required to operate a green light with the same perceived brightness as a red light.
- [104_Pc_6] In clear air experiments, the color of exit signs (when controlled for brightness) seemed to have a marginal impact on visibility.
- [104_At_2] Exit sign color may be more important for the initial attraction of the attention of the viewer than for visibility
- [104_Pc_7] Sign color should contrast with the background surface.
- [104_Co_1] Interpretation of the color of a sign (i.e., the meaning associated with it) is important for comprehension under a variety of clear and smoky conditions. The meanings of many colors are culturally dependent
- [104_Pc_8] The chevron style is the most visible for the arrow symbol. The chevron should be at least 60 mm high to be visible from a distance of 30 meters.
- [104_Pc_9] Placement on the floor or low on the wall may be beneficial given the lower smoke densities and lower ambient light levels at this height. However, smoke densities may not always be lower at the floor, and low-level signs may also be more prone to damage. Low level exit signs may therefore be of most benefit as supplemental rather than replacement signage.
- [104_Pc_10] Increased maintenance and the selection of long-life illumination sources improve the reliability of exit signs.

 # 105. Ouellette MJ. (1988). Exit signs in smoke: Design parameters for greater visibility. *Lighting Research and Technology, 20*(4), 155-160.

L		Pc	At	Co	Cr	Ps	Ac

Key influences on visibility of exit signs in smoke were identified from several smoke and lighting scenarios. Three types of signs were studied: red trans-illuminated letters/opaque background, opaque letters/red trans-illuminated background, and red trans-illuminated letters/white trans-illuminated background. The threshold values of illuminance at the rear of each sign (which follows the intensity of the light source) and of luminance at the position of the viewer were determined by reducing the back illumination to find the point at which the sign could no longer be read. Twelve participants with ages ranging from 22 to 48 years viewed the signs through a smoke-filled chamber. The exit signs were examined under smoke conditions with optical density per meter of path length of 0.5 m^{-1} and 1.5 m^{-1} and in ambient lighting of 0 lx or 0.55 lx.

The visibility of all signs examined in this study could be improved by exceeding the recommended luminance of the sign. An increase of back illumination by a factor of 50 compensated for a factor of 10 increase in the smoke density. An increase of back illumination by a factor of 1.6 resulted in the same visibility in ambient illumination of 0.55 lx as had been achieved with the lights off. Turning off the ambient illumination improved readability. There was a significant interaction between sign type and smoke density. The trans-illuminated sign was less affected by changes in smoke density and required the lowest levels of illumination.

Key findings:
- [105_Pc_1] The luminance at the position of the viewer is a more reliable indicator of visibility than the illuminance of the light source
- [105_Pc_2] An increase of back illumination of an exit sign by a factor of 1.6 resulted in the same visibility in ambient illumination of 0.55 lx as had been achieved with the lights off.
- [105_Pc_3] An increase of back illumination of an exit sign by a factor of 50 compensated for a factor of 10 increase in the smoke density.
- [105_Pc_4] Ambient lighting can affect the readability of signage.

106. Paulsen T. (1994). The effect of escape route information on mobility and wayfinding under smoke logged conditions. *Proceedings of the 4th International Symposium on Fire Safety Science,* 693–704.

F		Pc	At	Co	Cr	Ps	Ac

Paulsen examined the effectiveness of several visual and tactile wayfinding systems. The study involved 79 participants (46 males, 33 females) with ages ranging from 19 to 25 years of age. All participants were unfamiliar with the conditions, the wayfinding systems and the experimental procedure. The participants were asked to travel a route 29 meters long that included a corridor, doors and a stairwell, using non-irritant smoke to produce smoke conditions with a visibility distance of about 2.5 meters. The wayfinding systems are shown in Table 20, along with the percentage of participants who selected the correct route and the average evacuation time for each system.

Table 20: Decision-making success and evacuation times for a variety of wayguidance systems.

Wayguidance System	Configuration	% Route Success	Evac. Time (s)
Discontinuous Visual	Mean luminance 250 cd/m²; signs were placed to indicate directional change.	66 % (8/12)	65.6
Semi-Continuous Visual	Cold cathode tube with text, symbols and pictograms attached. Monochromatic green and white with mean luminance of 250 cd/m² and 440 cd/m².	73 % (11/15)	30.5
Continuous Tactile	Notched rail whose notches indicated direction to an exit and changes in the direction of movement.	93 % (13/14)	91.7
Continuous Visual with High Luminance	Line of incandescent lamps assembled into modules, located at floor level. Each module had a luminance of 5.5 cd/m².	100 % (9/9)	45.9
Continuous Visual with Low Luminance	Photoluminescent material (PLM) products with mean illuminance of 130 lx and luminance of 0.3 cd/m² when fully charged. Positioned at floor and head level.	93 % (14/15)	51.7
Combined Visual/Tactile	Combined semi-continuous visual system and continuous tactile system.	100 % (14/14)	44.9

Discontinuous and semi-continuous visual wayfinding systems were less successful than the othersystems in guiding route selection. Significant differences were found in evacuation times, with participants using the tactile system traveling significantly slower than those using any other system. Thus although the tactile system resulted in a good outcome, the participants required a long time to achieve it. Although route selection was successful only 73 % of the time, the semi-continuous system produced significantly shorter evacuation times than any other system.

After the experiment, the participants reported their opinions on the wayfinding systems using a four point scale (1: very poor; 4: very good). The semi-continuous visual, continuous visual with high luminance, and combined tactile and visual signage wayfinding systems were preferred.

Key findings:
- [106_Pc_1] [106_Ac_1] Two continuous visual wayfinding systems (incandescent and PLM) and a combination of visual semi-continuous and tactile systems outperformed other wayfinding systems in terms of route adoption and evacuation times.
- [106_Pc_2] [106_Ac_2] The route decision-making of a visual wayfinding system was improved through the addition of a tactile component.
- [106_Pc_3] [106_Ac_3] The slow evacuation time for a tactile wayfinding system was improved by adding a visual component.

Key recommendations:
- [106_Pc_4] [106_Ac_4] A wayfinding system with continuous markings is superior to traditional exit signs.
- [106_Pc_5] [106_Ac_5] Continuous tactile wayfinding systems are recommended for optical density exceeding 1.5, visual continuous systems are recommended for optical density between 0.1 and 1.5, and traditional exit signs are recommended only for optical density under 0.1.

---

107. Perry RW, Lindell MK. (2003, June). Understanding citizen response to disasters with implications for terrorism. *Journal of Contingencies and Crisis Management, 11*(2), 49-60.

C		Pc	At	Co	Cr	Ps	Ac

The need to coordinate emergency management and planning among law enforcement, military, and policy officials in addition to emergency managers requires education of all parties on the behavior of humans under the stress of a natural or technological disaster. Expectations about disaster syndrome (shock and passivity), panic, and fear responses among the citizenry are particularly ill-matched with the observed behavior. Studies indicate that people tend to act in what they believe to be their best interest, given the limited information available. Shock reactions, panic flight, and anti-social behavior such as looting are relatively rare in disasters. Instead, people behave responsibly and altruistically, helping themselves and their neighbors to their best ability. Those with shock symptoms are still able to act responsibly and follow instructions from emergency response officials. Myths about irrational and anti-social behavior lead to counterproductive approaches by authorities. The fear of panic has been used to justify withholding information, although it is known that people are more reluctant to comply with instructions when the warning is vague or incomplete. In fact, the lack of communication is a factor that contributes to panic. Almost all disasters are accompanied by convergence behavior, in which the stricken community becomes the focus of personal and material aid. A rapid influx of volunteers and resources can lead to difficulties with communication and logistics.

Authorities dealing with a disaster should understand that citizens confronted with disaster are rational, responsible, and capable of action. They should expect fear (not panic) as a normal human reaction to extreme circumstances, which reduces the ability to reason through complex problems. They should expect citizens to take action. They should also expect that citizens will be compliant with government authorities, especially in rapid-onset disasters. Thus emergency managers should address the fears of unfamiliar hazards by disseminating information about the hazard, its potential consequences, and appropriate actions to minimize negative effects. These include both constructive actions that citizens can do themselves (and why) and what is being done by experts on their behalf. Such messages will both increase compliance and discourage actions that are not effective. Trust provides a window of opportunity (and a responsibility) for emergency authorities to impact the outcome of the disaster. During disaster recovery, crisis therapy should be provided to mitigate negative long-term psychological effects. Communication of the disaster experience helps to provide closure and a transition to normal life.

For long-term events such as terrorist attacks involving WMD, the information may be vague and accumulated slowly over time. Compliance cannot be assumed during these events. Effective public response on evacuation, sheltering in place, and other protective measures requires pre-event risk communication to the public, incident management planning and first responder training.

Key findings:
- [107_Ac_1] Disaster studies indicate that people tend to act in what they believe to be their best interest, given the limited information available. Shock reactions, panic flight, and anti-social behavior such as looting are relatively rare.

- [107_Cr_1] People are more reluctant to comply with instructions when the warning is vague or incomplete. Information should not be withheld out of a fear of panic. In fact, the lack of communication is a factor that contributes to panic.
- [107_Ac_2] Citizens confronted with disaster are rational, responsible, and capable of action. Emergency authorities should expect fear (not panic), action, and compliance from the victims and structure messages accordingly.
- [107_Co_1] Fear (not panic) is a normal human reaction to extreme circumstances, which reduces the ability to reason through complex problems.
- [107_Cr_2] People tend to be compliant and trustful of authorities during a rapid-onset disaster. This provides a window of opportunity (and a responsibility) for emergency managers to impact the outcome of the disaster.

Key recommendations:

- [107_Ac_3] Emergency managers should address the fears of unfamiliar hazards by disseminating information about the hazard, its potential consequences, and appropriate actions to minimize negative effects. These include both constructive actions that citizens can do themselves (and why) and what is being done by experts on their behalf. Such messages will both increase compliance and discourage actions that are not effective.
- [107_Cr_3] After an event, communication that explains the disaster and the disaster experience helps to provide closure and a transition to normal life.

 # 108. Phillips BD, Morrow BH. (2007, August). Social science research needs: Focus on vulnerable populations, forecasting, and warnings. *Natural Hazards Review, 8*(3), 61-68.

  C

Pc	At	Co	Cr	Ps	Ac

This article summarizes research on the experiences of vulnerable populations during natural disasters, with a focus on improving risk communication and planning. Vulnerable populations are defined as "groups historically disadvantaged by socioeconomic status; patterns of discrimination and exclusion, or both; a lack of political representation; or cultural distancing." People in these groups are often marginalized, and are less likely to receive warnings and to respond appropriately in emergency situations. Furthermore, these groups are not mutually exclusive, and interacting disadvantages may magnify the difficulties.

Vulnerable populations may be based on age, race and ethnicity, socioeconomic status, gender, physical disability (e.g., mobility, hearing, and vision), language, work, isolation, and other factors. Socioeconomically disadvantaged families may lack the ability to receive forecasts and warnings, a safe home environment and the transportation to evacuate. Non-native English speakers may lack access to information in their native language or a good translation of the message. Deaf populations may not be well served by the media despite regulations to provide messages live during emergencies through closed captioning and interpreters. Lip reading is impossible when television experts turn their backs to the camera or are not visible on screen. The interpretation of warning messages depends on the sociocultural context of the receiver, which is not well understood by those outside of the community. Gender roles in family and work influence vulnerability. Women appear to be more responsive to warnings, but may find themselves and their children in more dangerous situations. Elderly people may have less access to social networks and technology, and may find it more difficult to evacuate due to mobility problems. Children may be on their own or in settings that are unprepared to handle the emergency. Conversely, schools may provide an avenue to provide information to other family members. Cultural groups outside of the mainstream may be reluctant to trust warnings until they have confirmed the information with members of their own communities, thus delaying their response. Tourists are unfamiliar with the environment, available resources, and possibly the language. Those who have survived previous events such as hurricanes may be more reluctant to evacuate. Interactions among disadvantages, such as age, poverty, disability, and isolations, multiply the risk.

Effective risk communication requires the warning message to be received, understood, believed, and acted upon. In order for instructions to be followed, people must believe that the action is necessary and possible. People often seek confirmation from their own social networks. Warning messages should therefore target social networks in order to assist vulnerable populations in receiving the message. Research needs identified in this article include studies on how deaf people, elderly people, children, and other populations receive and disseminate information; on which programs are successful in motivating and supporting vulnerable people; on how to effectively use social networks; and on the effects of structural and situational variables on behavior. Communities must engage agencies, businesses, and community groups in connecting with individuals at risk. Students in media, emergency management, and meteorology must be made aware of these needs in their educational programs.

Key recommendations:

- [108_Pc_1] [108_Ac_1] People in vulnerable populations are often marginalized, and are less likely to receive warnings and to respond appropriately in emergency situations. Vulnerable populations include those based on age, race and ethnicity, socioeconomic status, gender, physical disability (e.g., mobility, hearing, and vision), language, work, and isolation.
- [108_Pc_2] [108_Ac_2] Socioeconomically disadvantaged families may lack the ability to receive forecasts and warnings, a safe home environment and the transportation to evacuate.
- [108_Co_1] Non-native English speakers may lack access to information in their native language or a good translation of the message.
- [108_Pc_3] Deaf populations may not be well served by the media despite regulations to provide messages live during emergencies through closed captioning and interpreters. Lip reading is impossible when television experts turn their backs to the camera or are not visible on screen.
- [108_Co_2] The interpretation of warning messages depends on the sociocultural context of the receiver, which is not well understood by those outside of the community.
- [108_Ac_3] Gender roles in family and work influence vulnerability. Women appear to be more responsive to warnings, but may find themselves and their children in more dangerous situations.
- [108_Pc_4] [108_Ac_4] Elderly people may have less access to social networks and technology, and may find it more difficult to evacuate due to mobility problems.
- [108_Pc_5] [108_Ac_5] Children may be on their own or in settings that are unprepared to handle the emergency. Conversely, schools may provide an avenue to provide information to other family members.
- [108_Cr_1] Cultural groups outside of the mainstream may be reluctant to trust warnings until they have confirmed the information with members of their own communities, thus delaying their response.
- [108_Pc_6] [108_Co_3] [108_Ac_6] Tourists are unfamiliar with the environment, available resources, and possibly the language.
- [108_Ps_1] Those who have survived previous events such as hurricanes may be more reluctant to evacuate.
- [108_Pc_7] [108_Ac_7] Interactions among disadvantages, such as age, poverty, disability, and isolations, multiply the risk.

Key recommendations:

- [108_Pc_8] [108_Co_4] [108_Cr_2] Messages should be targeted for the social networks of vulnerable populations and developed by people that understand those communities.
- [108_Pc_9] Research is needed on how vulnerable populations receive and disseminate information.

 # 109. Proulx G. (2001, May). Occupant behaviour and evacuation. *Proceedings of the 9th International Fire Protection Symposium,* Munich, Germany, 219-232.

F | Pc | At | Co | Cr | Ps | Ac |

Building fire systems often assume certain occupant behavior that may or may not reflect how occupants actually behave in a fire. Occupants often ignore fire alarms in large public buildings, possibly out of concern about overreacting to a false alarm or a situation that is already under control. This delays evacuation and other protective measures. Occupant behavior depends on the occupant, the building, and the fire. Commitment to an activity (eating, watching a movie, waiting in line) can delay response. Other factors include social roles, the presence of smoke and other cues, time to investigate, collecting children and valuables.

Audible alarm features that affect behavior include the alarm type, audibility, and location, the number of nuisance alarms that reduce the credibility of the alarm system, and the voice communication system. In fire drill studies, an apartment building was judged to have a good audible fire alarm system when over 80 % of residents reported that the fire alarm in their apartment was loud enough. In these buildings, the mean delay time to start evacuation was about three minutes, compared to about nine minutes in the two buildings where over 20 % of residents judged the alarm signal to be too quiet. For 1000 occupants in three office buildings, the mean time to start evacuation was 50 seconds. Despite training, occupants spent time on other tasks before leaving their offices. Staff actions were credited for rapid evacuation in a retail store and got transit passengers moving in an underground station. In the absence of staff instruction, passengers did not start to evacuate after activation of the fire alarm. Voice communication informing passengers of the type of incident, its location, and what to do was as successful as staff in provoking action, with a delay time of only 15 seconds after the voice message. From fire victims, a delay time of 10 minutes after hearing the alarm was reported in a nighttime highrise fire. In a nighttime fire where the alarm was inaudible, evacuation did not start for 10 minutes to 30 minutes. In another case with alarm sounders in every unit, delay times were five minutes, with occupants waiting for the voice communication system.

To reduce the delay time before evacuation, a recognizable fire alarm signal (the standard T-3 evacuation signal as described in ISO 8201) should be installed. Attention-getting changes should be made to the environment, such as shutting off the movie, turning off the background music, and/or turning up the lights. A voice communication system should be installed in large public buildings. Voice alarms should be given as soon as the emergency has been identified in order to reduce the delay time. The message should be simple, direct, and truthful. It should give the location of the fire and what is expected of occupants, including specific directions to the exit. A live voice should be used since recorded messages can be ineffective or dangerous. A live voice can be updated with new information, can convey the appropriate urgency, and is perceived as more credible and reliable. Closed-circuit TVs installed for security purposes can be used to monitor the situation. Staff must be well-trained, since they will be perceived as reliable sources of information. Calculation of movement should include the willingness of occupants to try to travel through smoke, which is potentially lethal and slows movement considerably.

Key findings:

- [109_Pc_1] In buildings where over 80 % of residents reported that the fire alarm in their apartment was loud enough, the mean delay time to start evacuation was about three minutes, compared to about nine minutes when this was not the case.
- [109_Cr_1] The credibility of the fire alarm system is reduced by false alarms, test alarms and prank alarms.
- [109_Ac_1] Voice communication informing passengers of the type of incident, its location, and what to do was as successful as staff in provoking action, with a delay time of only 15 seconds after the voice message.
- [109_Pc_2] In a nighttime fire where the alarm was inaudible, evacuation did not start for 10 minutes to 30 minutes.

Key recommendations:
- [109_At_1] To reduce the delay time before evacuation, a recognizable fire alarm signal (the standard T-3 evacuation signal as described in ISO 8201) should be installed.
- [109_At_2] Attention-getting changes should be made to the environment during an alarm, such as shutting off the movie, turning off the background music, and/or turning up the lights.
- [109_Co_1] A voice communication system should be installed in large public buildings.
- [109_Co_2] [109_Ac_2] Voice alarms should be given as soon as the emergency has been identified in order to reduce the delay time.
- [109_Ps_1] [109_Ac_3] The voice message alarm in an emergency should be simple, direct, and truthful. It should give the location of the fire and what is expected of occupants, including specific directions to the exit.
- [109_Cr_2] [109_Ps_2] A live voice should be used since recorded messages can be ineffective or dangerous. A live voice can be updated with new information, can convey the appropriate urgency, and is perceived as more credible and reliable.
- [109_Ac_4] Closed-circuit TVs installed for security purposes can be used to monitor the situation.
- [109_Cr_3] Staff must be well-trained, since they will be perceived as reliable sources of information.

 # 110. Proulx G, Creak J, Kyle BR. (2000). A field study on photoluminescent signage used to guide building occupants to exit in complete darkness. *Proceedings of the International Ergonomics Association/Human Factors and Ergonomics Society 44th Annual Meeting,* SanDiego, CA, 4-407–4-410; also (Report No. NRCC-43959), National Research Council of Canada, Ottawa, Canada.

Bénichou N, Proulx G. (2007). Evaluation and comparison of different installations of photoluminescent marking in stairwells of a highrise building. *Proceedings of Interflam 2007, 11th International Fire Science & Engineering Conference,* Royal Holloway College, University of London, UK, 183-194; also (Report No. NRCC-49320), National Research Council of Canada, Ottawa, Canada.

Proulx G, Kyle BR, Creak J. (2000). Effectiveness of a photoluminescent wayguidance system. *Fire Technology, 36*(4), 236–248; also (Report No NRCC-44216), National Research Council of Canada, Ottawa, Canada.

E	F		Pc	At	Co	Cr	Ps	Ac

Proulx reviewed several studies in addition to performing her own trials. Work by Webber and Aizlewood demonstrated that although electrically powered systems had a higher degree of visibility, high-mounted luminaries reduced contrast and produced unhelpful scattered light. They also found that photoluminescent material (PLM) systems did not produce this 'veiling' effect. Jensen found during smoke tests involving 84 subjects that PLM performed well even at smoke densities (optical density per meter of path length) of 1.47 m^{-1} to 1.49 m^{-1}. Subjects could not move at smoke densities above this level, regardless of the guidance system.

Proulx conducted an unannounced fire drill to test a range of different wayguidance systems in a clear environment (i.e., in the absence of smoke). Data was collected from four stairwells, each employing a different wayguidance system: reduced emergency lighting (57 lx), full lighting (245 lx), PLM without other lighting, and PLM with lighting reduced to 74 lx. The lack of control of the number of people using each staircase was found to influence the results from this experiment, as did the three fire-fighters ascending the stair containing the PLM system. To address this, the results from the experiment were compared to travel speeds expected from the population density measured in the stairwell, using the Nelson and Mowrer formulation from the SFPE Handbook (3rd edition, 2002). This enabled an estimate of the potential impact of the population density on travel speed. The use of reduced emergency lighting resulted in mean travel speeds in accordance with the calculated value (0.70 m/s and 0.72 m/s respectively). The use of full lighting resulted in lower than expected travel speeds (0.61 m/s and 0.70 m/s respectively). The PLM alone resulted in higher than expected travel speeds (0.57 m/s and 0.49 m/s respectively). Finally, the use of PLM with elevated lighting resulted in lower than expected travel speeds (0.72 m/s and 0.79 m/s respectively).

Key findings
- [110_Ac_1] Simple sign designs are more effective than signs with complex designs, especially those presenting a lot of information.
- [110_Ac_2] Occupants are willing to enter stairwells even where the signage system is unfamiliar.
- [110_Pc_1] PLM should not be considered equivalent to wayguidance systems
- [110_Pc_2] PLM may be considered as functionally equivalent to emergency lighting.
- [110_Pc_3] PLM can be used to reduce the levels of emergency lighting.

- [110_Pc_4] The use of PLM in stairways may reduce the veiling effect of ambient lighting when smoke is present.
- [110_Pc_5] PLM produces favorable results when compared to emergency lighting.

111. Qi, DS. (1998, April). An inquiry into language-switching in second language composing processes. *The Canadian Modern Language Review, 54*(3), 413-435.

L		Pc	At	Co	Cr	Ps	Ac
				▓			

A native Mandarin speaker with a very high level of English proficiency was asked to perform a series of three composing tasks in English, each with a hard and an easy version. One task was performed in each of six sessions. The first set of tasks was to write two compositions: a letter to a friend about a topic of her choice (easy) and a persuasive essay on a topic of the author's choice (hard). The second was to solve two sets of five mathematical problems: problems that only required a single, basic operation (easy) and problems that required multiple operations and more complex operations (hard). The third was to carry out two sets of translations from Chinese to English (with the difficulty determined by vocabulary and structure). The subject provided think-aloud narratives to explain her thinking. When a task involved a demand that the subject thought would overload her memory or when she was unsure of the proper word, she would switch to Chinese. On the more difficult tasks, there was a higher percentage of thinking in her native language.

Key findings:
- [111_Co_1] When facing memory overload or when unsure of the proper word, a non-native speaker with a very high level of English proficiency reverted to speaking in her native language.
- [111_Co_2] When asked to carry out difficult tasks, a non-native speaker with a very high level of English proficiency reverted to thinking in her native language.

Key recommendations:
- [111_Co_3] Messages should use simple words that are easily understood by non-native speakers, and the messages should avoid memory overload.

| # **112**. Rea MS, Ouellette MJ, Clark FRS. (1985). Design considerations for egress signs based upon visibility through smoke. *General Proceedings of AIA Research & Design 85, Architectural Applications of Design and Technology Research,* American Institute of Architects, Washington, DC, 295-297; also (Report No. NRCC 25185), National Research Council of Canada, Ottawa, Canada. |

B		Pc	At	Co	Cr	Ps	Ac
		▓					

Photometric and psychophysical observations were made with thirteen exit signs. First the luminance of each sign was measured in four smokeless, dark environments. Then the exit signs were placed in a smoke chamber to which cosmetic oil smoke was added, simulating credible environmental conditions for smoke and reducing sign visibility. Sixteen volunteers made threshold observations of each sign through the smoke according to two criteria: detectability and readability. The smoke densities (optical density per meter path length) were recorded at the points at which each sign became unreadable or undetectable. Observations were made both with and without ambient illumination within the smoke chamber. As shown in Figure 8 below, the visibility of a sign depended to a large extent on its brightness, and less smoke was required to make the sign illegible than to make it undetectable.

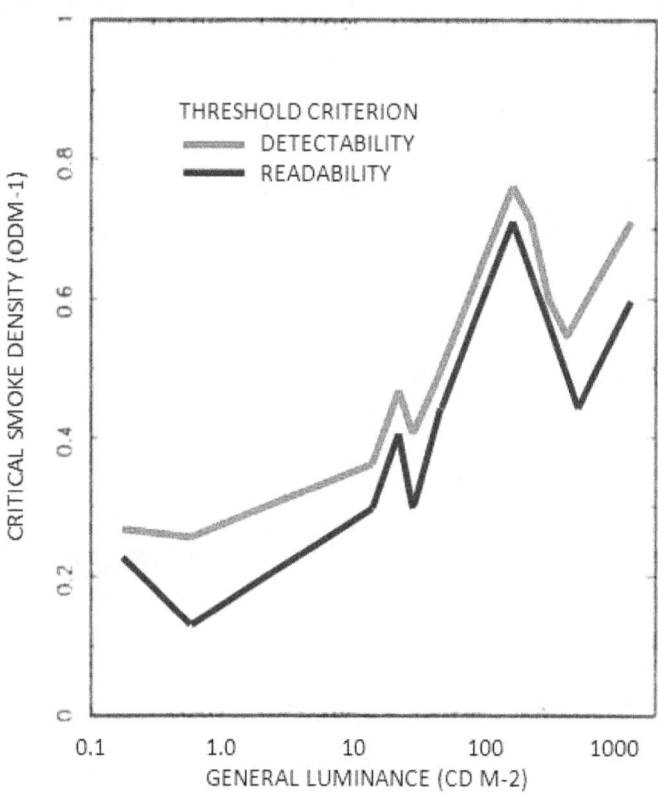

Figure 8: Readability and detectability thresholds (in terms of smoke density) as a function of sign luminance.

The effect of ambient illumination in the room on the visibility of the exit signs in smoke is shown in Figure 9. The exit signs were found to be more visible through smoke with the room lights off than with room lights on.

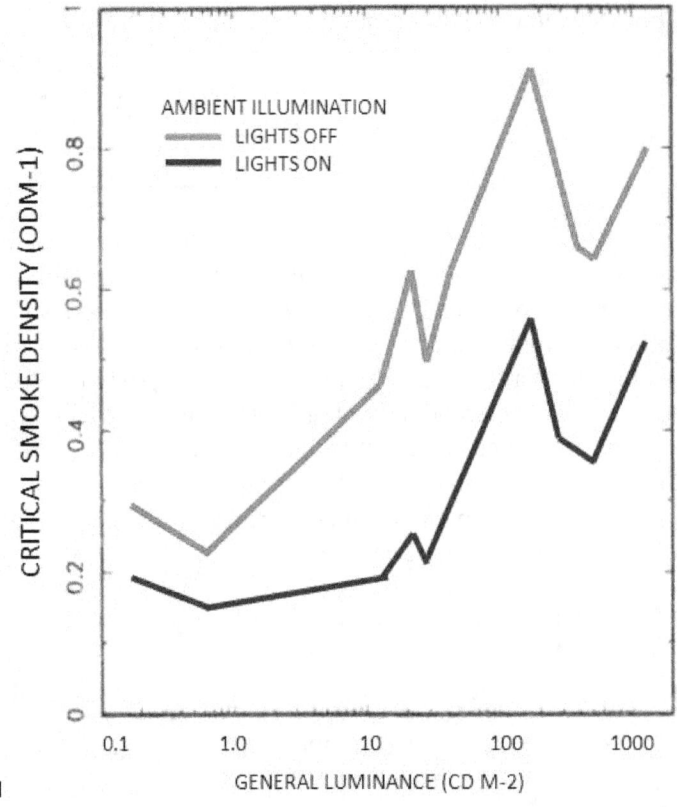

Figure 9: Detectability threshold (in terms of smoke density) as a function of sign luminance, with and without ambient illumination.

The authors make the following recommendations (based on the assumption that buildings employ conventional, 'static' signs that operate in the same way under all building conditions):
- Exit signs should be sufficiently bright that they are visible through smoke.
- The visibility of exit signs would be improved if scattered light from light fixtures and any other luminous sources was reduced. This can be accomplished by appropriately locating exit signs with respect to other light fixtures.
- The readability of exit signs can be improved by limiting unnecessary sources of scatter from the sign itself.
- Typically, translucent green materials are brighter to more people than translucent red materials for the same light source. Green signs are therefore likely to be more visible through smoke than red signs.
- Use color to aid occupants in discriminating between exit signs and other luminous sources. Other strategies may be needed to aid discrimination in environments where many colored light sources are used.

Key findings:
- [112_Pc_1] Smoke densities at detectability and readability thresholds have been measured for standard exit signs.

- [112_Pc_2] Exit signs are more visible through smoke with the room lights off than with room lights on.

Key recommendations:
- [112_Pc_3] Exit signs should be sufficiently bright that they are visible through smoke.
- [112_Pc_4] The visibility of exit signs would be improved if scattered light from light fixtures and any other luminous sources was reduced. This can be accomplished by appropriately locating exit signs with respect to other light fixtures.
- [112_Pc_5] The readability of exit signs can be improved by limiting unnecessary sources of scatter from the sign itself.
- [112_Pc_6] Typically, translucent green materials are brighter to more people than translucent red materials for the same light source. Green signs are therefore likely to be more visible through smoke than red signs.
- [112_Pc_7] Use color to aid occupants in discriminating between exit signs and other luminous sources.
- [112_Pc_8] Strategies other than color may be needed to aid discrimination of exit signs in environments where many colored light sources are used.

 # 113. Rogers GO. (1985, September). *Human components of emergency warning: Implications for planning and management* (Planning Report, FEMA Cooperative Agreement EMW-K-1024 with University of Pittsburgh). Washington, DC: Federal Emergency Management Agency.

C		Pc	At	Co	Cr	Ps	Ac

This article discussed how emergency messages could be the most effective, given the human elements in dissemination. These involve the decision-making processes of emergency officials in whether and when to broadcast warnings and the general public in what actions to take. Human elements in warning dissemination include: 1) sending the message, 2) receiving the message and evaluating its credibility, 3) evaluating the message for pertinence, 4) perceiving the threat, 5) selecting appropriate behavior, and 6) implementing the behavior.

Analysis of existing research on emergency warnings suggests the following:
1) For national emergencies, a Federal authority should alert Americans to the nature of the threat and local officials should notify the public of specific actions to be taken by local residents. This makes best use of the findings that Federal authorities have access to early warning systems, and local officials have greater credibility.
2) Warning messages should focus on simple protective or avoidance measures, since people naturally behave in familiar patterns during emergencies.
3) Take advantage of the public's willingness to participate in the warning process. In-home warning devices may be useful.
4) Make emergency warnings as specific as possible, including the nature of the threat, the area and likelihood of impact, and what simple actions can be taken alone and with others to ameliorate the situation. Nonspecific warnings encourage nonchalance in the presence of a real threat. Standard, generic warnings only lead people to seek more information since people will seek to reduce uncertainty.
5) Emergency preparedness and planning combined with drills and practice associate emergency responses with more routine and familiar behavior. Personal history matters. Enhanced adaptive response is seen with survivors of previous disasters, are more likely to act or obey authorities quickly rather than seeking additional information.
6) An all-clear signal is needed to prevent premature return to a hazardous area.
7) Issue warning messages as early as possible, instructing local officials to transmit the warning without delay.
8) Repetition of a warning message that includes the actions that should be taken strengthens adaptive responses to the hazard. People will seek confirmation from multiple sources, so consistency is important.
9) Encouraging people to contact their neighbors with information will reduce the overloading of telephone systems, warn individuals that might not have received the warning, and help to confirm the warning for those who have received the message. People are less likely to seek confirmation if with family.
10) Warning messages should be followed with a short list of easily recognized indicators of danger.
11) Warnings need to be understandable and in language with a common meaning. Messages should use simple, everyday language with declarative sentences. Officials who speak other languages should be available to help non-native speakers.

12) Emergency officials must seek to affect the perception of the hazard in the minds of the public, since perception and not the actual hazard is what determines people's response. Older people must be convinced that they are able to respond. Individual responses to any message will vary based on past experience, source of warning, and interpretation.

Key findings / recommendations:

- [113_Cr_1] Because they have access to early warning systems, a Federal authority should alert Americans to the nature of the threat in a national emergency.
- [113_Cr_2] Because they have greater credibility with local residents than Federal authorities, local officials should notify the public of specific actions to be taken.
- [113_Ac_1] Warning messages should focus on simple protective or avoidance measures, since people naturally behave in familiar patterns during emergencies.
- [113_Pc_1] Take advantage of the public's willingness to participate in the warning process, including encouraging the installation of in-home warning devices.
- [113_Ps_1] Make emergency warnings as specific as possible, including the nature of the threat, the area and likelihood of impact, and what simple actions can be taken alone and with others to ameliorate the situation. Nonspecific warnings encourage nonchalance in the presence of a real threat. Since people seek to reduce uncertainty, standard generic warnings only lead people to seek more information.
- [113_Ac_2] Emergency preparedness and planning combined with drills and practice associate emergency responses with more routine and familiar behavior. Personal history matters. Enhanced adaptive response is seen with survivors of previous disasters, are more likely to act or obey authorities quickly rather than seeking additional information.
- [113_Ac_3] An all-clear signal is needed to prevent premature return to a hazardous area.
- [113_Ac_4] Issue warning messages as early as possible, instructing local officials to transmit the warning without delay.
- [113_Co_1] [113_Cr_3] Repetition of a warning message that includes the actions that should be taken strengthens adaptive responses to the hazard.
- [113_Cr_4] People will seek confirmation from multiple sources, so consistency is important.
- [113_At_1] [113_Cr_5] Encouraging people to contact their neighbors with information will reduce the overloading of telephone systems, warn individuals that might not have received the warning, and help to confirm the warning for those who have received the message. People are less likely to seek confirmation if with family.
- [113_Ps_2] Warning messages should be followed with a short list of easily recognized indicators of danger.
- [113_Co_2] Warnings need to be understandable and in language with a common meaning. Messages should use simple, everyday language with declarative sentences. Officials who speak other languages should be available to help non-native speakers.
- [113_Ps_3] Emergency officials must seek to affect the perception of the hazard in the minds of the public, since perception and not the actual hazard is what determines people's response. Older people must be convinced that they are able to respond. Individual responses to any message will vary based on past experience, source of warning, and interpretation.

114. Rousseau GK, Lamson N, Rogers WA. (1998). Designing warnings to compensate for age-related changes in perceptual and cognitive abilities. *Psychology & Marketing, 15*(7), 643-662.

P		Pc	At	Co	Cr	Ps	Ac
		▓		▓			

The researcher states that by 2025, 25 % of the United States population will be over the age of 60 years old. It is therefore important to make sure that warning signs are effective for all ages. For a sign to function, it needs to be noticed, encoded, understood and then complied with. If there are failures at any of these levels, the warning sign will not be effective. As people age, it becomes harder to discriminate between different colors, especially those with shorter wavelengths. The elderly have trouble discriminating colors close in hue, especially those in the violet, blue and green range. Older people also have trouble distinguishing between adjacent areas that differ in light intensities and patterns that vary in width. To compensate for this, backgrounds and text should have high contrast. Older people become more susceptible to glare; therefore, lighting should be controlled to minimize the effects of glare. Older adults have trouble perceiving flashing stimuli in the range of 10 Hz and 45 Hz.

The elderly have difficulty reading small fonts. Fonts should therefore not go below a 12 pt to 14 pt type. In order to allow older people to notice and comprehend a sign, the minimum vertical type size should be at least 6.7 point and no more than 39 characters per inch. A stroke width ratio of 1:6 to 1:8 for black letters on a white background and 1:8 to 1:10 ratios for white type on a black background is recommended. For older adults, a sans serif font should be used with a medium or bold weight typeface.

Signs or labels should avoid clutter or irrelevant information to increase the ability for people to select and process the key relevant information. Cognitive abilities generally affect the comprehension and compliance stages of warning signals. Working memory, which is the ability to keep information in active in memory, declines with age. This is important in warnings or signs that require multiple steps. To reduce the age related differences, older adults benefit from environments or contexts that provide cues for remembering information. Warning designs should have simple sentences, with the information given explicitly and not left to the reader to infer its meaning. Older people have difficulty remembering time-based tasks, so it is better to present instructions in terms of event-based tasks (e.g., eat after breakfast vs. eat every eight hours).

Key findings:
- [114_Pc_1] Older adults have trouble with colors in the violet, blue and green range.

Key recommendations:
- [114_Pc_2] Text on signs should have a high contrast from the background and glare should be minimized.
- [114_Pc_3] Font should be at a minimum 12 point to 14 point and be sans-serif.
- [114_Pc_4] The minimum vertical type size of 6.7 point and no more than 39 characters per inch should be allowed if older people need to comprehend the sign.
- [114_Pc_5] A stroke width ratio of 1:6 to 1:8 for black letters on a white background and 1:8 to 1:10 ratios for white type on a black background is recommended.
- [114_Pc_6] A ratio of 1:100 of letter size to reading distance on signs with a minimum letter size of 15 mm is recommended.

- [114_Co_1] Signs or labels should avoid clutter or irrelevant information to increase the ability for people to select and process the key relevant information.
- [114_Co_2] Cognitive abilities generally affect the comprehension and compliance stages of warning signals. Working memory, which is the ability to keep information in active in memory, declines with age. This is important in warnings or signs that require multiple steps.
- [114_Co_3] To reduce the age related differences, older adults benefit from environments or contexts that provide cues for remembering information.
- [114_Co_4] Warning designs should have simple sentences, with the information given explicitly and not left to the reader to infer its meaning.
- [114_Co_5] Older people have difficulty remembering time-based tasks, so it is better to present instructions in terms of event-based tasks (e.g., eat after breakfast vs. eat every eight hours).

 # 115. Sanders MM, McCormick EJ. (1993). Chapter 4: Text, Graphics, Symbols and Codes. In *Human Factors in Engineering and Design* (7th ed.) (pp. 91-128). New York: McGraw-Hill.

E

Pc	At	Co	Cr	Ps	Ac

This chapter on human factors considerations for text and graphics begins with a discussion of the properties of the human eye. *Accommodation* is the ability of the lens of the eye to focus light rays on the retina that allows us to see details of an object. *Visual acuity* is the ability to discriminate fine detail and depends largely on the accommodation of the eyes. The most common measurement of acuity is the minimum separable acuity, which refers to the smallest space between the parts of a target that the eye can detect. This can also be expressed as a ratio called Snellen acuity. A Snellen acuity of 20/30 means that a person can read from a distance of 20 feet what a normal person can read from a distance of 30 feet. The *visual angle* (*VA*) is the angle (in degrees or minutes of arc) that a viewed object subtends at the eye. For angles less than 10°, the formula for visual angle is VA (arc minutes) = $3438 \times H/D$, where H is the height of the detail and D is the distance from the eye (both must be in the same units). Normal vision is considered to be a visual acuity of 20/20 = 1.0, or the ability to resolve points with a visual angle of one arc minute. *Spatial frequency* is the number of bars per unit distance in a grating. Humans are most sensitive to spatial frequencies between two and four cycles per degree. The more detailed the picture, the higher its spatial frequency. When the contrast between a picture and the background is low, the object must be made larger to accommodate. Under conditions of high illumination, acuity increases with exposure time up to 100 milliseconds to 200 milliseconds and then levels off. Visual acuity and contrast sensitivity decline with age. At age 75, the average visual acuity has declined to about 0.6. For visual displays, the design should be capable of being seen clearly and should help the viewer to correctly perceive the meaning of the display.

For text, the ratio of the thickness of the stroke to the height of the letter is considered the width-to-height ratio. Irradiation causes white features on a black background to appear to spread into dark areas, increasing the apparent thickness of the stroke. Black on white letters should therefore have higher width-to-height ratios than white on black ones. For printed material, the width-to-height ratio should be 1:6 to 1:8 for black on white and 1:8 to 1:10 for white on black. When there is less illumination, thicker letters are easier to read. When there is low illumination, letters should be bold type with a low stroke width-to-height ratio. A formula has been developed to determine the stroke width (Ws) for letters to be read at a specific distance d by a person with a given Snellen acuity score S: $Ws = (1.45 \times 10^{-5}) \, d / S$. Lowercase is easier to read than all capital letters. Close set type can be read faster than regular spaced type.

Reading on a computer screen is different than reading from a hardcopy. It has been recommended that letter sizes have a minimum visual angle of (11 or 12) arc minutes, which converts to 0.06 inches to 0.07 inches for lower case letters. For computer screens, it is recommended to avoid using color pairs of red and blue and to a lesser extent, red and green or blue and green on a dark background.

For symbols, these basic principles are applicable to signs in certain contexts: *Figure to Ground:* Clear and stable figure to ground articulation is essential. *Figure Boundaries*: A solid shape is preferable to a line boundary. *Closure:* A closed figure should be used unless there is a reason for the figure to be discontinuous. *Simplicity:* Figures should be as simple as possible, with only necessary features. *Unity:* Symbols should be as unified as possible.

Key findings:
- [115_Pc_1] Visual acuity and contrast sensitivity decline with age. At age 75, the average visual acuity has declined to about 0.6, from normal vision of 20/20 = 1.0 in young adults.
- [115_ Pc _2] For printed material, the ratio of the thickness of the stroke to the height of the letter should be 1:6 to 1:8 for black on white and 1:8 to 1:10 for white on black.
- [115_ Pc _3] The stroke width (Ws) for letters to be read at a specific distance d by a person with a given Snellen acuity score S is given by $Ws = (1.45 \times 10^{-5})\, d / S$.
- [115_ Pc _4] Lowercase letters are easier to read than all uppercase letters.
- [115_ Pc _5] Close set type can be read faster than regular type.
- [115_ Pc _6] On computer screens, letter sizes should have a minimum visual angle of (11 or 12) arc minutes, which converts to 0.06 inches to 0.07 inches for lower case letters.
- [115_ Pc _7] Clear and stable figure to ground articulation, which differentiates a figure from the background using texture, is essential.
- [115_Co_1] A solid shape is preferable to a line boundary for the boundary of a figure.
- [115_Co_2] A closed figure should be used unless there is a reason for the figure to be discontinuous.
- [115_Co_3] Figures should be as simple as possible, with only necessary features.
- [115_Co_4] Symbols in a set of related signs should be as unified as possible.

116. Schieber F, Kline DW. (1994). Age differences in the legibility of symbol highway signs as a function of luminance and glare level: A preliminary report. *Proceedings of the Human Factors and Ergonomics Society 38th Annual Meeting*, 133-136.

E		Pc	At	Co	Cr	Ps	Ac

The ability of older adults to see well is largely dependent on the ambient level of lighting. Age-related losses in visual sensitivity are less noticeable during daylight (given the luminance levels present), but can be severe with a glare source at nighttime. These conditions usually occur with road signs while driving at night.

In this study, highway signs were evaluated by older people during the daytime (high photopic luminance), nighttime (low photopic luminance) and nighttime luminance with a narrow angle glare source. This experiment used three groups of participants: 12 people with ages from 18 to 25, 12 people with ages from 40 to 55 and 18 people with ages from 65 to 79. Eighteen symbol highway signs that represented the range of legibility distances found today were tested. The images were placed in the center of a computer monitor screen with a white background. Under the nighttime and the nighttime with glare conditions, the luminance of the screen was 5 cd/m². The glare source was at an illuminance level of 8 lx at the eye, which is approximately the intensity of a pair of headlights 30 meters away. The sign first appeared small on the screen and then gradually got larger. The participants were asked to describe the sign details that they could detect at different sizes. The results, reproduced in Table 21, showed that reducing the luminance from daytime to nighttime had a small effect on the legibility distance for the young and middle aged adults, and that adding the glare source had no significant impact. For the older adults, both reducing the luminance from daytime to nighttime and adding the glare source had large effects on the legibility distance.

Table 21: Relative reduction in highway sign legibility by age group under three lighting conditions, normalized to average young adult daytime performance.

	Lighting condition		
Age Group	Daytime	Nighttime	Nighttime + Glare
Young	1.00	0.70	0.69
Middle-Aged	0.88	0.60	0.60
Old	0.80	0.46	0.34

Key findings:
- [116_Pc_1] Under daylight viewing conditions, the legibility distance for older adults to read highway signs is 80 % of that for young adults.
- [116_Pc_2] Nighttime reduces the distance at which people are able to read highway signs. Relative to the value for young adults under daylight viewing conditions, the legibility distance is reduced to 70 % for young adults and to 46 % for older adults.
- [116_Pc_3] The presence of glare reduces the distance at which older people are able to read highway signs at night by about 26 %. The legibility distance is 34 % that for young adults under daylight viewing conditions

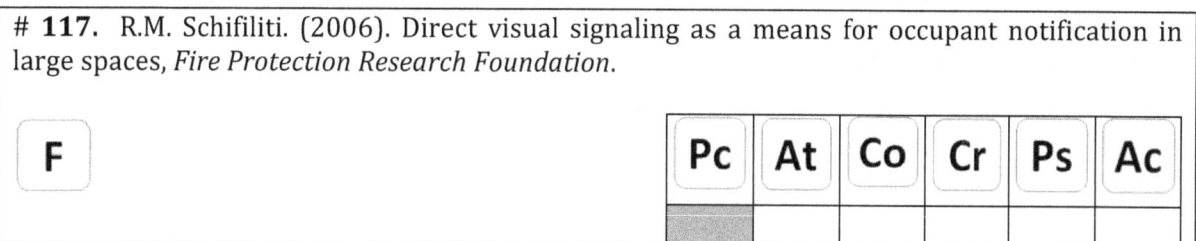

 # 117. R.M. Schifiliti. (2006). Direct visual signaling as a means for occupant notification in large spaces, *Fire Protection Research Foundation.*

F			Pc	At	Co	Cr	Ps	Ac

Requirements for visual signaling described in NFPA 72 are based on occupants being alerted by indirect signalling effects. These requirements were based on tests in relatively small, enclosed spaces. This article reports on a study of visual signaling devices in large well-lit spaces. Tests were carried out in three different warehouse stores: two Home Depot stores in MA and a Wal-Mart store in FL. Participants and visual signals are described in Table 22. Lighting levels for all three locations were within the range for general circulation and merchandise areas, as stipulated in the Illuminating Engineering Society of North America guidelines for High Activity Areas.

Table 22: Warehouse store evacuation experiments with visual signaling devices.

Year	Location	Participants	Notification
2005	Home Depot Store, MA	13 (three with hearing impairment, two with corrective devices)	Audible signals disabled. Strobes located 23 feet above the floor.
2005	Home Depot Store, MA	12 (two with hearing impairment, one with a corrective device)	Strobes located 21.5 feet above the floor.
2005	Wal-Mart Store, FL	22	Strobes located 15 feet to 20 feet above the floor. Horns disabled after initial chirp.

Occupant alerting was achieved through a combination of direct and indirect signalling from the installed lighting systems. However, blind-spots were noted that need to be addressed to ensure comprehensive coverage, which is especially important for the hearing impaired. Participants were surveyed after each event to better understand their experiences. They were asked if they could indirectly see a strobe – from a reflection, for example – without necessarily seeing the strobe directly. As can be seen in

Figure **10**, the majority of participants could indirectly see a strobe in all of the three test cases.

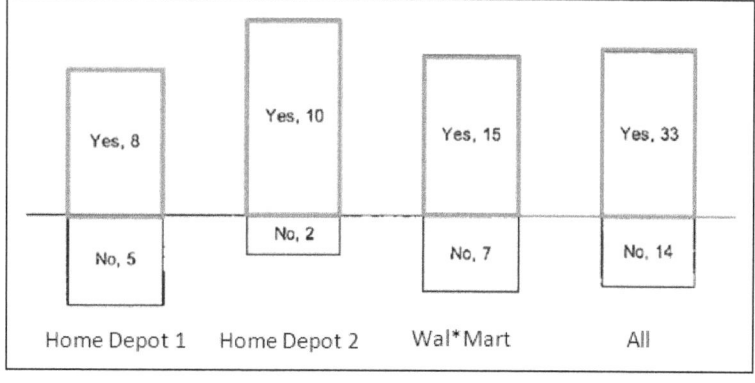

Figure 10: Reported indirect visibility of strobes

Participants were asked if they could directly see a strobe without deliberately seeking one out or looking at the ceiling. As can be seen in
Figure **11**, the majority of participants could directly see a strobe in all of the three test cases.

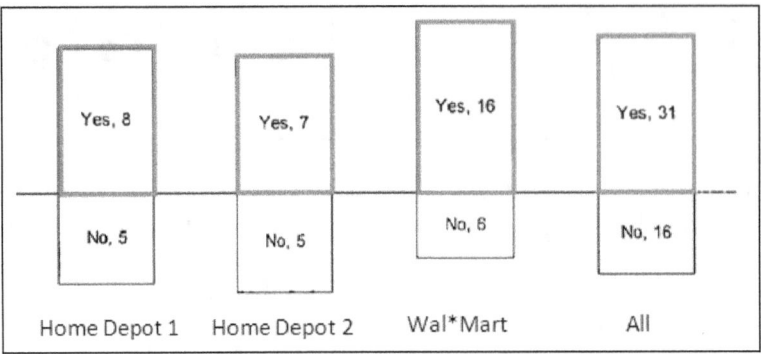

Figure 11: Reported direct visibility of strobes

Participants were asked to report the number of strobes that they could see during the event. It is apparent from Figure 12 that all participants could see at least one strobe, with most participants being able to see between three and six strobes during the event.

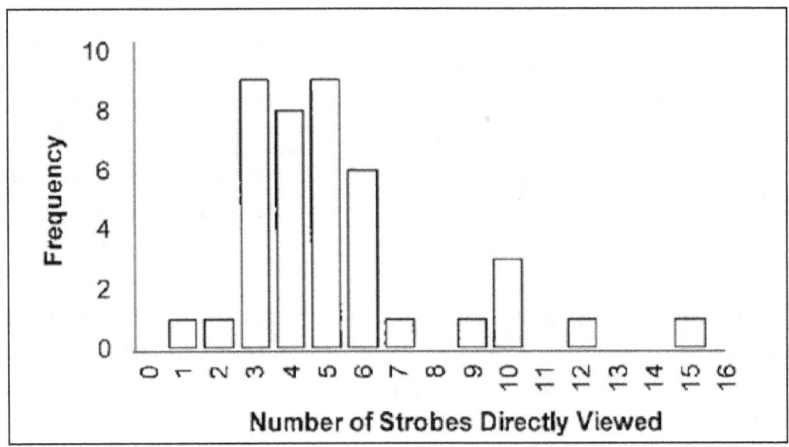

Figure 12: Number of strobes viewed directly

Perhaps most importantly, the participants were asked whether there were blind spots from which strobes could not be seen. It is apparent from Figure 13 that the number of blind spots differed between the three cases (as would be expected given structural, procedural and population differences); however, blind spots existed in all cases.

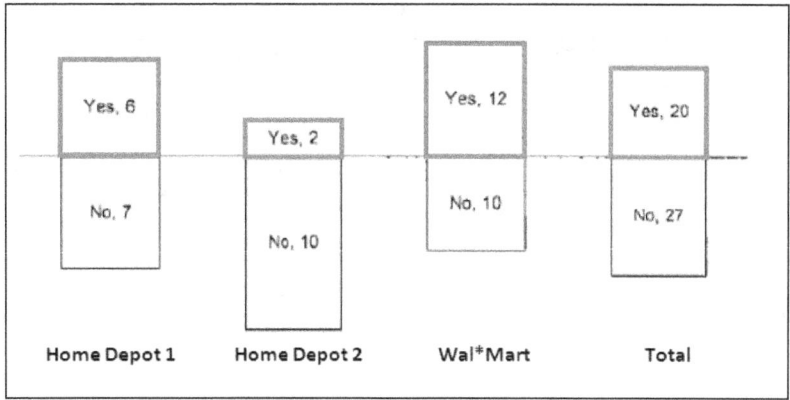

Figure 13: Existence of locations where strobes could not be seen.

The following key points were made:

- 'Strobe lights are effective for both direct and direct viewing even if not located directly over an aisle, provided there is sufficient penetration to the aisle.
- A design with strobe lights over every aisle is more effective than one where strobes serve several aisles.
- Aisles focused the occupant's vision and improved direct signalling effects.'

However, the effectiveness of strobes (especially in stores such as those examined) may be susceptible to frequent changes in the configuration of the space and not just upon the performance of the devices themselves.

Key findings:
- [117_Pc_1] Strobes can be seen both directly and indirectly – from reflective surfaces, for example.
- [117_Pc_2] Blind spots were found in each of three large warehouse stores where strobe alarms could not be seen either directly or indirectly.
- [117_Pc_3] The existence of blind spots in facilities with visual signaling alarm devices depends on a number of factors, including structural, procedural and population, not necessarily related to strobe performance.

 # 118. Schmidt JK, Kysor KP. (1987). Designing airline passenger safety cards. *Proceedings of the Human Factors Society 31st Annual Meeting,* 51-55.

E		Pc	At	Co	Cr	Ps	Ac

The purpose of this study was to identify characteristics on airplane passenger safety cards that are easy to comprehend.

In the first part of the experiment, 25 participants were asked to rank passenger safety cards in order of their supposed effectiveness in portraying the correct safety information. The cards were taken from all of the 33 major airlines in the United States. The results showed that the graphic cards that were slightly larger, less wordy, and more colorful were preferred by the participants. The lower ranked cards were smaller, wordier, and less colorful.

The second part of the experiment had 25 participants. Each participant was given a different safety card. They were asked to study the safety card for three minutes and were told that they would be tested on the information in the safety card. After the test, they were given back their cards and were asked to correct their answers in red pen if they found any mistakes. Participants were found to prefer cards that displayed words integrated with diagrams. They also liked the cards that presented pictures in a pictogram sequence with little text. Illustrations with words were remembered better than photographs with words.

Key findings:
- [118_Co_1] Larger, less wordy, and more colorful airplane passenger safety cards are preferred.
- [118_Co_2] Airplane passenger safety cards that display words integrated with diagrams or present pictures in pictogram sequence with little text are preferred.
- [118_Co_3] Illustrations with words are remembered better than photographs with words.

119. Schroeder DG, Hancock HE, Rogers WA, Fisk AD. (2001). Phrase generation and symbol comprehension for 40 safety symbols. *Proceedings of the Human Factors and Ergonomics Society 45th Annual Meeting,* Santa Monica, CA, 1479-1480.

E		Pc	At	Co	Cr	Ps	Ac

Symbols have the potential to replace textual messages for warnings. However, it is important to design the symbol so that all who need to be warned about the product will comprehend the symbol in a consistent and reliable manner. This test studied safety symbol comprehension by having participants generate phrases related to particular safety symbols. Participants were given a packet with symbols and were asked to write down phrases that each symbol prompted. It was emphasized to the participants that there were no right or wrong answers. Fifty two younger adults (ages 18 to 23) and 52 older adults (ages 64 to 75) participated in the study. Younger adults were given 45 seconds to generate a phrase while older adults were given 60 seconds.

Younger adults scored significantly higher for every symbol type than older adults, and also generated a lot more phrases per symbol than older adults did.

Key findings:
- [119_Co_1] Younger adults comprehend standard safety symbols better than older adults.

 # 120. Sharlin E, Watson B, Sutphen S, Liu L, Lederer R, Frazer J. (2009). A tangible user interface for assessing cognitive mapping ability. *International Journal of Human-Computer Studies, 67*(3), 269-278.

P		Pc	At	Co	Cr	Ps	Ac

Wayfinding is the skill that allows people to find their way from one destination to another and then back again. The definition from psychology is "an overall mental image or representation of the space and layout of a setting" [Arthur P & Passini R. 1992. *Wayfinding-People, Signs, and Architecture*, New York: McGraw-Hill], which is used to guide movement throughout a space. A major component in wayfinding is cognitive mapping, which describes the mental map employed by an individual that is relied upon during wayfinding. The Cognitive Map Probe (CMP) is an automated tool that measures cognitive mapping ability. The procedure is for participants to view a drive through a neighborhood on a large computer screen and then recreate the neighborhood using 3D blocks on a special Segal tabletop input surface. The CMP then scores the participant based on the 3D model. People acquire cognitive maps by direct physical contact with an environment, such as by looking around an area or by tapping an area with a cane.

In this experiment, the number of buildings was varied in the virtual CMP environment to determine whether wayfinding abilities are adversely affected by an increase in the number of buildings. Twenty people participated in the study, of whom 10 were under 55 years old and 10 were between the ages of 55 and 81. Participants each performed 10 trials using the CMP, viewing the same neighborhoods in the same order with the same number of buildings.

The results showed that the older participants built less accurate maps than the younger participants. As the number of buildings increased, the mapping performance of participants deteriorated.

Key findings:
- [120_Co_1] Older adults create less accurate cognitive maps than younger adults.
- [120_Co_2] An increased number of map details results in poorer performance in cognitive mapping.

121. Shieh, K-K, Huang S-M. (2003). Factors affecting preference ratings of prohibitive symbols. *Applied Ergonomics, 34*(6), 581-587.

E		Pc	At	Co	Cr	Ps	Ac

The symbol of a circle with a line through it typically indicates that access is prohibited. The purpose of this study was to investigate the effects on comprehension of two components of prohibitive symbols: the representation of the prohibitive symbol and the circle-slash component. The pictorial solidity, size, and the direction of elongation (DE) of the representation and the slash orientation and thickness of the circle-slash component were studied. Six commonly used symbols in Taiwan were used to help examine the different features of the circle-slash prohibitive symbol.

Twenty-four undergraduate students from the Oriental Institute of Technology participated in the study. Seventy-two designs showing variations of a pictorial with a red circle-slash on top of the symbol were constructed on cards and shown to the students. At the beginning of the procedure, the test participants were told the meanings of the prohibitive symbols. Then they were asked to rate each card and sort the cards into ten groups based on their preference.

The results showed that solid pictorials were preferred over outline pictorials. This may be because outline figures contain more lines, angles, and segments than solid figures and are thus more complicated visually. The results indicated that pictorial size should be at least 75 % of the inner diameter of the circle of the prohibitive symbol. This is consistent with previous data and with evidence that larger pictures are usually easier to recognize [Sanders MM & McCormick EJ. 1993. *Human Factors in Engineering Design.* New York: McGraw Hill, pp. 121-125]. A slash that comprised 25% of the area within the circle was rated the highest. Slash thicknesses that exceeded this value, reducing the available space for the pictorial and covering more of it, resulted in lower ratings. The orientation of the slash at an angle of either 45°or 135° was preferred unless the picture behind the prohibitive symbol was obscured, in which case preference was given to angles that obscured the picture less.

Key findings:
- [121_Co_1] Solid pictorials are preferred to outline pictorials, likely because outline figures contain more lines, angles, and segments than solid figures and are thus more complicated visually.
- [121_Co_2] For prohibitive symbols, the pictorial size should be at least 75 % of the inner diameter of the circle.
- [121_Co_3] For prohibitive symbols, a slash that comprised 25 % of the area within the circle was rated the highest. Slash thicknesses that exceeded this value, reducing the available space for the pictorial and covering more of it, resulted in lower ratings.

 # 122. Siegel JT, Burgoon JK. (2002). Expectancy theory approaches to prevention: Violating adolescent expectations to increase the effectiveness of public service announcements. In DC William, M Burgoon (Eds.), *Mass Media and Drug Prevention: Classic and Contemporary Theories and Research* (pp. 163-186). New Jersey: Lawrence Erlbaum Associates, Inc.

M		Pc	At	Co	Cr	Ps	Ac

Expectancy theory describes behavioral choices as arising from the expectations of what the results of the choices will be. This chapter suggests that this class of sociological and psychological theories may provide useful insights into designing effective public service announcements. Some theories in this family are described here.

- Expectancy Value Theory seeks to explain and predict the attitudes of individuals. The three basic components are: 1) individuals respond to new information about an object or action by developing or modifying a belief about it, 2) individuals assign a value to each attribute on which the belief is based, and 3) beliefs and values are used to calculate an expectancy, or attitude, about his/her interactions with the object or action.
- Expectation States Theory (EST) addresses how observed differences between groups affect attitudes, perceptions, and behavior. While primarily applied to gender inequality, it can be applied to study any group.
- Expectancy Violations Theory (EVT) seeks to explain reactions to unexpected behavior. It is based on three primary assumptions: 1) an individual seeks to reward others and avoid punishing them, 2) behavior violations call attention to the violator and the relationship with the individual, and 3) the evaluation of the violation is based on the interpretation of the violation in light of the values and needs of the individual. The response to communication depends on the relationship of the communicator to the listener.
- Language Expectancy Theory (LET) addresses the use of language to persuade. It hypothesizes that cultural and societal expectations about language use affect the receptiveness of individuals to persuasive messages. In particular, there are normative expectations about the appropriate use of fear and intensity in persuasion. Credible sources have a wider range of language strategies to choose from, while less credible sources and women are more persuasive when they stick with low intensity appeals. Men may be considered weak if they do not use intense language.

The chapter discusses these three theories about attitudes (EST, EVT, and LET) as they apply to the production of more effective public service announcements (PSAs) on adolescent marijuana use. EST contributes the idea that attitudes toward similar previous public service campaigns will influence the response to the next one. EVT suggests that behavior that violates expectations can be used to attract attention to the PSA. From LET, males are usually more persuasive using intense, compliance-gaining messages while females are more persuasive in using low intensity, unaggressive messages. Fear arousal that is irrelevant to the message makes listeners more receptive to low intensity messages.

Key recommendations:
- [122_Cr_1] Attitudes toward previous false alarms and emergency drills affect behavior during the next emergency.
- [122_At_1] Messages that violate expectations attract attention. Warnings should therefore

be distinct from normal ambient sounds and given in a different voice than that used for normal announcements.

- [122_Cr_2] The response to communication depends on the relationship of the communicator to the listener.
- [122_At_2] [122_Cr_3] Credible sources have a wider range of language strategies to choose from, including more intense language that may attract more attention.
- [122_Cr_4] [122_Ac_1] A male voice is more persuasive for presenting specific instructions that need to be followed.
- [122_Cr_5] Fear arousal makes listeners more receptive to low intensity messages, which are most effectively delivered by women.
- [122_Ps_1] Unwarranted fear can be avoided by including specific details about the risks for each individual in a warning message.

 # 123. Silver NC, Gammella DS, Barlow AS, Wolgalter MS. (1993). Connoted Strength of Signal Words by Elderly and Non-Native English Speakers. *Proceedings of the Human Factors and Ergonomics Society 37th Annual Meeting*, 516-519.

E		Pc	At	Co	Cr	Ps	Ac

Currently standards on safety messaging suggest using the words DANGER, WARNING, and CAUTION to indicate the highest to lowest levels of hazard. Most experiments supporting this ordering used college students to evaluate the danger words. The purpose of this study was to determine whether the words DEADLY, DANGER, WARNING, CAUTION, CAREFUL, NOTICE, ATTENTION and NOTE were rated similarly for meaning by elderly and non-native English speaking populations as by college students and grade-school children. This study recruited 233 participants, with 98 living in retirement homes and 135 non-native English speakers. Forty-three words from a previous study were used. The words were less than 10 letters each and were rated above 4.0 ('understandable') on a scale of comprehensibility. The participants rated the 43 terms based on the question: "How careful would you be after seeing this term?" They rated the words on a scale from zero to eight, where zero meant that they would not be careful at all and eight meant they would be extremely careful after seeing the word. The results showed that the elderly and non-native speakers tended to rank the signal words in the same order as college students. The only difference was that the word ATTENTION had greater carefulness ratings than the word CAREFUL among the elderly and non-native English speakers. DANGER also had a greater carefulness rating than DEADLY among the non-native English speakers. The article suggests that if non-English speakers are likely to use a dangerous product, words that are more familiar to non-native English speakers, like STOP and DON'T would be better warning words. If important safety terms were incorporated into early English-training curricula, then it would allow non-native English speakers to be aware of danger.

Key findings:
- [123_Co_1] Elderly people, non-native English speakers, college age students, and grade school children generally rate hazard words in the same order by implied meaning.
- [123_Co_2] While the word CAREFUL implies a higher level of carefulness than ATTENTION for college students (and in current standards), the two words were rated in the opposite order by elderly and non-native English populations.
- [123_Co_3] While the word DEADLY implies a higher level of carefulness than DANGER for college students (and in current standards), the two words were rated in the opposite order by non-native English speakers.

Key recommendations:
- [123_Co_4] Use more familiar words like STOP and DON'T on dangerous products that are likely to be used by non-English speakers.

 # 124. Singer JP, Lerner ND. (2005, March). *Countdown Pedestrian Signals: A Comparison of Alternative Pedestrian Change Interval Displays.* Federal Highway Administration.

T

Pc	At	Co	Cr	Ps	Ac

In 2003, the conventional pedestrian signals for crossing the street were WALK (a white outline of a person telling pedestrians to begin crossing the street), Flashing DON'T WALK (an orange hand flashing telling pedestrians that they should not begin crossing the street but should continue crossing if they have already started) and the steady DON'T WALK (a steady orange hand which means that people should not cross the street). Previous research in Chicago showed that the Flashing DON'T WALK sign was comprehended at a very low rate (below 50 % in some cases), and that this sign can cause pedestrians to cross the street in an unsafe manner, sometimes staying in the intersection at the median or returning to the original curb. Many recent studies have looked at the effectiveness of countdown pedestrian signals (CPS). Standard CPS gives the same information as the conventional pedestrian signals with the orange hand but also include a countdown timer that shows pedestrians how much time they have left to cross the street. The pedestrian change interval begins at the time when pedestrians should not begin crossing the street but should continue crossing if they have already started. This research looks at the effects of the flashing hand during the pedestrian change interval of a CPS by comparing the standard CPS that shows both a flashing hand and the countdown with an experimental CPS that shows the countdown alone. Two studies were performed to look at comprehension of pedestrian signals including the CPS and at their effect on pedestrian behavior.

The first study compared the comprehension of the conventional pedestrian signal with flashing hand only, the standard CPS with flashing hand and countdown, and the experimental CPS with the flashing hand removed. Forty five people of various ages and gender participated. Participants were shown a photograph of a crosswalk with a large pedestrian signal on the top of the photograph and a pedestrian crossing the street. Five scenarios were shown in the photograph: 1) the pedestrian arrives at the curb when the signal changes from WALK to the pedestrian change interval; 2) the pedestrian starts crossing during the WALK phase but the signal changes to the pedestrian change interval and the countdown begins at 16; 3) the pedestrian change interval is active when the participant reaches the curb and is at seven seconds when they reach the curb; 4) the pedestrian change interval begins soon after the pedestrian starts walking; and 5) the countdown signal is at zero when the participant reaches the curb. Participants were shown the photographs in a small group setting and were each asked to write down what the pedestrian in the photograph is supposed to do.

The results showed that for the pedestrian change interval, 14 % of the pedestrians believed it was safe to cross with the conventional signal but 50 % believed it was safe to cross for the standard CPS and 75 % for the experimental CPS. The participants felt that the CPS allowed participants to decide if it was safe or not to cross while the conventional sign did not. Participants preferred having the countdown timer present over the flashing hand alone but did not have a clear preference for the standard or the experimental CPS. The results indicate that participants are more likely to believe that they can cross during the pedestrian change interval with a CPS than with the standard flashing hand signal, removing a significant amount of confusion. Older people and females tended to be more cautious when deciding whether to cross or not, while young males were more likely to believe that they could begin crossing. Older people benefited most from the removal of confusion by the experimental CPS because they tend to walk slower and are more likely to find themselves in the middle of the road during the pedestrian change interval. Without the flashing hand, they understood that they were able to complete the crossing.

The second experiment looked at the differences between the effects of a standard CPS and an experimental CPS on pedestrian behavior in an actual crosswalk. The crosswalk was filmed during the morning, midday and afternoon. Pedestrians were initially observed with a standard CPS positioned at the crossway; later the standard CPS was replaced with the experimental CPS. A video recording system filmed people crossing the street. The video was later reviewed to note the signal phase when each person came to the curb, whether the person crossed when they arrived at the curb or waited for the next WALK phase, the signal phase at the other side of the street, the estimated gender and age, and any key events that occurred. The results showed that the experimental CPS did not have a significant effect on whether or not pedestrians began to cross during the pedestrian change interval. With the experimental CPS, the pedestrians that arrived at the curb early in the pedestrian change interval were less likely to cross the street, but the ones that arrived late were more likely to cross.

The overall results indicated that the experimental CPS may lead to less confusion than the standard CPS, especially for older people. The results do not provide dramatic evidence in favor of the experimental CPS over the standard CPS, although older pedestrians may benefit from the removal of the flashing hand.

> Key findings:
> - [124_Co_1] A pedestrian signal with countdown only (no flashing hand) resulted in the least confusion about whether the pedestrian was allowed to be in the crosswalk. The standard orange flashing hand signal resulted in the most confusion.
> - [124_Ac_1] Older people and women are more likely to show caution when crossing the street.
> - [124_Ac_2] People prefer countdown pedestrian signals to a flashing hand signal with no countdown because the additional information provides more control over the decision.

125. Smith-JacksonTL, Wogalter MS. (2000). Users' hazard perceptions of warning components: An examination of colors and symbols. *Proceedings of the 14th Triennial Conference of the IEA/HFES*, 6, 55-68.

E		Pc	At	Co	Cr	Ps	Ac

It is important to design product safety and product labels to include signal words, symbols, and colors that communicate levels of risk consistently across cultures. This study was designed to explore cross-cultural hazard perceptions by comparing the hazard perceptions of colors and symbols by monolingual English-speakers to those by Spanish speakers. Participants were asked to rank colors and symbols on a hazard rating scale. The ten colors were RED, YELLOW, BLACK, ORANGE, MAGENTA, BLUE, BROWN, GREEN, WHITE, and GREY. Six symbols were also presented: SKULL, SHOCK, PROHIBITION, MR. YUK, ALERT, and ASTERISK.

For both the Spanish speakers and English speakers, RED was given the highest hazard rating, and BLACK and YELLOW did not differ significantly. English speakers perceived YELLOW to be the second highest hazard color while Spanish speakers rated ORANGE as the second highest color. Both groups rated the SKULL symbol significantly higher than all the other symbols. The colors RED, BLACK, ORANGE, and YELLOW mean hazard across cultures within the US. The SKULL, SHOCK, and PROHIBITION symbols were also given higher hazard ratings for both groups.

Key finding:
- [125_Ps_1] Red, black, orange, and yellow connote hazard across different cultures (Spanish-speakers and English-speakers) in the United States.
- [125_Ps_2] The SKULL, SHOCK, and PROHIBITION symbols communicate higher hazard across both Spanish-speaking and English-speaking populations in the United States.

 # 126. Steinberg FS, Horwitz EK. (1986, March). The effect of induced anxiety on the denotative and interpretive content of second language speech. *TESOL Quarterly, 20*(1), 131-136.

L		Pc	At	Co	Cr	Ps	Ac

Anxiety adversely affects the proficiency of speakers of a second language. Previous studies have looked at how anxiety impairs fluency and grammatical structure. This study investigates the effect of anxiety on the content of oral communication by students. It was hypothesized that nonanxious second language students may attempt to communicate more complicated topics than their anxious peers, thus appearing to have a lower grasp of the language than they actually do.

The subjects were 20 Spanish-speaking young adults in an intensive English as a Second Language (ESL) course at the University of Texas at Austin. All students were studying at the low-intermediate level, and they were divided into high or low oral proficiency groups based on teacher evaluations. Subjects from each group were randomly assigned to two test conditions. For the first test condition, anxiety was generated through the use of a video camera, formal (cold) interactions between the researcher and student, and emphasizing that the performance in the experiment was indicative of ability and that good performance was crucial to its success. For the second condition, students were put at ease as much as possible by a comfortable chair, the absence of a video camera, casual conversation before the experiment, and supportive comments, including emphasizing that the experiment was meant to be an enjoyable experience. All subjects were informed that they would be taped by an audio recorder, and the same researcher gave all interviews. The task was to look at three pictures from the standard Thematic Apperception Test (TAT) one at a time and describe orally elements in each picture, what was depicted, and what the subject thought was happening. Basic vocabulary was available from a written list or from the researcher by request.

Audio recordings were reviewed by three native speaking ESL teachers, who evaluated the proportions of denotative (limited to elements clearly depicted in the pictures) and interpretive (referring to elements not specifically depicted) information in each interview. The students responding under anxiety conditions were found to provide descriptions that were significantly more denotative and less interpretative. Although they may still have been affected, high proficiency students did not perceive themselves as anxious under the anxiety conditions.

Key findings:
- [126_Co_1] Students of English as a second language are more limited in their ability to communicate subjective information when they are anxious than when they are relaxed and comfortable.
- [126_Co_2] Higher proficiency non-native English speakers feel less anxiety than lower proficiency speakers when asked to speak in a high pressure environment.

 # 127. Stout C, Heppner CA, Brick K. (2004, December). *Emergency preparedness and emergency communication access: Lessons learned since 9/11 and recommendations.* Deaf and Hard of Hearing Consumer Advocacy Network (DHHCAN) and Northern Virginia Resource Center for Deaf and Hard of Hearing Persons (NVRC).
http://tap.gallaudet.edu/Emergency/Nov05Conference/EmergencyReports/DHHCANEmergencyReport.pdf

D

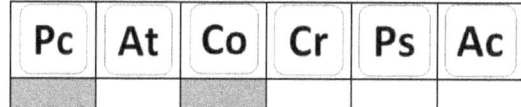

Pc	At	Co	Cr	Ps	Ac

The events of September 11, 2001 and its aftermath exposed weaknesses in emergency preparedness for building occupants, especially for Americans who are deaf, hard of hearing, late-deafened, and deaf blind. In emergencies, sirens, calls or yells from colleagues or emergency personnel, knocks on doors, phone calls, radios, and public address systems can be useless for these populations at times when emergency information is most important. Television captions, captioned news video on websites, and the availability of text messaging can't be depended on. This report provides recommendations to increase communications options and redundancy in an emergency, with emphasis on developing new communication technology and requiring more effective and wide-reaching dissemination. Little attention is provided here to the content of emergency messages. Relevant recommendations include:

- Television broadcasters should:
    - Provide accurate, real-time, and verbatim captioning of all news broadcasts, with contrasting background to ensure readability
    - Provide captioning at the same time that audio information begins
    - Identify new emergency information in some way
    - Locate sign language interpreters onscreen during broadcasts
- Televisions should be installed with a clearly-designated button at the front of all televisions and on any remote control to allow for instant caption activation.
- The FCC should increase enforcement of regulations requiring that emergency captions not be blocked, that emergency captions not block other captions, and that captions not block sign language interpretations.
- Televisions in public places should have captions turned on at all times.
- A captioned radio system should be developed to send emergency text messages to radio
- Text messaging systems should be interoperable to provide multiple communications channels in an emergency.
- Text devices should be built with backlit displays and keys, should permit the adjustment of font size, and should be compatible with Braille and larger print displays.
- All audio announcements should be broadcast with a simultaneous display of text (for Transportation and Transportation System information sources, although this recommendation could also apply to areas in buildings). Also ensure that these textual displays provide appropriate information, including the nature of the emergency and instructions on what to do.
- Internet providers should provide accurate, realtime, and verbatim captioning of all live streaming news and news updates.

More recommendations on this topic are presented in the online document.

Key recommendations:
- [127_Pc_1] [127_Co_1] Television broadcasters and internet providers should provide accurate, real-time, and verbatim captioning of all news broadcasts, with contrasting background to ensure readability and nothing blocking the message.
- [127_Pc_2] [127_Co_2] Captioning of news broadcasts should begin immediately as soon as audio information begins.
- [127_Pc_3] [127_Co_3] Important new emergency information should be identified in captions as well as in the news broadcast.
- [127_Pc_4] Sign language interpreters should be positioned onscreen during emergency broadcasts and not blocked by captions or other visual material.
- [127_Pc_5] Televisions in public places should have captions turned on at all times.
- [127_Pc_6] Text messaging systems should be interoperable in order to provide multiple communications channels in an emergency.
- [127_Pc_7] [127_Co_4] All audio announcements should be broadcast with a simultaneous display of text, which should provide appropriate information, including the nature of the emergency and instructions on what to do.

128. Subervi F. (2010, December). *An Achilles heel in emergency communications: The deplorable policies and practices pertaining to non English speaking populations* (Report). San Marcos, TX: Center for the Study of Latino Media & Markets School of Journalism & Mass Communication, Texas State University.

F

Pc | At | Co | Cr | Ps | Ac

This two-year project reviewed the current state of emergency communication policies and practices for alerting Spanish-speaking populations in Central Texas. The field research phase involved the collection of data on demographics, media, and government and media websites, and interviews with government emergency response officials, media representatives, and community leaders. The second phase was a forum at Texas State University with 80 participants from public and private sectors. The project was based on two assumptions: that the safety of a community in a crisis depends on well-informed citizens, and that the safety of an individual may be at risk if others in the vicinity are not well-informed.

The Federal Emergency Management Agency (FEMA) has recently adopted a digital protocol (the Common Alerting Protocol – CAP) for emergency warnings to replace the Emergency Alert System (EAS) currently in use. Although the CAP standard allows for Spanish-language messaging, its effectiveness requires support of this functionality by government agencies and the media. Major gaps and deficiencies exist in the capabilities of both government agencies and broadcast media to adequately communicate with Spanish speakers before and during an emergency.

Data collection: Demographic data about Hispanics in states, counties, and cities was easily obtained from U.S. Bureau of the Census records. A list of Spanish-language news and information programs was developed from internet sites and radio stations, when possible. Spanish-language versions of government emergency communications policies and procedures were difficult to obtain. Written emergency policies and procedures from Spanish-language broadcast media were nearly impossible to obtain.

Interviews: Government officials recognize the challenge of communicating with non-English speakers during an emergency. Although the need to reach this community is understood, bilingual employees are not always available to emergency managers when needed. Officials were not fully aware of the limited capabilities of local Spanish-language media in disseminating emergency information in Spanish. Spanish-language reporters do not take advantage of press conferences and other informational opportunities offered by emergency managers (in English). The corporate owners of Spanish-language radio stations often do not support local news operations, which cost money and are not required by FCC regulations. Barriers to emergency broadcasts on Spanish-language television stations include a lack of authorization to preempt syndicated network programming, in addition to limited local news staff and technical equipment for live coverage. Some Spanish-language television broadcasters accept the responsibility for notifying their community in an emergency situation; others view emergency communication in Spanish as the responsibility of the government. Leaders in Spanish-speaking communities agreed on the need for attention on emergency communication policies and practices from both government and media. English-speaking business and community leaders were unaware of the needs of the Spanish-speaking population but supportive of improvements. Problem areas with emergency communications with the general public include: coordination between state and federal agencies, the interface between government and media, the lack of regulations requiring state and local governments to provide warnings to broadcasters, and the lack of regulations requiring rebroadcast of EAS signals. The Common Alerting System (CAP) by itself will not deliver the comprehensive information needed by people in an emergency.

Forum: The Latinos a Salvo forum, sponsored by the Center for the Study of Latino Media and Markets at Texas State University, was attended on November 5, 2010 by 80 participants from public and private sectors, including state, county, and city emergency response and government agencies, the Red Cross, community leaders, academic researchers, Spanish-language television stations, and other broadcasting organizations. Presentations were made on journalist training for emergency coverage, government interaction with ethnic media, social and mobile technologies, and multilingual emergency alert announcements. These were followed by small group discussions divided by government emergency agencies/offices and first responder organizations; media; community leaders and insurance agencies; and social media.

Findings:
- Government agencies are in general not properly staffed to produce and disseminate messages in multiple languages. Not all information on websites is translated, and content in languages other than English may be difficult to find.
- Spanish-language radio stations rarely have the capability to provide emergency broadcasts, especially after daylight operating hours. Although the substantial population of native Spanish speakers in the Austin and San Antonio areas is served by several Spanish-language radio stations, news and information programs (and the staff to support them) were found to be rare. Access to emergency news and information is not assured, especially outside of metropolitan areas. Signal strength of Spanish-language radio stations may be reduced at night, and staff members are generally not available to cover nighttime emergency broadcasts.
- Few Spanish-language media websites contain local news and information. The capability to update the websites with information about developing emergency situations is lacking. Up-to-date lists of emergency contact persons are rare.
- Regulations on emergency message content for broadcast media are lacking (in Spanish or English).

Recommendations:
- A training program for government representatives dealing with ethnic media, including crisis communication plans.
- A training program in emergency communications for journalists.
- Attention for crisis situations affecting local communities with little or no access to broadcast/cable media news and emergency information.
- Adequate multilingual staffing, including volunteers, for governmental agencies to handle any emergency at any time.
- Regulations that require broadcasters to transmit emergency alerts in the operating language of the licensed station, at any time of day or night.
- Paid staff on call at broadcast stations to provide emergency broadcasts at any time.
- Arrangements for English-language broadcast stations to air multilingual messages if ethnic stations are knocked off the air.

- Adoption of a multilanguage emergency alert system being developed by New America Media, a national nonprofit organization for ethnic news organizations, for quick alerting capability.
- Multicultural and multilanguage course material to be included in the training of emergency and media professionals and students.
- Government agencies and first responder organizations saw a need to work more closely with their constituents, including using regulation to enforce non-English emergency communication, studying how crisis information is exchanged with the community, using bilingual preparedness materials and social media, developing trust, and partnering with organizations within non-English speaking communities.
- Spanish-language media representatives suggested a larger role for ethnic-oriented print media, development of disaster or emergency plans for every broadcast station and print medium, and partnerships with government and emergency response agencies. They saw a need for automated weather alert systems, Spanish-speaking spokespersons at government agencies, and notification on crime and terrorist threats. Because Spanish-speaking communities have limited computer and social media access, emergency materials should be disseminated in print.
- Community leaders recommended neighborhood alert systems to organize members of the community to reach isolated people and those distrustful of government-based alerts; neighborhood radio, siren, and text messaging systems; grassroots efforts; and business involvement.
- Strategies for social media include sending multiple language audio and/or text messages to cell phones with brief guidance on actions to be taken; requiring smart phone owners to opt-out (not opt-in) to emergency messaging systems; targeting women and children for dissemination in the home; engaging businesses and public relations companies; and using traditional media to send breaking news feeds through social media. People need tools, knowledge, and skills to act proactively. A cultural connection is necessary for prompt and efficient response. Community-based information hubs, training sessions, and crisis communication plans are needed.

Key findings:
- [128_Pc_1] Barriers to emergency broadcasts on Spanish-language television stations include a lack of authorization to preempt syndicated network programming, in addition to limited local news staff and technical equipment for live coverage.
- [128_Pc_2] [128_Co_1] [128_Cr_1] Government agencies are in general not properly staffed to produce and disseminate messages in multiple languages. Not all information on websites is translated, and content in languages other than English may be difficult to find.
- [128_Pc_3] Signal strength of Spanish-language radio stations may be reduced at night, and staff members are generally not available to cover nighttime emergency broadcasts. Spanish speakers outside of metropolitan areas are particularly affected.
- [128_Pc_4] [128_Ps_1] It would be useful for Spanish-language media websites to contain local news, information about developing emergency situations, and up-to-date lists of emergency contact persons.
- [128_Ps_2] Regulations on emergency message content for broadcast media are lacking.

Key recommendations:
- [128_Co_2] [128_Cr_2] A training program is recommended for government representatives dealing with ethnic media, including crisis communication plans.
- [128_Ps_3] A training program in emergency communications for journalists is recommended.

- [128_Pc_5] Attention is needed to planning for crisis situations affecting local communities with little or no access to broadcast/cable media news and emergency information.
- [128_Co_3] Adequate multilingual staffing, including volunteers, is needed for governmental agencies to handle any emergency at any time.
- [128_Pc_6] Regulations are recommended that require broadcasters to transmit emergency alerts in the operating language of the licensed station, at any time of day or night.
- [128_Pc_7] Paid staff should be on call at broadcast stations to provide emergency broadcasts at any time.
- [128_Pc_8] [128_Co_4] Arrangements should be made for English-language broadcast stations to air multilingual messages if ethnic stations are knocked off the air.
- [128_Pc_9] [128_At_1] [128_Co_5] A multilanguage emergency alert system is being developed by a national nonprofit organization for ethnic news organizations.
- [128_Co_6] [128_Cr_3] Multicultural and multilanguage course material should be included in the training of emergency and media professionals and students.
- [128_Co_7] [128_Cr_4] Government agencies and first responder organizations need to develop bilingual preparedness materials.
- [128_Cr_5] Government agencies and first responder organizations need to develop trust and partner with organizations within non-English speaking communities.
- [128_Pc_10] Because Spanish-speaking communities have limited computer and social media access, emergency materials should also be disseminated in print. The Spanish language print media would make a good partner.
- [128_Cr_6] [128_Ps_4] Community leaders recommended neighborhood alert systems to organize members of the community to reach isolated people and those distrustful of government-based alerts
- [128_Pc_11] [128_At_2] Neighborhood radio, siren, and text messaging systems were recommended by community leaders.
- [128_Pc_12] [128_Ps_5] [128_Ac_1] Strategies for social media include sending multiple language audio and/or text messages to cell phones with brief guidance on actions to be taken.
- [128_Ps_6] Women and children are good targets for disseminating emergency information to households.

| # **129**. Tang CH, Lin CY, Hsu YM. (2008). Exploratory research on reading cognition and escape-route planning using building evacuation plan diagrams. *Applied Ergonomics 39*(2), 209-217. |

E		Pc	At	Co	Cr	Ps	Ac
				▓			▓

This study examined how effective evacuation plan diagrams are at conveying meaning. The study looked at how long it takes for individuals to read the evacuation plan diagram (the time it takes to locate the exit) and how long it takes them to plan an emergency route (the time it takes to find their current location and exit). These were seen as key elements in measuring the explanatory power of these diagrams.

The evacuation plan diagram of the fourth floor of a department store in Taiwan was selected for the study. The time was measured from the point at which participants started to read the evacuation plan diagram to the point at which each participant finished planning his or her route. The time to plan an escape route was found to be 1.1 to two times longer than the time to read the diagram. Furthermore, lay people required two to three times more time to read the diagram and plan the escape route than those with an architectural background. Reducing the amount of time required to read and interpret the diagram could therefore make escape route planning significantly more efficient.

Evacuation plans are a form of architectural drawing, with symbolic meanings understood by professionals. For the layperson, the symbols may be less familiar. Additional interpretation, especially under stress, may increase the likelihood of misinterpretation of the intended plan leading to mistakes. The inclusion of certain characteristics would improve the design of evacuation plan diagrams, including: comprehensibility, ease of being conventionally understood, clarity, ability to provide directions and safety concepts. Additional recommendations for improving the effectiveness of the diagrams are to decrease the number of technical symbols and to use education and training to increase the general public's ability to interpret the diagrams. Metaphorical and abstract symbols should be eliminated.

Key findings:
- [129_Co_1] The number of technical symbols in evacuation plan diagrams should be reduced. Metaphorical and abstract symbols should be eliminated.
- [129_Co_2] [129_Ac_1] The general public's ability to interpret evacuation plan diagrams should be improved through education and training.
- [129_Co_3] Evacuation plan diagrams should have the following characteristics: comprehensibility, ease of being conventionally understood, clarity, ability to provide directions, and safety concepts.

130. Thomas IR, Bruck D. (2008). Strobe lights, pillow shakers and bed shakers as smoke alarm signals. *Fire Safety Science-Proceedings of the Ninth International Symposium,* Karlsruhe, Germany, 415-423.

Bruck D, Thomas I. (2007, June). *Waking effectiveness of alarms (auditory, visual and tactile) for adults who are hard of hearing* (Optimizing Fire Alarm Notification for High Risk Groups Research Project Report). Quincy, Massachusetts: The Fire Protection Research Foundation.

F		Pc	At	Co	Cr	Ps	Ac

The research in this study focused on technologies to wake up people that are hard of hearing and those that are intoxicated with alcohol. Seventy people participated in this experiment, of which 38 aged 18 to 77 years (16 males, 22 females) were hard of hearing and 32 young adults aged 18 to 26 years (15 males, 17 females) were alcohol impaired. The strobes were presented at an intensity of 177 cd, 210 cd, or 420 cd. A commercial pillow shaker and a commercial bed shaker were modified to vibrate at any of five intensity levels in a T-3 pulsing pattern. Two auditory signals, a 520 Hz square wave and a 3100 Hz pure tone, were tested. The participants went to sleep in a sleeping facility, where a sleep technician monitored their EEG output. Once the participants were in deep sleep, the different signals were presented in 30 second intervals. Each signal continued for three minutes or until the participant woke up and pressed a bedside button three times.

The strobe lights at the 177 cd level (compliant with the 2002 NFPA 72 standard) only woke a quarter of the deaf and intoxicated participants. The 520 square wave auditory signal was the most effective of all signals in waking people up. The results showed that strobe lights alone are ineffective in waking up people; 42 % of the hard of hearing and 32% of intoxicated participants did not wake up when using the highest intensity. For the hard of hearing group, bed shakers were much more effective than strobe lights, with only 3 % of people sleeping through pillow shakers and 11 % sleeping through bed shakers at the highest intensity level (higher than that originally set for these products). For the alcohol impaired, bed shakers were not as effective as had previously been found for sober adults. At the highest intensity level, 32 % slept though the pillow shaker and 25 % slept through the bed shaker, compared to reported successful awakenings of 70 % to 100 % for sober adults.

Key findings:
- [130_Pc_1] [130_At_1] Strobe lights alone are not able to attract sufficient attention to overcome sleep.
- [130_Pc_2] [130_At_2] Pillow and bed shakers at high intensity are much more effective than strobe lights at waking individuals who are hard of hearing.
- [130_Pc_3] [130_At_3] Although they are effective at waking sober adults, pillow and bed shakers are not effective at waking those impaired by alcohol.
- [130_Pc_4] [130_At_4] A 520 square wave auditory signal at 75 dBA or above is more effective than either strobe lights or bed shakers at waking both hard of hearing and alcohol impaired individuals.

131. Thorley P, Hellier E, Edworthy J. (2001). Habituation effects in visual warnings. In MA Hanson (Ed.) *Contemporary Ergonomics 2001*, Taylor and Francis, London, UK, 223-228.

E		Pc	At	Co	Cr	Ps	Ac

There has been relatively little research regarding the effectiveness of warnings upon a target population over a period of time. Thorley, Hellier and Edworthy looked at the effects of the presence of a warning sign on behavior over time. The study took place in a lecture room where 220 undergraduates were observed entering and exiting the room. Three signs (A, B and C) were created for the experiment to show different levels of hazard. Signs A and B were medium hazard signs, while sign C was a high hazard sign. Signs A and B had the signal words STOP (red font) or CAUTION (yellow font), listed the hazard (DEFECTIVE DOORS), and instructions to avoid the hazard (PLEASE USE OTHER DOORS in black font). Sign C used the signal word DANGER (red font) and had the same hazard warning instructions as signs A and B. Over a period of eight sessions, door behavior was observed into and out of the lecture room. Initially, the number of people using the doors without the presence of warning signs was observed in order to establish a baseline. Then, they put sign A on a door inside and outside the lecture theater and recorded the people who still tried to use the defective door. Participant's behavior was considered either compliant or non-compliant. After five sessions, Sign A replaced Sign B and the later Sign C replaced Sign B.

The results showed that compliant behavior decreased as time went on with repeated exposure to a warning. There was no significant difference when replacing Sign B with Sign A. When Sign B was replaced with sign C, compliant behavior increased. The results follow the characteristics of habituation (Thompson and Spencer, 1966) where repeated exposure to a stimulus decreases the response strength.

Key findings:
- [131_Ac_1] People are more likely to comply with a warning sign when the word "DANGER" is used compared to "STOP" or "CAUTION."

132. Timmons RP. (2009). *Sensory overload as a factor in crisis decision-making and communications by emergency first responders.* PhD dissertation, The University of Texas at Dallas.
Timmons RP. (2007, February). Interoperability: Stop Blaming the Radio. *Homeland Security Affairs, III*(1), 1-17.

F

Pc	At	Co	Cr	Ps	Ac

In large-scale disasters, many first responders attempt to use the available communications channels at the same time. The intensity of the environment, the large number of users, and the patching together of systems from multiple organizations leads to many failures to successfully transmit a message to the intended party. Review of transcripts after the incident reveals that the chatter is nearly constant, and that a portion of the content is unintelligible. A modification of first responder radio use habits at large-scale emergency scenes can help to overcome shortcomings of the technology.

Adoption of new procedures to deal with large-scale disasters will be a challenge due to psychological and physiological factors, including the tendency of people to fall back on old habits when under stress. Psychological factors that affect the communications of first responders at large-scale disasters include: sensory overload from the large amount of information at the scene; cognitive bias, in which information is filtered through preconceptions; speech center deficit, reflecting the effect of hormones on the character of the human voice; and suppressed emotions, which are expressed through other stress reactions of the human body. Communications are also hampered by personal protective equipment (PPE).

In a study of training exercises, 4.9 % of radio messages needed to be repeated. Another 11.9 % were unacknowledged (more than half of these were directed to the incident commander), and 2.6 % were judged as a questionable use of the emergency radio. In total, communications could be improved for almost one-fifth of messages.

Review of radio transmissions during critical incidents found a pattern of use. Messages prefaced by a keyword such as 'urgent', 'priority message', or 'emergency traffic' received greater attention than those delivered in an urgent tone. Repetitious and superfluous requests for certain information such as staging areas and command post locations could be reduced by broadcasting a summary of this information before it was requested.

The dissertation work of the author was directed at understanding the causes for missed radio messages by first responders during an emergency response. The main interest was on the effects of sensory overload during a crisis on the ability to receive or make the best use of messages. This was investigated through case studies of actual fire department incidents and data collected during realistic fireground training exercises.

Most radio communications failures (missed or unanswered messages and messages requiring repeating) were found to be due to inattention caused by the distraction of other conversations, confusion over radio channel selection, unintelligible messages, and background noise. No apparent reason for failure was found in 38 % of cases. Physiological or psychological stress provides a possible explanation for these missed messages. Evidence of sensory overload in the incident command environment included ignoring one message when another, more urgent-sounding message followed; failing to complete interrupted messages; and forgetting message content in the face of distractions. Emotional stress can cause a physiological hearing deficit, preventing some messages from being heard. The narrowing of attention in challenging environments can result in inattentional deafness, in which unexpected features are simply not noticed.

Message senders can transpose numbers or words without realizing it when their attention is fully focused elsewhere, as in one case in which a firefighter reported he was "putting fire on the water."

The messages that were responded to with the most accuracy were stated clearly and avoided multiple interpretations. Overly wordy messages could cause confusion. Calm, controlled voices were most easily understood and promoted calmness in others at the scene. Message redundancy and ensuring that multiple people (e.g., the incident commander and an aide) are listening to the messages can help overcome sensory overload issues. Training and experience promotes optimal radio use and teaches first responders about their own reactions to stress.

Recommended changes to procedures include prioritization of radio transmissions (life safety first), standardized nomenclature, alternatives to radio such as face-to-face communications, the use of staging areas before deployment, and emphasizing the quality of radio messages over the quantity. A crisis communications plan can be developed, which could be triggered by a prediction of communications overloading using factors such as the number of victims, the area involved, and the type of attack.

Key findings:
- [132_Co_1] Psychological factors that affect the communications of first responders at large-scale disasters include sensory overload, cognitive bias, speech center deficit, and suppressed emotions.
- [132_Pc_1] [132_At_1] Emotional stress can cause a physiological hearing deficit, preventing some messages from being heard.
- [132_Pc_2] [132_At_2] The narrowing of attention in challenging environments can result in inattentional deafness, in which unexpected features are simply not noticed.
- [132_At_3] [132_Co_2] Message senders can transpose numbers or words without realizing it when their attention is focused elsewhere and not on the message itself.
- [132_Co_3] The messages that were responded to with the most accuracy were stated clearly and avoided multiple interpretations.
- [132_Co_4] Overly wordy messages could cause confusion.
- [132_Co_5] Calm, controlled voices were most easily understood and promoted calmness in others at a disaster scene.

Key recommendations:
- [132_Pc_3] Message redundancy and ensuring that multiple people are listening can help overcome sensory overload issues.
- [132_Co_6] [132_Ps_1] Training and experience promotes optimal radio use and teaches first responders about their own reactions to stress.

133. UL 1971. (2008). Standard for Signaling Devices for the Hearing Impaired.

S					Pc	At	Co	Cr	Ps	Ac

The UL 1971 standard addresses visual signaling devices that communicate using illumination, including strobes and other flashing devices. The standard covers the construction of the device enclosure; cover; ventilation openings; corrosion protection; insulating materials; mounting parts; operating mechanisms; and wiring, cables, connections, and circuit boards. The standard also provides requirements for the performance of the system by specifying a series of tests to measure operating currents at a range of voltage ratings (both maximum and minimum ratings). Additionally, installation and operating instructions are included. However, the UL 1971 standard does not specify the means by which the device should be used to disseminate information; for example, the ways in which the lights must flash to provide a specific message to occupants.

Key findings:
- [133_Co_1] The UL 1971 Standard for Signaling Devices for the Hearing Impaired specifies the mechanical and electrical properties of these devices but does not specify how the device is to be used; for example, the ways in which the lights must flash to provide a specific message to occupants.

 # 134. Ullman BR, Dudek CL, Trout ND, Schoeneman SK. (2005 October). AMBER alert, disaster response and evacuation, planned special events, adverse weather and environmental conditions, and other messages for display on dynamic message signs (Texas Department of Transportation Report No. FHWA/TX-06/0-4023-4). Texas Transportation Institute, College Station, TX.

Pc	**At**	**Co**	**Cr**	**Ps**	**Ac**

Occasions for which messages may be displayed for drivers on dynamic message signs (DMS) include America's Missing: Broadcast Emergency Response (AMBER) alerts, disaster response and evacuation (e.g., flooding, hurricanes, or terrorist attacks), sports events and other planned special events, and adverse weather and environmental conditions. Some issues have been noted with current DMS practice, including problematic driving behavior and complaints from drivers. This report describes recommendations from focus groups and human factors laboratory studies for improving the effectiveness of DMS messages.

DMS messages must be carefully designed. Information must be communicated in a short time period to people who are performing a cognitively engaging task. Consistency over time and from region to region is important for driver comprehension, traffic flow, and credibility. Comprehension is improved when DMS messages are brief, use simple words, use well-understood abbreviations, standardize the word order, and standardize the order of message lines. Message elements in an unexpected location or order can result in driver confusion and increase the time to read the message.

Important design factors for DMS include message load, message length, and message familiarity. The message load is the number of informational units in a message, and the message length is the total number of words or characters. DMS messages may be divided into multiple frames that are displayed sequentially. Previous studies have found that the average driver can process no more than eight words of four to eight characters each when driving at high speeds. For drivers to be able to recall all information when making a decision, there should be no more than two units on a single line, and no more than three units on a single frame. The full message should be no longer than four units of information long, and four message lines should be divided into two frames of two lines each.

The most efficient way to specify the location of incidents, numbers, and other information was investigated using focus groups with seven to 10 participants each (54 total) and human factors laboratory studies with a total of 192 participants in six Texas cities. The most effective words and phrases were identified for describing the type of incident, its location, and what action to take. Letters were easier to remember than numbers in license plate numbers displayed on the DMS, and a radio station to tune to was easier to remember than a license plate. Participants listed, rated, and ranked information needs for different types of incidents. In the case of a terrorist attack, people identified what to do, evacuation routes, the areas affected, safe areas, and the type of attack as critical information to display on the DMS. Some participants were shown a message on a computer screen for eight seconds (the amount of time a driver usually has to read a sign) and were asked to recall the message. As expected, the comprehension increased as the viewing time increased.

Key findings:

- [134_Co_1] [134_Ac_1] Consistency in the design of DMS highway messages, over time and from region to region, is important for driver comprehension and traffic flow.
- [134_Cr_1] Consistency in the design of DMS highway messages is important to maintain the credibility of the agency responsible for the signs.
- [134_Co_2] Comprehension is improved when DMS messages are brief, use simple words and well-understood abbreviations, standardize the word order, and standardize the order of message lines.
- [134_Co_3] Message elements in an unexpected location or order can result in driver confusion and increase the time to read the message.
- [134_Co_4] The average driver can process no more than eight words of four to eight characters each when driving at high speeds. For drivers to be able to recall all information when making a decision, there should be no more than two informational units on a single line, no more than three units on a single frame, and no more than four units in the full message.
- [134_Co_5] There should be no more than four DMS message lines total, which should be divided into two frames of two lines each.
- [134_Ps_1] Testing identified the most effective words and phrases for describing the type of incident, its location, and what action to take.
- [134_At_1] Letters were easier to remember than numbers in license plate numbers displayed on the DMS.
- [134_At_2] A radio station to tune to was easier to remember than a license plate.
- [134_Ps_2] In the case of a terrorist attack, people identified what to do, evacuation routes, the areas affected, safe areas, and the type of attack as critical information to display on the DMS.
- [134_Co_6] Message comprehension increases as viewing time increases.

 # 135. Ullman BR, Trout ND, Dudek CL. (2008, March). *Use of graphics and symbols on dynamic message signs: Technical report* (Texas Department of Transportation Report No. FHWA/TX-08/0-5256-1). Texas Transportation Institute, College Station, TX.

T		Pc	At	Co	Cr	Ps	Ac

Previously, dynamic message signs (DMS) could not use symbols because the technology to display them on these roadway signs was not available. Now that technology has advanced, the use of color and symbols in DMS needs to be reevaluated. The purposes of this project were to evaluate how easily drivers can comprehend graphic information, quantify comprehension of graphic elements, and assess the information load placed on drivers by symbols and graphics.

Several issues must be kept in mind in considering the use of graphic elements in DMS. Maintaining the credibility of these signs requires drivers to be confident that they can obtain from them the information they need. DMS signs are particularly important for communicating unexpected circumstances, such as traffic incidents, roadwork, and special events, that all require decisions to be made by drivers. Useful information includes the type of problem, its location, the action to take, lane closures, consequences of the problem, the audience for the message, and the reason for taking the action. Drivers must be able to read and comprehend a message in only a few seconds (eight seconds or less) at highway speeds, possibly while engaged in a complex driving task. A decision may need to be made quickly.

Well-designed symbols can be advantageous because they are quicker to read and comprehend than words, can be recognized from farther away, and are more recognizable under poor environmental conditions such as darkness and weather. They are also more easily understood by drivers who have trouble comprehending word messages, such as non-English speakers. Recent changes include hardware technologies, such as full-matrix, that make it possible to display graphic features on DMS, and the increasing use of symbols on static traffic signs.

A recent study on the comprehension of symbols on static regulatory and warning signs showed that drivers understand the majority of these signs but not all of them. It is therefore important for the DMS designers to pick appropriate symbols that are easily understood. In Europe, symbols displayed on DMS, known as pictograms, are widely used to inform drivers of travel situations. Studies of Europeans drivers have revealed that while many pictograms (including those for roadwork, congestion, queue, and slippery road conditions) are well-understood, others (including fog, restricted lanes, and crash) are not. A study in the U.S. indicated that correct interpretation of pictograms deteriorated rapidly with longer distances from the sign and with darkness, and that the symbols were often associated with potential danger.

A graphical route information panel (GRIP) is a sign that graphically portrays a route and uses color to highlight changes in conditions (e.g., congestion) along the route. Studies in Texas in the early 1970's indicated that drivers prefer color to distinguish between normal and abnormal conditions. GRIP signs are used with success in Japan and Australia and are being tried in the Netherlands. Advantages include the ability to provide more information, more complex information, the location of an incident, relevant information for drivers with different destinations, and information for those who don't speak English. However, drivers who are uncomfortable with maps are not as capable at using GRIPs.

Focus groups were established in four Texas cities (with a total of 38 frequent drivers) and in a fifth city with 11 long-haul truckers (including three non-English speakers). They reviewed graphics for five travel scenarios. The color red on an interchange map, intended to communicate slow or stopped traffic, was misinterpreted by some as indicating closure. Information on delay time, cross streets at the start of congestion, and alternate routes was suggested. Ramps and cross streets were considered helpful information by some and too much detail by others. A three dimensional perspective of the area did not improve comprehension. An accident symbol showing two cars was better understood than a single car overturned.

A human factors laboratory study was performed on potential DMS graphics to determine the most promising candidates. The topics for study were Graphical Interchange Information, GRIP displays (showing regional maps), HOV lane identification, adjacent toll lanes, and truck routes. Candidate signs were displayed on a computer monitor for 160 participants to measure comprehension times and accuracy. Although GRIP displays were able to communicate a large amount of information, study results showed that viewing times were significantly higher than for text or other types of graphics. Research on possible overloading of driver comprehension capability was recommended. Graphical interchange information was able to communicate congestion on specific lanes and provided faster and more accurate comprehension for non-native English speakers. The use of graphics makes it easier to communicate unusual scenarios on a roadway: i.e., high occupancy lanes or cash-only toll lanes. Without graphics, scenarios like these would require many words to explain using text. Congestion in lanes can be communicated using red for stop-and-go conditions, yellow for slow, and green for normal. Arrows on a traffic lane indicate that the lane is open beyond the point of congestion. Display of delay times was not found to be necessary.

Key findings:
- [135_Cr_1] Maintaining the credibility of DMS roadway signs requires drivers to be confident that they can obtain from them the information they need.
- [135_Ps_1] DMS signs are particularly important for communicating unexpected circumstances, such as traffic incidents, roadwork, and special events, that require decisions to be made by drivers.
- [135_Ps_2] Graphics on DMS roadway signs may be able to communicate more complex information regarding the type of problem, its location, the action to take, lane closures, consequences of the problem, the audience for the message, and the reason for taking the action.
- [135_Co_1] Evaluation of the use of graphics on DMS roadway signs must take into account that drivers must be able to read and comprehend a message in only a few seconds (eight seconds or less) at highway speeds, possibly while engaged in a complex driving task.
- [135_Pc_1] [135_Co_2] Well-designed symbols can be advantageous because they are quicker to read and comprehend than words, can be recognized from farther away, and are more recognizable under poor environmental conditions such as darkness and weather.
- [135_Co_3] New hardware technologies and the increasing use of symbols on static traffic signs encourage the use of graphic features on DMS roadway signs.
- [135_Co_4] A study of pictograms on DMS roadway signs showed that correct interpretation deteriorated rapidly with distance and in darkness,
- [135_Ps_3] Pictogram symbols on DMS roadway signs may be (rightly or wrongly) associated with potential danger.

- [135_Co_5] A three dimensional map perspective did not improve comprehension of a graphical DMS roadway sign.
- [135_Co_6] An accident symbol showing two cars was better understood than a single car overturned.
- [135_Co_7] Although a graphical route information panel (GRIP) was able to communicate a large amount of information, study results showed that viewing times were significantly higher than for text or other types of graphics.
- [135_Ps_4] Graphical interchange information was able to communicate congestion on specific lanes.
- [135_Co_8] Graphical information on DMS roadway signs provided faster and more accurate comprehension for non-native English speakers.
- [135_Co_9] The use of graphics makes it easier to communicate unusual scenarios on a roadway, such as high occupancy lanes or cash-only toll lanes, that would require many words to explain using text.
- [135_Co_10] Congestion in lanes can be communicated using red for stop-and-go conditions, yellow for slow, and green for normal.
- [135_Co_11] Arrows on a traffic lane indicate that the lane is open beyond the point of congestion.
- [135_Co_12] The use of graphics to display traffic information requires testing on drivers to determine the optimal amount of information that can be displayed without overloading the driver.

136. Ullman GL, Dudek CL, Ullman BR. (2005, September). *Development of a field guide for portable changeable message sign use in work zones* (Texas Department of Transportation Report No. FHWA/TX-06/0-4748-2). Texas Transportation Institute, College Station, TX.

T

Pc At Co Cr Ps Ac

Portable Changeable Message Signs (PCMS) have been found to attract more attention than static signs for controlling traffic in work zones. They are also far more versatile, since they can be programmed to display a range of messages. However, improper use can destroy their credibility, create confusion, and promote improper driving behavior. Research-based guidelines for PCMS message design using human factors and traffic engineering principles have been developed, but they are not always followed by personnel in the field.

The objectives of this research project were to identify and prioritize areas still in need of research for PCMS use in work zones, and to develop guidance for field personnel. Key research gaps were published in a separate document (Ullman et al. 2005, review #138 in this annotated bibliography). This report presents a summary of current guidance on PCMS design and use that has been condensed to a single sheet for use by field personnel.

At the national level, basic guidance on PCMS messaging is provided by the Manual on Uniform Traffic Control Devices (MUTCD) by the Federal Highway Administration (FHWA). The message should be simple, brief, and clear. To maintain credibility of the PCMS communication mode, messages should not be vague or overly simple. The PCMS format should not have the appearance of an advertising display and should display only traffic-related information. Messages are limited to one or phases, each with up to three lines with no more than eight characters per line. Five general recommendations for message content are:
1. Present one thought per phase;
2. For messages that can be displayed on a single phase, present the problem on top, the location on the second line, and the recommended action on the bottom;
3. Make the message as brief as possible;
4. Use an additional PCMS for messages longer than two phases; and
5. Use easily understood abbreviations.

Other guidelines in this manual include positioning of the PCMS for sign visibility of half a mile under ideal conditions and for legibility at 650 feet, separation of multiple PCMS by at least 1000 feet, and adjustable display rates (three seconds per phase minimum) to allow each driver to read the message twice. Messages should be steady while they are displayed, not scrolled or flashed. Recommendations on positioning the PCMS upstream of decision points are found in The Portable Changeable Message Sign Handbook PCMS, another FHWA document. Specific wordings for thirty traffic scenarios were established in the mid-1990s.

In Texas, the Dynamic Message Sign Message Design and Display Manual is a comprehensive (nearly 500 pages) guidance document developed by the Texas Department of Transportation (TxDOT) for both full-size DMS and PCMS. This manual includes a highly specific 48-step process for designing messages based on the concept of a "base message set." The base message consists of seven elements: the roadwork descriptor, the location, lanes closed, the effect on travel, the action to take, the audience, and the reason to take the action. Since all of this information is unlikely to fit on the DMS or PCMS, methods are provided for condensing or prioritizing the message. Exact wordings of a variety of messages are provided. More PCMS guidance is provided in the TxDOT construction standard plans, including that the words "DANGER" or "CAUTION" should not be used in PCMS for road work (although there are exceptions). Other states have also developed guidance documents, some of which are described here.

The report concludes that available guidance documents are fairly generic, require some degree of training in human factors or traffic engineering, and are not organized or formatted for quick retrieval of information. Field personnel, who need to make quick decisions on message content in the field, do not necessarily have the required background nor the time to dig for the recommendations to fit a specific situation. A single page field guide with the most critical information has been developed.

Key recommendations:
- [136_At_1] Portable Changeable Message Signs (PCMS) are versatile, since they attract attention and can be programmed to display a wide range of messages.
- [136_Cr_1] [136_Ac_1] Improper use of highway signage (PCMS) can destroy their credibility, create confusion, and promote improper driving behavior.
- [136_Cr_2] Research-based guidelines for message design are not helpful unless they are in a form that can be used by field personnel that need to make quick decisions.
- [136_Co_1] Safety messages (PCMS) should be simple, brief, and clear. Messages are limited to one or phases, each with up to three lines with no more than eight characters per line.
- [136_Cr_3] To maintain credibility of the PCMS communication mode, messages should not be vague or overly simple.
- [136_Cr_4] [136_Ps_1] The PCMS format should not have the appearance of an advertising display and should display only traffic-related information.
- [136_Co_2] Present one thought per phase for PCMS roadway signs.
- [136_Co_3] [136_Ps_2] For messages that can be displayed on a single phase of a PCMS roadway sign, present the problem on top, the location on the second line, and the recommended action on the bottom
- [136_Co_4] If a message can't be displayed in two phases, additional PCMS should be used.
- [136_Pc_1] Positioning considerations for PCMS roadway signs include the distances needed for sign visibility, for legibility, and for separation of multiple PCMS.
- [136_Cr_5] [136_Ps_3] [136_Ac_2] Roadway signs need to be positioned far enough upstream of the decision point to provide adequate warning but close enough that the message is relevant.
- [136_At_2] Adjustable display rates (three seconds per phase minimum) are needed to allow each driver to read a PCMS message twice.
- [136_Ps_4] The base message for roadway signs consists of seven elements: the type of roadwork, the location, lanes closed, the effect on travel, the action to take, the audience, and the reason to take the action. If not all information fits on the sign, methods are provided for condensing or prioritizing the message.
- [136_Cr_6] [136_Ps_5] The words "DANGER" and "CAUTION" should not be used in PCMS for road work (although there are exceptions).

137. Ullman GL, Ullman BR, Dudek CL, Trout ND. (2004, July). *Legibility distances of smaller character light-emitting diode (LED) dynamic message signs for arterial roadways* (City of Dallas Transportation Management Systems Project No. 404940). Texas Transportation Institute, College Station, TX.

T		Pc	At	Co	Cr	Ps	Ac

Previous human factors research on DMS has resulted in design principles that define the maximum amount of information to present, the most important information desired by drivers, the ordering of message elements, and appropriate abbreviations, among other design factors. This assumes that drivers have an adequate amount of time to read the message on the DMS.

The legibility of any sign depends on the height of the characters and the distance of the reader from the sign. The maximum distance at which a dynamic message sign (DMS) is legible translates into the amount of time available for a driver to read the sign while traveling on the roadway. Previous research found that drivers require two seconds to read each unit of information on a DMS and can process no more than four units per message at high speed (five units is acceptable at low speed). The required DMS viewing time is therefore up to eight seconds (or 10 seconds at low speed). If less time is available, drivers may slow down, focus less attention from their driving, and contribute to traffic problems. Knowledge of the relationship between character height on a DMS and legibility distance (which differs under daytime and nighttime conditions and is not necessarily proportional) will enable the generation of a chart that relates character height and driving speed to the number of units that drivers can be expected to read and comprehend.

The standard size of characters on a DMS is 18 inches, which was established based on driving speeds for a high-speed roadway. The legibility distance of drivers from a DMS using 18 inch characters is about 800 feet during the day and 600 feet at night, based on the highway industry standard of accommodating 85 % of drivers. The purpose of this work was to investigate the legibility of smaller characters that might be used on lower speed (arterial) roadways, to recommend an appropriate character height for this application, and to develop recommendations for reducing the length of typical DMS messages in order to reduce reading time and legibility distance in general.

The first part of this study investigated the legibility distances for DMS displaying nine inch and 10.6 inch characters based on light emitting diodes (LEDs) under daytime and nighttime conditions. Two portable DMS, displaying words with nine inch and 10.6 inch characters respectively, were set up in a parking lot in Dallas, Texas. Each character was created by a 5 x 7 pixel matrix that had been shown to provide maximum legibility distances for larger character sizes. The 60 participants in the study were asked to drive a designated course at forty miles per hour, half of them during the day and half at night. Each DMS in the pathway displayed a three-letter word. When participants were able to identify the word, they were asked to say it out loud. If they did not correctly identify the word, they were asked to try again until they got it right. Participants passed each of the two DMS three times each, with different three-letter words displayed each time. They then finished by driving toward a four unit display on one of the DMS, on which questions were asked regarding the information content.

The results showed that the 10.6 inch characters could be seen about 100 feet sooner than the nine inch characters. A strong linear relationship was found between character height and legibility distance, although it was not directly proportional – a 50 % change in character height does not result in a 50 % change in legibility distance. Defining the legibility index as the legibility distance divided by the character height, the legibility index in daylight was 45 feet per inch of character height for 18 inch characters and only 25 feet per inch of character height for nine inch characters. Drivers traveling above 30 mph are able to read no more than two units of information from nine inch characters on DMS during the day. At night, drivers can read no more than one unit for nine inch characters, even at 30 mph, and no more than two units for 10.6 inch characters. Since most messages communicate three units or more, using nine inch and 10.6 inch characters are not recommended for arterial roads. If cars are traveling under 40 mph, DMS with 12 inch characters can be used to display up to four units of information during the day, but only up to three units at night,

Neither character height succeeded in achieving an 85 % correct response on the questions asked about the comprehension of the four unit sign. Comprehension results at nighttime (33 % correct for nine inch characters and 40 % correct for 10 inch characters) were much worse than those in the daytime (53 % correct for nine inch characters and 73 % correct for 10 inch characters).

The display time for each of two phases delivering four unit messages should be proportional to the number of units in each phase. If there are two units of information on each phase, the display time for each phase should be equal, and if one phase has two units and the other has one unit, the display phase should be twice as long for the first phase than for the second.

After the experiment, the researchers recommend that nine inch and 10.6 inch characters not be used for DMS and they should continue using the 12 inch characters.

Key findings:
- [137_Co_1] Drivers require two seconds to read each unit of information on a DMS and can process no more than four units per message at highway speeds or five units at low speed. The required DMS viewing time is therefore up to eight seconds or 10 seconds, respectively.
- [137_Ac_1] If less time than two seconds per informational unit is available for reading a DMS, drivers may slow down, focus less attention from their driving, and contribute to traffic problems.
- [137_Pc_1] The legibility distance of drivers from a DMS using 18 inch characters is about 800 feet during the day and 600 feet at night, based on the highway industry standard of accommodating 85 % of drivers.
- [137_Pc_2] At night, drivers can read no more than one unit for nine inch characters, even at 30 mph, and no more than two units for 10.6 inch characters. Since most messages communicate three units or more, using nine inch and 10.6 inch characters are not recommended for arterial roads.
- [137_Co_2] If cars are traveling under 40 mph, DMS with 12 inch characters can be used to display up to four units of information during the day, but only up to three units at night,

138. Ullman GL, Ullman BR, Dudek CL, Williams A, Pesti G. (2005, July). *Advanced notification messages and use of sequential portable changeable message signs in work zones* (Texas Department of Transportation Report No. FHWA/TX-05/0-4748-1). Texas Transportation Institute, College Station, TX.

Pc	At	Co	Cr	Ps	Ac

Portable changeable message signs (PCMS) are commonly used in Texas for work zones to control traffic, since they can provide more information than a static sign. Improper usage of PCMS can ruin the credibility of the signs or confuse motorists. To be effective, they must provide an important message that can be read very quickly.

The researchers performed a case study in six Texas districts in which the current usage of PCMS was compared to human factors principles, including simplicity of words, brevity, standardized word order and message line order, and understandable abbreviations. The message on many of the PCMS (40 %) continued across more than two phases. This exceeds guideline recommendations based on previous research. Information units were often split improperly across phases, display of calendar dates was inconsistent, and inappropriate abbreviations were used. Previous research has determined, for example, that the term ALT is not an acceptable abbreviation for alternate.

Motorist comprehension of dates was tested through a laptop laboratory study. Drivers were better able to comprehend dates when the month was given as an abbreviation (AUG 25) rather than in numerical form (8/25). Regardless of the format, however, only about two-thirds to three-fourths of drivers were able to correctly interpret whether the PCMS message was applicable to them on their current trip or a later one.

In order to convey longer messages, the use of two PCMS in sequence has been recommended. A driver simulator study tested whether drivers are able to piece together two PCMS messages displayed sequentially along a road into a single message, in comparison with the same information displayed on a single large dynamic message sign (DMS) in one or two phases. Significantly lower comprehension and recall rates were found for sequential PCMS displaying five units of information than for a large DMS. Neither PCMS nor DMS were able to convey five units of information with comprehension rates that were acceptable for highway applications (at least 85 %). For four units of information, comprehension rates for two sequential PCMS were comparable to those for a large single-phase DMS and acceptable for highway use.

Key findings:
- [138_Co_1] Driver comprehension of dates on work zone PCMS was improved by giving the month as an abbreviation (AUG 25) rather than in numerical form (8/25).
- [138_Ps_1] Regardless of the format for the date, only about two-thirds to three-quarters of drivers were able to correctly interpret whether the PCMS message is applicable to them on their current trip or a later one.
- [138_Co_2] For four units of information delivered on work zone message boards, comprehension rates for two sequential PCMS were comparable to those for a large single-phase DMS and acceptable for highway use.

- [138_Co_3] Significantly lower comprehension and recall rates were found for sequential PCMS displaying five units of information compared with a large DMS.
- [138_Co_4] Neither portable changeable message signs (PCMS) nor dynamic message signs (DMS) were able to convey five units of information with comprehension rates that are acceptable for highway applications.

139. U.S. Department of Health and Human Services' Office of Minority Health. (2011, February). *Guidance for integrating culturally diverse communities into planning for and responding to emergencies: A toolkit.* Recommendations of the National Consensus Panel on Emergency Preparedness and Cultural Diversity. Washington, DC: DHHS/OMH (Author).

C		Pc	At	Co	Cr	Ps	Ac

As evidenced by Hurricane Katrina and the H1N1 Pandemic Influenza of 2009-2010, racially and ethnically diverse communities remain vulnerable to disproportionate losses of life and property in disasters. Federal, state, and community organizations are working to improve preparedness and response capability to handle these populations. The National Consensus Panel on Emergency Preparedness and Cultural Diversity was created by the Office of Minority Health (OMH) of the U.S. Department of Health and Human Services (DHHS) in 2007, and consists of experts from agencies at all levels of government, emergency management organizations, health-related professional organizations, and racial/ethnic, immigrant, and limited English proficiency (LEP) communities. This document provides guidance from the National Consensus Panel on engaging and integrating diverse communities into emergency planning.

The goals of the toolkit are to highlight the importance of including communities of diverse race, ethnicity, culture, and language in emergency programs and planning, to incorporate state-of-the-art research and practices in efforts to include these populations, and to provide information and initiative areas for informing program and policy decisions. The information in the toolkit complements those available from other resources on emergency planning. Many of these resources are provided in the document as references and internet links.

Effective emergency preparedness and response depends on the engagement of diverse communities. Eight guiding principles for public health and emergency management professionals begin with the identification of at-risk populations. Trust with the members of these populations can be built in partnership with community representatives, who can provide insight into the design of risk communication strategies. Different communities obtain information from different sources, including social networks, ethnic media, and places of worship. The level of trust in various messengers and communication channels and the degree to which information is sought from these sources must be assessed. The specific vulnerability of each community and the availability of resources to carry out various types of emergency response, such as evacuation, isolation, and quarantine, must be understood, along with the attitudes of community members to these practices. The design, execution, and evaluation of drills and exercises must include these diverse populations. Information, resources, and actions must be coordinated among all organizations and diverse communities in order to maximize compliance with preparedness practices. Building and maintaining trust is of critical importance. Funding streams to promote preparedness of members of diverse communities must be identified and sustained.

Effective risk communication requires the engagement of communication representatives in designing and evaluating educational materials, emergency messages, and dissemination practices, including language resources and accessibility. Messages must be culturally appropriate and accurately translated. Resources to assist with this process are listed in the toolkit. Trusted members of the community and channels of communication (formal and informal) are critical to disseminate and reinforce the messages, making sure that messages are received, understood, and followed. Multiple modes of communication are required. A guide is available for preparing a database of trusted sources, interpreters, and effective communication modes.

Key findings:
- [139_Ac_1] Racially and ethnically diverse communities remain vulnerable to disproportionate losses of life and property in disasters.

Key recommendations:
- [139_Ac_2] Effective emergency preparedness and response depends on the engagement of diverse communities.
- [139_Pc_1] Emergency management professionals need to identify at-risk populations.
- [139_Cr_1] Partnerships are needed with community representatives in order to build trust and to obtain insight into the design of risk communication strategies and other emergency preparedness tools.
- [139_Pc_2] [139_Co_1] [139_Cr_2] [139_Ps_1] Different communities obtain information from different sources, including social networks, ethnic media, and places of worship.
- [139_Cr_3] [139_Ps_2] The level of trust in various messengers and communication channels and the degree to which information is sought from these sources must be assessed for each diverse community.
- [139_Ac_3] The specific vulnerability of each community and the availability of resources to carry out various types of emergency response, such as evacuation, isolation, and quarantine, must be understood.
- [139_Cr_4] The attitudes of community members to emergency response practices must be understood.
- [139_Pc_3] [139_Co_2] [139_Cr_5] [139_Ps_3] Effective risk communication requires the engagement of communication representatives in designing and evaluating educational materials, emergency messages, and dissemination practices, including language resources and accessibility.
- [139_Co_3] Messages must be culturally appropriate and accurately translated.
- [139_Cr_6] [139_Ps_4] Trusted members of the community and channels of communication (formal and informal) are critical to disseminate and reinforce the messages, making sure that messages are received, understood, and followed.
- [139_Pc_4] [139_At_1] Multiple modes of communication are required to reach diverse communities.
- [139_Co_4] [139_Cr_7] A database of trusted sources, interpreters, and effective communication modes for diverse populations should be prepared.
- [139_Ps_5] The design, execution, and evaluation of drills and exercises must include member of diverse populations.
- [139_Cr_8] Building and maintaining trust is of critical importance.

140. U.S. Department of Justice. (2010, September). *2010 ADA Standards for Accessible Design.* Washington, DC: DOJ (Author).
http://www.ada.gov/2010ADAstandards_index.htm

S			Pc	At	Co	Cr	Ps	Ac

The U.S. Department of Justice (DOJ) has published revised regulations for Titles II and III of the Americans with Disabilities Act of 1990 (ADA), which will take effect on March 15, 2012. The regulations in the 2010 ADA Standards for Accessible Design apply to facilities covered by the ADA, including public accommodations, commercial facilities, and state and local government facilities. The intent is to remove barriers for individuals with disabilities using a range of structural and design features. The 2004 ADA Accessibility Guidelines (ADAAG) document is incorporated into the 2010 standard. Chapter 7 of this document, entitled Communication Elements and Features, includes the design of alarms and warnings to ensure that they reach individuals with disabilities as well as other building occupants. The scopes of requirements for fire alarm systems (in section 215), signs (216), telephones (217), assistive listening systems (219), and two-way communication systems (230) are in Chapter 2. Elevator signs are covered in section 407. The 2010 ADA standard includes the directives summarized below.

Fire alarms: Fire alarm systems must incorporate permanently installed audible and visible alarms complying with NFPA 72, although medical care facilities are permitted to design fire alarm systems in accordance with industry practice. Visible alarms must be located within the space they serve in order to ensure that the signal is visible to the target population. Although non-fire facility alarm systems do not need to comply with the regulations in Chapter 7, the alarm signal should be distinguishable from fire alarms. Every effort should be made to safeguard everyone in all emergency situations (such as tornados).

Elevators: Elevator control buttons must be identified both tactilely and visually, with raised characters and Braille positioned directly to the left of the control button, and using standard tactile symbols for emergency stop, alarm, door open/close, main entry floor, and phone buttons. The emergency two-way communication device must be identified and its use explained by adjacent tactile symbols and characters. Verbal messages must be provided at sound levels 10 dB minimum above ambient, without exceeding 80 dB measured at the source. The frequency range of the annunciator is between 300 Hz and 3000 Hz.

Signs: Raised characters must be duplicated in Braille. They must be uppercase and plain (sans serif, not italic or decorative) and raised at least 0.8 mm above their background. Font proportions and character and line spacings are restricted, and character height must be between 16 mm and 51 mm. The raised characters must be separated from raised borders and decorations by at least 9.5 mm. Uppercase letters in Braille are allowed only before the first word of sentences, in proper nouns and names, and as individual letters of the alphabet, initials, and acronyms. The document provides detailed guidance on the configuration of the Braille system. All tactile characters must be located between 1.22 meters and 1.525 meters above the ground surface, and in specified locations by single and double doors.

Visual characters and their background must have a non-glare finish and as high contrast as possible (color and texture can help). Lighting shadows and surface glare should be minimized.

Characters in a message can be uppercase, lowercase or a combination, and must be conventional in form. Font proportions, stroke thickness, and character and line spacings are restricted. The minimum allowable character height ranges from 16 mm to over 75 mm depending on the viewing distance, defined as the horizontal distance between the character and an obstruction that prevents closer approach. Visual characters must be located at least 1.015 meters above the floor.

Pictograms, such as restroom signs, must be at least 150 mm high, with no characters or Braille within the pictogram field. Pictograms and their fields require a non-glare finish and high contrast in order to improve their legibility for persons with low vision. Lighting shadows and surface glare should be minimized. Text descriptors must be located directly below the pictogram field.

Accessibility symbols (for physical accessibility, TTY, volume control telephones, and assistive listening systems) must be of standard design, with a non-glare finish and high contrast.

Telephones: Public telephones with volume control must provide an adjustable gain up to 20 dB or more. Incremental volume control requires at least one intermediate step of at least 12 dB of gain. An automatic volume reset is required.

Detectable warnings: Truncated domes that are detectable through shoes and wheels can be used to provide a warning of dangerous surfaces, such as near the edge of boarding platforms. The acceptable size, height, and spacing of the domes are defined. Detectable warning surfaces must contrast visually with adjacent walking surfaces. For boarding platforms, the surface must be at least 610 mm wide and extend the full length of the platform.

Assistive listening systems: Assembly areas that require assistive listening systems can choose from hard-wired and wireless systems. A professional sound engineer is required to match the system to the space for maximum intelligibility. The sound pressure level capability must be between 110 dB and 118 dB, with a dynamic range of 50 dB on the volume control. The signal-to-noise ratio for internally generated noise must be at least 18 dB, and peak clipping is not allowed to exceed 18 dB relative to the peaks of speech.

Two-way communication systems: Both audible and visual signals are required. A light can be used to indicate visually that assistance is on the way. Signs should be provided to indicate the meaning of visual signals. Devices that do not require handsets are encouraged, since they are easier to use by people who have a limited reach. The system must support both voice and TTY communication.

Key findings:
- [140_Pc_1] [140_At_1] The ADA regulations require the use of both audible and visual signals for fire alarm systems.
- [140_Pc_2] The ADA regulations require elevator control buttons to be identified both tactilely and visually, with raised characters and Braille positioned directly to the left of the control button and using standard tactile symbols for emergency stop, alarm, door open/close, main entry floor, and phone buttons.
- [140_Pc_3] [140_At_2] The ADA regulations require verbal messages in elevators to be provided at sound levels 10 dB minimum above ambient, without exceeding 80 dB measured at the source.
- [140_Pc_4] The frequency range of speech messages in elevators is required by ADA regulations to be 300 Hz to 3000 Hz.
- [140_Pc_5] The ADA regulations require raised characters to be uppercase, in a plain and conventional font, and raised at least 0.8 mm above their background.
- [140_Pc_6] The ADA regulations require visual messages to have a non-glare finish and high

---

contrast between the image and background in order to improve their legibility for persons with low vision. Lighting shadows and surface glare should be minimized.

- [140_Pc_7] The ADA regulations require tactile characters to be located between 1.22 meters and 1.525 meters above the floor and visual characters to be located at a height of at least 1.015 meters.
- [140_Co_1] The ADA regulations require text descriptors for pictograms to be located directly below the pictogram field.
- [140_Pc_8] Public telephones with volume control are required by ADA regulations to provide an adjustable gain up to 20 dB or more. Incremental volume control requires at least one intermediate step of at least 12 dB of gain
- [140_Pc_9] Assistive listening systems are required by ADA regulations to have sound pressure level capability between 110 dB and 118 dB, with a dynamic range of 50 dB on the volume control. The signal-to-noise ratio for internally generated noise must be at least 18 dB, and peak clipping is not allowed to exceed 18 dB relative to the peaks of speech.

Key recommendations:

- [140_Pc_10] Visible alarms must be located within the space they serve in order to ensure that the signal is visible to the target population.
- [140_Co_2] Alarm signals should be distinguishable for different scenarios (e.g., fire, tornado) that require different responses.
- [140_Pc_11] Multiple means of communicating important messages should be used when feasible. The ADA regulations require the use of both audible and visual signals for fire alarm and two-way communication systems. The use of tactile information with visual information is required for signs in elevators, detectable warning surfaces on boarding platforms, and other critical applications.
- [140_Pc_12] Visual messages require a non-glare finish and high contrast between the image and background in order to improve their legibility for persons with low vision. Lighting shadows and surface glare should be minimized.
- [140_Pc_13] For legibility, character fonts in warning messages should be plain and conventional, and character size, character spacing, line spacing, and message height need to take into account viewing distance for visual messages and finger legibility for tactile messages.
- [140_Pc_14] The appropriate use of uppercase and lowercase letters for legible signs depends on the application.
- [140_Co_3] For two-way communication systems, a light can be used to indicate visually that assistance is on the way. Signs are needed to indicate the meaning of all visual signals.

141. Using Universal Symbols: Improving wayfinding through universal signage systems. (2007, March). *Environment of Care News*, 8-10.

E		Pc	At	Co	Cr	Ps	Ac

The Hablamos Juntos project, administered by the University of California- San Francisco, has conducted research into ways to develop practical solutions for written language and signage barriers in health care. The researchers developed 28 universal symbols and tested them at four different hospitals. The test required participants to find six locations within each hospital, with some of the existing signs replaced by the experimental symbols.

The test showed that participants walked faster to their destination when symbols were used in the signs than when they were guided by multilingual word signs. The study concluded that for successful wayfinding, the signs must be easily visible for people, consistent in height, and placed in easily observable locations. Signs should be placed in every location where a decision must be made and spaced so that successive signs are within view of each other. Sign clutter should be reduced to a minimum. Lighting should be sufficiently bright that the signs can be read from at least 25 feet away. If lighting is low, a white or light background should be used on the signs. The contrast between the letter or symbol sign content and the background should be at least 60 %.

Key findings:
- [141_Ac_1] Participants walked faster to their destination when symbols were used in wayfinding signs than when they were guided by multilingual signs.
- [141_Pc_1] Wayfinding signs should be consistent in height, type of location, and style.
- [141_Pc_2] Lighting should be sufficiently bright that wayfinding signs can be read from at least 25 feet away.
- [141_Pc_3] A white or light background should be used for signs in low lighting.
- [141_Pc_4] The contrast between sign content and the background should be at least 60 %.

142. Vukelich M, Whitaker LA. (1993). The effects of context on the comprehension of graphic symbols. *Proceedings of the Human Factors and Ergonomics Society 37th Annual Meeting-1993*, 511-515.

E

Graphic/pictorial symbols are simplified pictures of the objects they represent. They can be used instead of words when dealing with an illiterate or multilingual population because they can represent meaning without the use of language, thus overcoming language vulnerabilities. A successful symbol transmits the same meaning to everyone who encounters it, and that meaning is the same as that intended by the creator of the symbol. This is known as consistent comprehension.

In this study, comprehension of symbols by 70 English-speaking college students was assessed using a survey. The participants viewed 20 symbols, for each of which they were given full, partial or no context. Full contextual information was presented as a description underneath the symbol. Partial context consisted of a short sentence describing the environment in which the symbol would be seen. The participants first rated their degree of familiarity with each symbol on a scale from one to five and then were asked to describe the symbol.

The more familiar the participants were with a symbol, the more accurate were their responses. The only symbols that were correctly understood by 85 % of the participants or more were those for 'flammable' and 'fuel.' Context greatly increased the comprehension of a symbol.

Key findings:
- [142_Co_1] Familiarity with a symbol results in more accurate identification of its meaning.
- [142_Co_2] Context greatly increases the comprehension of a symbol.

# **143**. Waller RHW. (2007). Comparing typefaces for airport signs. *Information Design Journal 15*, 1-15.							

E		Pc	At	Co	Cr	Ps	Ac

At Heathrow airport in London, new wayfinding systems were to be implemented in a newly constructed terminal (Terminal 5). This research was conducted to see if the current sign standard at the rest of Heathrow Airport was suitable for wayfinding. The researchers tested the five typefaces shown in Table 23: the serif font currently used for BAA airport signs, Frutiger Bold and Roman (the current corporate typeface), Vialog (used in transport), and Stempel Garamond as a control font.

Table 23: Typefaces tested.

Hxhp	BAA SIGN
Hxhp	FRUTIGER BOLD
Hxhp	FRUTIGER ROMAN
Hxhp	VIALOG
Hxhp	GARAMOND

The research also looked at color combinations, comparing the current design of black letters on yellow background with alternative designs of white on black, white on grey and black on white. Three research methods were used:
- legibility testing – measured by the time of recognition for word displays
- qualitative consumer research - participant preferences and associations
- expert review – opinion survey of panel of experts

The legibility test measured the average time for 23 participants to recognize words displayed in each font and color combination. A word displayed on an otherwise blank computer screen was gradually enlarged until the participant signaled that he or she could read it by clicking the mouse.

In legibility tests arranged by font, reproduced in Table 24, the Frutiger bold font was found to be the most legible of those tested. Frutiger Roman and the BAA Sign typeface were tied for second place. The widest typeface was found to be the most legible and the narrowest the least legible. Looking at legibility by color combination in Table 25, contrast was identified as the most important factor, with black on white giving the best results. Significant differences between results were found at the group level, but not for the combined data.

Table 24: Legibility test results by font.

Typeface	Average Recognition Time (s)	STATISTICAL SIGNIFICANCE OF LEGIBILITY TIME COMPARISON				
		Frutiger Bold	Frutiger Roman	BAA Sign	Vialog	Garamond
Frutiger Bold	6.7					
Frutiger Roman	6.9	NOT SIG				
BAA Sign	7.0	SIG	NOT SIG			
Vialog	8.1	SIG	SIG	SIG		
Garamond	11.4	SIG	SIG	SIG	SIG	

Table 25: Legibility test results by color combination.

Color Combination	Average Recognition Time (s)	STATISTICAL SIGNIFICANCE OF LEGIBILITY TIME COMPARISON			
		Black/White	Black/Yellow	White/Grey	White/Black
Black on White	6.9				
Black on Yellow	7.1	NOT SIG			
White on Grey	7.1	NOT SIG	NOT SIG		
White on Black	7.4	NOT SIG	NOT SIG	NOT SIG	

The qualitative consumer research was performed through an online survey of 400 participants in the UK and Germany. In the survey, the participants were asked to describe the fonts using a bipolar scale for various characteristics. Descriptive results are in Table 26.

Table 26: Font comparison using qualitative consumer research.

Typeface	Respondent Description
Frutiger Bold	Efficient, dull, formal, and British. British participants identified it as straight-talking and welcoming. German participants found it bureaucratic and technological.
BAA Sign	Old-fashioned, welcoming, fairly human while efficient, and formal.
Vialog	Modern, efficient, and technological. British participants thought it straight-talking. German participants saw it as dull, formal and less modern.
Garamond Italic	Seen as lively, human, welcoming, old-fashioned. German respondents found it informal and straight-talking. British respondents were split on the formal-informal dimension.

In the expert review of these typefaces, the experts thought that the sturdy strokes and relatively tight letter and line spacings of the BAA Sign typeface might reduce legibility at a distance. A majority of the experts predicted that the BAA Sign would be less legible than Frutiger Bold by about 10 % to 20 %. The experts favored the yellow/black color combination.

Key findings:
- [143_Co_1] Contrast was identified as the most important legibility factor for color selection.
- [143_Co_2] A bold sans-serif font was found to be slightly more legible than the serifed font in use at airports.
- [143_Co_3] The font with the widest typeface was found to be the most legible.
- [143_Co_4] Opinions on the meaning of font styles (e.g., formality, freshness, directness) may vary by cultural background.

144. Wang J-H, Cao Y. (2005). Assessing Message Display Formats of Portable Variable Message Signs. *Transportation Research Record: Journal of the Transportation Research Board,* 1937, 113-119.

	Pc	At	Co	Cr	Ps	Ac
T						

Variable Message Signs (VMS) are traffic signs that can be composed of letters, symbols, or both. They have to be understood in a short time frame so that the appropriate responses can be made while driving. VMS can either display information discretely or sequentially. A discrete display shows all the information at once, while the sequential display can present a longer message by dividing it into multiple phases that are displayed sequentially. The maximum number of phases is limited by the available display time and by the ability of drivers to process the information. It has previously been recommended that a sequential display separate each message with a blank display for one second. Research has also shown that a sequential display can separate each message by an asterisk displayed for 0.5 seconds with a 0.25 seconds blank time before and after the asterisk. One-phase messages should be used whenever possible, and two-phase should only be used when information could not be kept to a single phase.

The previous study (Phase 1) found that green was the best font color for VMS and a 5×7 dot matrix was the best font size. The study found that older drivers responded faster than younger drivers but with lower accuracy. People 20 to 40 years old responded with 93 % accuracy, those 41 to 60 years old were about 85 % accurate, and those over 60 years old were about 80 % accurate. The Phase 2 study compared the effects of discrete and sequential displays. The discrete display contained a two line message and a three line message. The sequentially displayed messages showed a single line message followed by either a single line or a two line message. In the experiment, participants were put into a simulated driving environment. When the participant was able to read and understand a VMS sign, they pressed a button to signal comprehension. The results showed that the response time of younger drivers was shorter than that for older drivers, and that the response time of male drivers was shorter than that for female drivers. The response accuracy was higher for female than for male drivers. For both discrete and sequential VMS, fewer message lines resulted in a shorter response time. The participants were interviewed following the driving simulation, and 71.4 % of them preferred the discrete display.

Key findings:
- [144_Co_1] [144_Ac_1] The response time was shorter for simultaneously displayed (discrete) messages than for the same message displayed sequentially in multiple phases. Drivers preferred the discrete displays.
- [144_Co_2] [144_Ac_2] The response time was shorter for variable message signs (VMS) when there were fewer lines in the message.
- [144_Ac_3] Older drivers showed slower response times to variable message signs (VMS) and less response accuracy than younger drivers.
- [144_Ac_4] Female drivers showed slower response times to variable message signs (VMS) but higher response accuracy than male drivers.

145. Wang XH, Kapucu N. (2008). Public complacency under repeated emergency threats: Some empirical evidence. *Journal of Public Administration Research and Theory, 18*(1), 57-78.

C		Pc	At	Co	Cr	Ps	Ac

Public preparedness and emergency management decisions depend on assumptions about how much the public response is affected by alerts that warn of more serious events than actually occur (the "cry wolf" effect), and can be improved by better understanding. Do such occurrences cause the public to become more complacent over time and ignore warnings to take protective measures? This is a concern especially for low-probability high-impact events such as serious crimes, health care threats, terrorist attacks, and natural disasters. Florida was struck by four major hurricanes within a two month period in 2004. This provided an opportunity to study public complacency in the presence of repeated threats and warnings and to develop a theoretical framework for modeling its role in the public response. The decision-making model in the framework assumes individual decisions being made in a social context, using bounded rationality and risk communication theories.

Data on the response to the four Florida hurricanes was collected from emergency response managers using surveys, interviews, and documentation. The authors sent questionnaires to emergency managers/directors in Florida's 67 counties as well as four large cities in the state and the state government. At least 28 % of the emergency managers who responded believed that the people in their county were complacent during the hurricanes. The most common reason given for not evacuating was the belief that the individual was not at risk. The level of complacency decreased after the first and second hurricanes, but increased after the third one. The authors proposed two possible explanations for this change in complacency. First, since the population had recently practiced the skills and knowledge needed, they saw less of a need to prepare as far in advance. Second, the population now viewed hurricanes as less of a danger.

Key findings:
- [145_Cr_1] [145_Ps_1] In a set of repeated emergencies requiring evacuation (four hurricanes in two months), public complacency decreased after the first two and then increased.
- [145_Cr_2] Even if they are ultimately deemed to be unnecessary, the first one or two evacuations in response to a repeated threat may decrease complacency at the time of the subsequent evacuation.
- [145_Cr_3] Too many unnecessary evacuations in a short period of time may result in increased public complacency.
- [145_Ps_2] The most common reason given for not evacuating was the belief that the individual was not at risk.

Key recommendations:
- [145_Cr_4] Emergency response managers need to have an evidence-based understanding of the "cry wolf" effect in order to properly manage the possibility of public complacency.

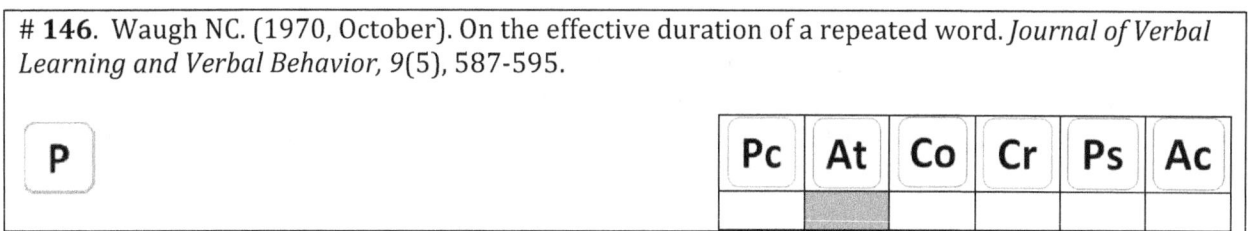

 # 146. Waugh NC. (1970, October). On the effective duration of a repeated word. *Journal of Verbal Learning and Verbal Behavior, 9*(5), 587-595.

The retention of information in verbal learning is known to depend on repetition, but the mechanics of this process are not well understood. Some previous studies have suggested that the likelihood of retaining information from a list is linearly dependent on the total time of presentation, independent of the distribution of repeated items within the list. Other studies have found that there is indeed a dependency on distribution, with well-distributed repetitions remembered better than massed ones. This study investigates whether the discrepancy between results may be explained by differing rates of presentation of repeated items.

In experiment one, twelve undergraduates at Tufts University listened to 20 lists, each containing 33 monosyllabic words, of which 19 words were unique and seven were repeated once at set lags (interval between repetitions). No word appeared in more than one list. Each list of words was spoken and recorded at two speeds (one or four seconds per word). The participants were divided into four groups. Each group listed to half of the lists spoken at one speed followed by the second half spoken at the other speed. After hearing an entire list, the subjects were instructed to write down every word that they could remember from the list within two minutes. At the fast speaking rate, all repeated words were about twice as likely to be recalled as words read only once, independent of lag. When the speaking rate was slower, more words were recalled, but the dependence on lag was very different. In this case, words that were repeated twice in a row were recalled only slightly more often than words spoken only once, and repeated words separated by more than three words were recalled about half again as often. The relationship between repetition and recall of words in a list thus depends on the rate at which the words are read. This suggests that with slower speech rates more attention is given to unique items at the expense of repeated ones, especially those that appear in close proximity.

In experiment two, twenty undergraduates at Tufts University each listened to 48 lists. Twelve lists each were generated of four word lengths: 34, 58, 82, and 106, in which six, 12, 18, and 24 of these words, respectively, were presented one to eight times. The same lists were generated for the control condition, in which the repeated words were randomly distributed throughout the list, and for the experimental condition, in which all repetitions of a given word were sequential. The speech rate was one second per word in all cases. Half of the subjects listened to the 48 control lists and the other half listened to the 48 experimental lists. As with Experiment I, subjects were given two minutes after each list to write down all of the words that they could remember. The relationship between the probability of recall and the number of presentations was well described by a linear least-squares fit in all cases but one. For distributed repetitions, the probability of recall was proportional to the number of presentations. Massed words yielded better recall for one to three presentations within the list, and distributed words yielded better recall for more than four presentations. In experiment three, 10 undergraduates at Tufts University listened to the 12 experimental lists of 58 words each from Experiment two, spoken at a speed of four seconds per word, plus word lists with the same number of words and pauses between words up to eight seconds. The length of the pause between words did not increase recall.

Key findings:
- [146_At_1] More items in a list are recalled at a slower speech rate (four seconds per word) than at a faster speech rate (one second per word).
- [146_At_2] At a faster speech rate (one second per word), the probability of recall of a word on a list is linearly dependent on the number of times it appears for at least seven repetitions.
- [146_At_3] At a speech rate of one second per word, consecutive repetition leads to better recall for two or fewer repetitions, and distributed repetitions are better for four or more repetitions
- [146_At_4] At a slower speech rate (four seconds per word), recall is improved by repeating the information at distributed intervals.
- [146_At_5] Word recall is independent of the pause time between words.

147. Webber GMB, Aizlewood CE. (1994). *Emergency wayfinding lighting systems in smoke* (BRE Information Paper IP 17/94). Building Research Establishment, Watford, UK, 1994.

F		Pc	At	Co	Cr	Ps	Ac

The research studied the effectiveness of different wayfinding systems in smoke-filled conditions. Sixteen participants were asked to move toward an exit at the end of a corridor 13.7 meters long. The corridor was filled with non-irritant, non-toxic, white smoke, with smoke density (optical density per meter of path length) ranging from 0.18 m⁻¹ to 3.7 m⁻¹. Ambient lighting was at an illuminance level of 0.3 lx. The visibility of three structural components (signs, doors, and wall) accentuated by elements of a variety of wayfinding systems was assessed in different smoke densities by measuring the distance from which the feature could be recognized.

The results, reproduced in Table 27, were categorized according to the structural component being examined and the wayfinding system being used:

Table 27: Visibility of structural components accentuated by features of wayfinding systems.

Structural component	Wayfinding system	Smoke Density (m^{-1})	Recognition Distance (m)
Signs	LED system with pictogram	1	3.5
		2	2
		3	1.5
	Photoluminescent pictogram	1	1.5
		2	0.75
		3	0.5
Doors	Miniature incandescent and LED systems (equivalent results)	1	3
		2	1.5
		3	1
	Photoluminescent signs	1	1.5
		2	1
		3	0.75
Wall/Floor Track	Triplet LED floor track	1	3
		1.8	2
	Miniature incandescent system	1	2.5
		1.4	2
	LED system	1	2.25
		1.8	1.5
	Photoluminescent system	0.7	2

The LED and miniature incandescent systems consisted of a set of small light sources, each light having a brightness of 0.1 cd. The components accentuated by these systems were recognized from farther away than photoluminescent systems. The presence of other emergency lighting was found to compromise the effectiveness of the wayfinding system.

Key findings
- [147_Pc_1] Ambient lighting compromises the effectiveness of emergency signage and wayfinding systems in smoke.
- [147_Pc_2] Powered (LED and miniature incandescent) wayfinding systems are visible in smoke at a significantly greater distance than photoluminescent systems.
- [147_Pc_3] Groups of small lights placed close together make an effective wayfinding system, enabling the accentuation of complex structural features and the use of dynamic configurations.

148. Wogalter MS. (2006). Purposes and scope of warnings. In MS Wogalter (Ed.), *Handbook of Warnings* (pp. 3-9). Mahwah, NJ: Lawrence Erlbaum Associates, Inc.

M		Pc	At	Co	Cr	Ps	Ac

Warnings are the third level of defense hazard, when engineering design that eliminates the hazard and physical guards against the hazard are impractical or ineffective. This chapter of the *Handbook of Warnings* provides an overview of several major warning-related concepts, including the purposes of warnings, the fundamentals of hazard control, the 'who, what, when and where of warnings', and a discussion of how to warn (i.e., the warning systems used to dissemination the information). The chapter contains a helpful table (1.1 on page 6) on the design guidelines of a successful warning system. The specifics associated with each of these guidelines are discussed in the many chapters included in this handbook.

The literature on visual warnings generally recommends a signal word panel (ANSI 2002 Z535) and a message panel, which can be text or symbols. Design considerations include using standard colors and meanings for signal words (e.g., DANGER, WARNING, CAUTION) and symbols. Red connotes the highest level of hazard. People do not distinguish a difference in hazard level between orange and yellow and between WARNING and CAUTION. The message panel consists of information about the hazard, its consequences, and instructions on how to deal with it. Text should be legible and high contrast, with a plain font, good use of white space, and bullet points to separate statements. Instructions should be explicit but brief. Words and phrases should be familiar and short, without multiple interpretations. Messages should be in multiple languages when appropriate. Clearly comprehensible symbols can augment or replace the text. Symbols attract attention and are beneficial for poor readers and non-English speakers. However, the meaning must be clear. The symbol should be bold and avoid irrelevant graphical details. It should be legible from a distance and after moderate degradation. Open and obvious hazards, such as scissors, cliffs, and naturally uneven terrain, do not necessarily require a warning. The hazards of modern technology, including airbags and household cleaners, are frequently not open and obvious. Priority in warning messages for multiple hazards, as demonstrated by ordering and conspicuousness, should go to hazards that are more severe, more likely to occur, unfamiliar, and of higher importance.

A warning message needs to consider the audience. It should be understandable by an audience member with the lowest level of skills, including consideration of sensory-perceptual difficulties, cognitive limitations, and literacy levels. Task analysis is a method for determining good potential locations for the warning.

Communication of warning messages may be visual, auditory, olfactory (as in the odor added to natural gas), or tactile (as in rumble strips for automobiles). A systems approach using multiple modalities can be used to reach people in different ways, at different times, and in different subgroups. Training may be used in addition to warning messages. People receive warnings from indirect channels such as colleagues and friends in addition to direct channels. The effectiveness of warning systems should be tested using participants from the target population.

Key recommendations:

- [148_Pc_1] [148_At_1] The literature on visual warnings generally recommends a signal word panel (DANGER, WARNING, CAUTION, etc.) and a message panel, which can be text or symbols.
- [148_Pc_2] [148_At_2] Good warning design includes using standard colors and meanings for signal words (e.g., DANGER, WARNING, CAUTION) and symbols.
- [148_Co_1] Text should be legible and high contrast, with a plain font, good use of white space, and bullet points to separate statements.
- [148_Co_2] Instructions should be explicit but brief.
- [148_Co_3] Words and phrases should be familiar and short, without multiple interpretations.
- [148_Co_4] Messages should be in multiple languages when appropriate.
- [148_Co_5] Clearly comprehensible symbols can augment or replace text in the message panel. Symbols attract attention and are beneficial for poor readers and non-English speakers. However, the meaning of the symbols must be clear.
- [148_At_3] [148_Co_6] A warning symbol should be bold and avoid irrelevant graphical details. It should be legible from a distance and after moderate degradation.
- [148_At_4] Priority in warning messages for multiple hazards, as demonstrated by ordering and conspicuousness, should go to hazards that are more severe, more likely to occur, unfamiliar, and of higher importance.
- [148_Co_7] A warning message needs to consider the audience and be understandable by an audience member with the lowest level of skills, including consideration of sensory-perceptual difficulties, cognitive limitations, and literacy levels.
- [148_Pc_3] Task analysis is a method for determining good potential locations for the warning.
- [148_Pc_4] [148_At_5] Communication of warning messages may be visual, auditory, olfactory, or tactile. A systems approach using multiple modalities can be used to reach people in different ways, at different times, and in different subgroups.
- [148_Ac_1] Training may be used in addition to warning messages.
- [148_Ps_1] People receive warnings from indirect channels such as colleagues and friends in addition to direct channels.
- [148_Ac_2] Warning system effectiveness should be tested using participants from the target population.

149. Wogalter MS, Racicot BM, Kalsher MJ, Simpson SN. (1993). Behavioral compliance to personalized warning signs and the role of perceived relevance. *Proceedings of the Human Factors and Ergonomics Society 37th Annual Meeting*, 950-954.

E			Pc	At	Co	Cr	Ps	Ac

In a previous study by the same authors (reference #150 in this annotated bibliography), a highly visible posted warning sign was found to result in significantly lower compliance than the same warning embedded in a set of instructions. One possible reason for the low compliance was that the participants did not believe that the sign was relevant to the tasks they were performing. Adding features intended to enhance the salience of the sign (i.e., the ability to note the sign in relation to its surroundings), such as strobe lights and pictorials, failed to increase compliance.

The purposes of this study were to examine ways to make a sign more relevant to the intended audience, and to see whether understanding that the sign is relevant increases compliance. A new sign that can display different messages was used. Questions addressed by the study include whether dynamic warning signs are more salient than a static warning message and whether placing the warning sign in a cluttered area reduces compliance.

The participants in this study were 156 undergraduate students. A warning sign was placed on the laboratory counter eight feet, 15 feet or 18 feet from where each participant worked on a laboratory task. The position of the sign at a distance of 18 feet (farthest from the participant) was less cluttered than the position at 15 feet. At the nearest position, the sign was located eight feet in front but slightly to the left of the participant and was in the most cluttered environment. Two warning signs were tested: the 'impersonal' warning sign said "CAUTION! IRRITANT! Use Mask & Gloves," and the 'personalized' sign said "[participant's name]! IRRITANT! Use Mask & Gloves." Compliance was measured by whether subjects wore the required safety gear indicated in the sign. After completing the assigned task, each participant was asked to complete a questionnaire asking whether he or she noticed the safety equipment and the safety sign.

The participants exposed to a personalized sign demonstrated more compliance than the participants exposed to the impersonal sign. Participants exposed to the personal warning sign were also more likely to recall seeing it when answering the questionnaire. This supports the suggestion that the impersonal nature of the highly visible posted sign (with strobe lights and pictorials) used in the previous study contributed to the relatively low level of compliance. The dynamic LED display did not produce an additional effect over a static LED display. The personalized sign that was located the farthest distance away but in the least cluttered area resulted in the greatest compliance; however, the explanations for this are not definitive.

Key findings:
- [149_Ps_1] People exposed to personalized signs (that included their names) demonstrated more compliance than people exposed to an impersonal sign.
- [149_Pc_1] [149_Ac_1] Safety signs using dynamic LED displays were not found to increase compliance over static LED displays.
- [149_Pc_2] [149_Ac_2] The personalized sign that was located the farthest distance away but in the least cluttered area resulted in the greatest compliance.

150. Wogalter MS, Kalsher MJ, Racicot BM. (1992). The influence of location and pictorials on behavioral compliance to warnings. *Proceedings of the Human Factors Society 36th Annual Meeting-1992*, 1028-1033.

The purposes of this experiment were to determine the relative efficacy of warnings placed in a set of instructions compared to warnings posted on a wall, and to examine whether adding pictorials would affect compliance. For the first part of the experiment, participants were asked to use a triple beam balance, beakers, and flasks to measure and mix several substances in the laboratory. A visual warning message told participants to don protective gear while performing the tasks. The form of the warning was varied; participants were given no warning, a posted warning sign, a warning located inside the task instructions, or both a warning in the instructions and on a posted sign. The warning sign posted on the wall was black print on a yellow background. The warning sign located inside the task instruction was 4 % of the size of the posted warning sign and of identical design except for the color of the background, which was white instead of yellow. The results showed that if the participants wore one piece of protective gear, they were very likely to wear the other piece. The warning located within the instructions resulted in greater compliance than the posted-sign warning.

In the second part of the experiment, participants were asked to use the same equipment to measure and mix substances. This time however, the participants were given either no warning, a posted sign with no pictorials, a posted sign with pictorials, a warning located inside the task instructions without pictorials, or a warning located within the task instructions with pictorials. This experiment again showed that a posted sign produces a lower rate of behavioral compliance than the same warning posted in the instructions. In this study, no significant benefit was demonstrated from the use of pictorials, but compliance was greater when the pictorials were presented in the within-instruction warning.

Key findings:
- [150_At_1] [150_Ac_1] Warnings presented in task instructions for a chemistry experiment resulted in greater compliance than warnings posted on the wall.
- [150_At_2] [150_Ac_2] Adding a pictorial to a written warning in instructions for a chemistry experiment was not found to increase compliance.

 # 151. Wogalter MS, Murray LA, Glover BL, Shaver EF. (2002). Comprehension of different types of prohibitive safety symbols with glance exposure. *Proceedings of the Human Factors and Ergonomics Society 46th Annual Meeting-2002*, Santa Monica, CA, 1753-1757.

		Pc	At	Co	Cr	Ps	Ac
E							

The red circle with a red slash overlay from the top left to the bottom right has become a symbol that represents prohibition. This study looked at the 'glance legibility' (comprehension after quick examination) of 16 pictorials that are commonly found in public and industrial places. There were 64 participants in this study. Pictorials with a red prohibition circle were shown on a computer screen to each participant. The area of red was 35 % of the total area inside the outer rim of the circle, with 65 % of the space taken up by the pictorial. Several designs were shown, with the red prohibition sign with the slash in front of the pictorial, the slash behind the pictorial, a partial broken slash, or a translucent slash. Each image was displayed for 50 milliseconds. After the image disappeared, the participants recorded their impression of the meaning of the symbol.

The results showed that comprehension was higher for less complex symbols. The symbols with the lowest comprehension scores contained more detail than the others. Comprehension scores were highest for the translucent slash and the slash behind the pictorial. The slash covering the pictorial resulted in intermediate comprehension scores, and the broken slash was the worst.

Key findings:
- [151_Co_1] The meaning of prohibition signs was more rapidly comprehended when the slash in the red circle symbol was translucent or located under the pictorial (i.e., more of the pictorial could be seen) than when the slash covered the pictorial or was broken.
- [151_Co_2] Comprehension of prohibition signs was most rapid when the symbol was simple.

152. Wogalter MS, Conzola VC, Smith-Jackson TL. (2002). Research-based guidelines for warning design and evaluation. *Applied Ergonomics, 33*(3), 219–230.

E		Pc	At	Co	Cr	Ps	Ac

This article provides a review of the empirical research about warning design and evaluation. For a sign to be used, it must be noticed and be as salient (prominent) as possible to capture the attention of individuals. The salience of a sign can be enhanced by using large bold print, high contrast, color, borders, pictorial symbols, and special effects, such as flashing lights. Bold type is preferred because of its greater contrast with most backgrounds, but the letters cannot be so wide that the details are obscured. Colored warning labels are perceived as more readable and more likely to indicate a hazard. Effective warnings consist of four message components:
1. Signal word to attract attention
2. Identification of the hazard,
3. Explanation of the consequences if exposed to hazard, and
4. Directives for avoiding the hazard.

The signal word to attract attention should indicate the level of hazard present, such as DANGER or WARNING. The hazard description should be specific and complete. The explanation of the consequences of the hazard should be explicit. The directions for avoiding the hazard should also be explicit. Presenting warning text as bullets in outline form is preferred to continuous flowing text. The warning should be located close to the hazard. Visual clutter in the neighborhood of the warning should be reduced. Pictorials facilitate warning comprehension and improve comprehensibility for people who do not understand the language written on the sign.

Demographic variables are also important to consider in the understanding of a warning sign. For older people, print must be large enough to compensate for reduced visual capabilities. To accommodate the large print, clearer layouts using spacing or bullets can be used. Familiarity is an important factor in warning habituation. The more often a warning is encountered, the less likely the person will notice it during following encounters. One way to reduce the negative effects of habituation is to change the content or appearance of a warning sign.

Key findings:
- [152_Pc_1] [152_At_1] To be noticed and used, a sign must be as salient as possible. Salience can be enhanced by using large bold print, high contrast, colors, borders, pictorial symbols and special effects, such as flashing lights.
- [152_Co_1] Colored warning labels are perceived as more readable.
- [152_Ps_1] Colored warning labels are perceived as more likely to indicate a hazard.
- [152_At_2] [152_Ps_2] [152_Ac_1] Effective warnings consist of four message components: a signal word to attract attention, identification of the hazard, explanation of the consequences if exposed to the hazard, and directives for avoiding the hazard.
- [152_At_3] [152_Ps_3] The signal word to attract attention should indicate the level of hazard present, such as DANGER or WARNING.
- [152_Co_2] [152_Cr_1] [152_Ps_4] The description of the hazard should be specific and complete.
- [152_Co_3] [152_Cr_2] [152_Ps_5] The explanation of the consequences of the hazard and

the directions for avoiding the hazard should be explicit.

- [152_At_4] Warning text should be in bullet form as opposed to continuous flowing text.
- [152_Ps_6] The visual warning should be located close to the hazard.
- [152_Ps_7] Visual clutter in the neighborhood of the warning should be reduced.
- [152_Co_4] Pictorials facilitate warning comprehension and improve comprehensibility for people who do not understand the language.
- [152_Co_5] For older people, print must be large enough to compensate for reduced visual capabilities. To accommodate the large print, clearer layouts using spacing or bullets can be used.
- [152_At_5] Familiarity is an important factor in warning habituation. The more often a warning is encountered, the less likely the person will notice it during following encounters. One way to reduce the negative effects of habituation is to change the content or appearance of a warning sign.

153. Wong LT, Lo KC. (2007). Experimental study on visibility of exit signs in buildings. *Building and Environment, 42*(4), 1836–1842.

B		Pc	At	Co	Cr	Ps	Ac

The visibility of exit signs was studied experimentally. Participants were placed at the end of an 18 meters long internal corridor and asked to view a sign placed at the opposite end of the corridor through a viewing window under clear conditions. Four factors affecting visibility were investigated: graphics, colors, lighting conditions and the age of the observers. Two lighting conditions were examined: normal lighting with floor illuminance and wall luminance of 115 lx and 2.8 cd/m² respectively, and emergency lighting conditionsfloor illuminance and wall luminance of with 5 lx and 0.35 cd/m² respectively. In each case, the sign content height was increased until the participant could detect the sign, identify it, and identify it with confidence.

Combinations of designations and designs were presented under normal and emergency lighting conditions to 30 observers (22 males and eight females, ranging between 18 and 50 years old with a mean age of 36) to determine their visibility. Designs included the English and Chinese words for EXIT, five directional indicators, and two exit symbols, presented using different colors. The sign heights were recorded at the point at which the signs were detected and identified, and again when the content could be identified with confidence. Results under normal lighting for a variety of sign designs are reproduced in Table 28.

Table 28: Sign content height for visibility as a function of sign design.

Sign	Height for Visibility (mm)	
	Identified	Confident
	40	55
	41	57
EXIT	30	46
出口	53	69
❮	38 (L) / 34 (R)	52 (L) / 49 (R)
❮	41 (L) / 45 (R)	63 (L) / 59 (R)
◀	42 (L) / 43 (R)	59 (L&R)
⬅	28 (L) / 27 (R)	42 (L&R)
⬅	32 (L) / 27 (R)	45 (L) / 42 (R)
⬅	34	48
❮	47	62
⬅	32	47
EXIT ◀	46	62
出口 ❮	35	51

344

The English word EXIT exhibited higher visibility (required smaller sign content height from the same viewing distance) than the two pictograms, which were more visible than the Chinese symbols for exit. Arrows with tails were more visible than arrows without tails. Although an exit sign of combined symbols would be recognized if any one of its symbols could be identified, its visibility was generally reduced when compared to a sign containing only one of these symbols.

Table 29 and Table 30 reproduce the sign content heights for visibility for three sign designs in different sign and background colors under normal and emergency lighting conditions, respectively.

Table 29: Sign content height for visibility as a function of sign design and sign/background colors under normal lighting conditions

		Height for Visibility (mm)					
Sign Color		Green	Red	Black			
Background Color					Green	Red	Black
→🏃 (sign)	Detected/Identified	20 / 31	21 / 32	20 / 30	21 / 36	21 / 38	21 / 37
	Confident	39	46	40	51	51	49
→EXIT出口	Detected/Identified	20 / 33	21 / 35	20 / 32	20 / 37	20 / 40	20 / 36
	Confident	45	46	47	52	55	50
→EXIT出口	Detected/Identified	20 / 32	21 / 36	20 / 34	20 / 35	20 / 40	20 / 35
	Confident	45	47	45	51	54	51

Table 30: Sign content height for visibility as a function of sign design and sign/background colors under emergency lighting conditions

		Height for Visibility (mm)					
Sign Color		Green	Red	Black			
Background Color					Green	Red	Black
→🏃 (sign)	Detected/Identified	32 / 53	36 / 56	33 / 52	31 / 57	33 / 60	33 / 55
	Confident	70	72	68	75	78	74
→EXIT出口	Detected/Identified	33 / 58	37 / 59	34 / 55	32 / 56	33 / 57	32 / 55
	Confident	75	78	74	80	79	76
→EXIT出口	Detected/Identified	31 / 53	35 / 57	33 / 51	31 / 55	32 / 58	32 / 54
	Confident	72	77	73	75	75	72

As expected, the sign height needed for correct identification was considerably greater under reduced illumination than under normal lighting conditions. Signs whose content was in color on a white background were more visible than white signs on colored backgrounds. Visibility of green and black signs is slightly better than visibility of red signs, although the statistical significance of this advantage is unclear.

The impact of age on sign performance is shown in Figure 14. Under both emergency and non-emergency conditions, larger sign content heights were required for comparable visibility as people got older.

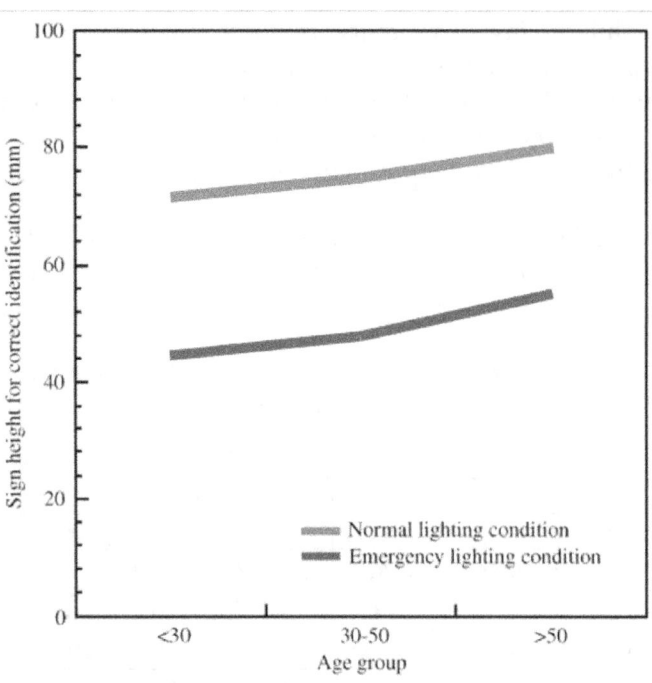

Figure 14: Relationship between performance and age.

Key findings:
- [153_Pc_1] The word EXIT exhibits higher visibility (requires smaller sign content height from the same viewing distance) than pictograms.
- [153_ Pc _2] Arrows with tails are more visible than arrows without tails.
- [153_ Pc _3] Visibility of an exit sign with combined symbols is slightly lower than that of a sign containing only one of these symbols.
- [153_ Pc _4] Green or black sign content is slightly more visible than red.
- [153_ Pc _5] Larger signs are needed for older adults than for younger adults.
- [153_ Pc _6] To be visible under emergency lighting conditions, exit sign content needs to be larger than under normal lighting conditions. For example, for people with normal eyesight, exit sign characters need to be 75 mm to 80 mm high to be confidently identified from 18 meters away under emergency lighting, compared to 50 mm to 55 mm high under normal lighting.

154. Wright MS, Cook GK, Webber GMB. (2001). The effects of smoke on people's walking speeds using overhead lighting and wayguidance provision. *Proceedings of the 2nd International Conference on Human Behavior in Fire,* MIT, Boston, MA, London, UK: Interscience Communications.

F		Pc	At	Co	Cr	Ps	Ac

In this study, walking speeds were recorded for each of 18 participants (23 to 63 years old) as they traversed a 29 meters long route consisting of a landing, a flight of stairs, and a long corridor ending at an exit door, and then returned along the same route. Walking speeds were tested in well-mixed air/smoke mixtures with optical density per meter of path length ranging from 0.5 m^{-1} to 1.5 m^{-1} using six different lighting systems, which are listed in Table 31 below.

Table 31: Lighting systems used in study of walking speed.

System Number and Type	Lighting System Details
1 – Overhead emergency lighting	Six overhead 4 W fluorescent bulbs provided a minimum of 0.68 lx of illuminance on the floor. A fluorescent sign was mounted at the exit door, and a green sign that was externally illuminated at 6 lx was located at the turn into the corridor. Both signs employed graphical symbols.
2 – Normal lighting	Six ceiling-mounted luminaires in the corridor and two incandescent luminaires in the stairwell produced a minimum illuminance of 19.6 lx. A fluorescent sign was mounted at the exit door, and a green sign externally illuminated at 80 lx was located at the turn into the corridor. Both signs employed graphical symbols.
3 – Electroluminescent (EL) wayguidance system:	A continuous green EL track was mounted on the wall one meter above the floor. On the track, clear plastic wedges indicated direction by touch. A vertical strip on the right side marked the final exit. Track luminance ranged from 18 cd/m^2 to 26 cd/m^2, with minimum illuminance along the route of 0.25 lx. Two internally illuminated signs marked the exit door and the turn into the corridor.
4 – LED wayguidance system	Wall mounted tracks had 35 mcd LED illuminators every 25 mm (pointing down) and green 35 mcd marker LEDs every 100 mm (pointing up). LEDs also marked stair nosings and outlined the doors. The minimum illuminance along the route was 0.25 lx. Signage consisted of a pictogram on the exit door and LEDs forming arrows pointing the way to the exit.
5 – LED wayguidance system	Same as (4) on landing and stairwell. The corridor was lined with a floor track that consisted of LEDs of 140 mcd in triplets spaced 200 mm apart that directed light upwards. The minimum illuminance was 0.16 lx, less than the recommended value of 0.2 lx. The exit door sign was an LED sign marked "EXIT." LEDs forming arrows provided directional guidance.
6 – Miniature incandescent (MI) wayguidance system:	A light strip with whitish lamps of mean intensity 100 mcd spaced 100 mm apart was mounted 18 cm from the floor. Tracks went down both sides of the corridor and stairwell and outlined the doorways. The minimum illuminance was 0.15 lx, less than the recommended value of 0.2 lx. A fluorescent exit sign was mounted at the exit door. No illuminated signs directed the way to the exit door.

People walked significantly more slowly in smoke with overhead lighting systems than with LED wayguidance systems. In the long corridor, walking speeds for a wayguidance system that illuminated the surroundings at lower than recommended levels produced faster speeds than when normal overhead lighting conditions were employed (by more than a factor of three). System (4) resulted in the fastest walking speeds in all cases. Systems (4), (5), and (6) produced the fastest walking speeds on the landing, averaging about 0.4 m/s. System (4) produced the fastest walking speeds on the stair (with a mean speed of about 0.43 m/s). System (3) produced the second fastest stair speeds, averaging 0.4 m/s. Systems (4) and (5) produced the fastest walk speeds in the stairwell, of between 0.6 and 0.7 m/s on average. Systems (4), (5), and (6) produced the fastest corridor walking speeds, averaging between 0.82 and 0.85 m/s.

Key findings:
- [154_Pc_1] [154_Ac_1] Wayguidance systems with lighting tracks mounted on the wall resulted in significantly faster walking speeds in smoke than overhead lighting. This was true for stairwells, stair landings, and corridors.
- [154_Pc_2] [154_Ac_2] Walking speeds for a wayguidance system that illuminated at lower than recommended levels were found to exceed speeds under normal overhead lighting by more than a factor of three in a long corridor.
- [154_Ac_3] Directional information incorporated into wayguidance systems may contribute significantly to their effectiveness and compensate for lower levels of illumination. Outlining the exit door may also improve performance.
- [154_Pc_3] Wayguidance systems can highlight architectural features such as steps and doorframes, giving people more confidence when entering a stairwell and approaching an exit door.

155. Wright MS, Cook GK, Webber GMB. (2001). Visibility of four exit signs and two exit door markings in smoke as gauged by twenty people (XiTlight Report). In Bright K, Cook G. (2010). *The Colour, Light and Contrast Manual: Designing and Managing Inclusive Built Environments,* John Wiley & Sons, Ltd.

During an emergency, it is important for building occupants to be able to navigate around areas of the building with which they are unfamiliar. This study tests the ability of people to find their way around an unfamiliar structure when walking through smoke. Although escape routes are designed to keep out smoke, smoke can still make its way into the stairwells. In this experiment, 30 participants ranging in age from 19 to 63 were asked to walk along a 13 meter hallway filled with smoke until they could see a lit element that was being tested. Once the element could be seen they stopped and the distance from the element to the person was measured. Six elements were tested; four signs and two wayguidance strips at the side of the door. Two of the signs were internally illuminated fire safety signs with a green background and white symbols. The other two signs were LED signs with green LEDs forming the symbols. One of the door marker strips used green LEDs and the other was electroluminescent. The people in the study all had normal eye sight. The smoke density (optical density per meter of path length) was between 0.8 m^{-1} to 1.4 m^{-1}. The hallway was lighted by wayguiding strips along the floor with no lighting from above, emergency overhead lighting (1 lx to 2 lx on the floor in clean air) or normal overhead lighting (130 lx to 190 lx on the floor in clean air). Under conditions of no overhead lighting and emergency overhead lighting the visibility distances were similar, and greater than visibility distances under normal lighting.

No differences were found in visibility distance between the electroluminescent sign and the electroluminescent door markings. The electroluminescent sign and door markings could penetrate a lower smoke density than the other four elements. Ambient lighting conditions allowed the illuminated element to be seen from a greater distance. In moderately thick smoke (smoke density around 0.8 m^{-1} to 1.4 m^{-1}), the viewing distances were reduced by about one meter to two meters. The fluorescent sign with a green and white stencil and mean luminance over 1000 cd/m^2 performed the same as the LED signs with 100 LEDs each having luminance of 140 mcd or 260 mcd. The electroluminescent sign (50 cd/m^2) did not penetrate the smoke as effectively as the signs with LEDs of 140 mcd or more.

Key findings:
- [155_Pc_1] In moderate smoke of smoke density around 0.8 m^{-1} to 1.4 m^{-1}, the viewing distance of signs is reduced by about one meters to two meters.
- [155_Pc_2] A fluorescent sign with a green and white stencil and mean luminance over 1000 cd/m^2 performed the same as LED signs with 100 LEDs each having luminance of 140 mcd or 260 mcd.
- [155_Pc_3] Fluorescent signs and LED signs perform better in smoke (longer visibility distances) than electroluminescent signs.

 # 156. Xie H, Filippidis L, Gwynne S, Galea ER, Blackshields D, Lawrence PJ. (2007). Signage legibility distances as a function of observation angle. *Journal of Fire Protection Engineering, 17*(1), 41-64.

Filippidis L, Galea ER, Gwynne S, Lawrence PJ. (2006, February). Representing the influence of signage on evacuation behavior within an evacuation model. *Journal of Fire Protection Engineering, 16*(1), 37-73.

F		Pc	At	Co	Cr	Ps	Ac

This study examined the effect of observation angle on legibility in order to determine the distance from which a sign could be read given the angle of approach. Three sign designs were considered: two plastic designs, (1) with small lettering and (2) with large lettering, and one photoluminescent design (3) with small lettering. Forty eight participants (29 males and 19 females) approached the sign (on an individual basis) along a 39 meter corridor. The corridor was lit, and there was no smoke present. The three signs were pivoted to simulate the approach of the volunteer from different directions; i.e., producing a different viewing angle between the participant and the sign. The study found that the angle of approach influenced the distance from which the sign could be seen, in the manner shown in Figure 15. Although the type of sign employed was a factor, the impact of the angle of observation appeared to be consistent for all signs.

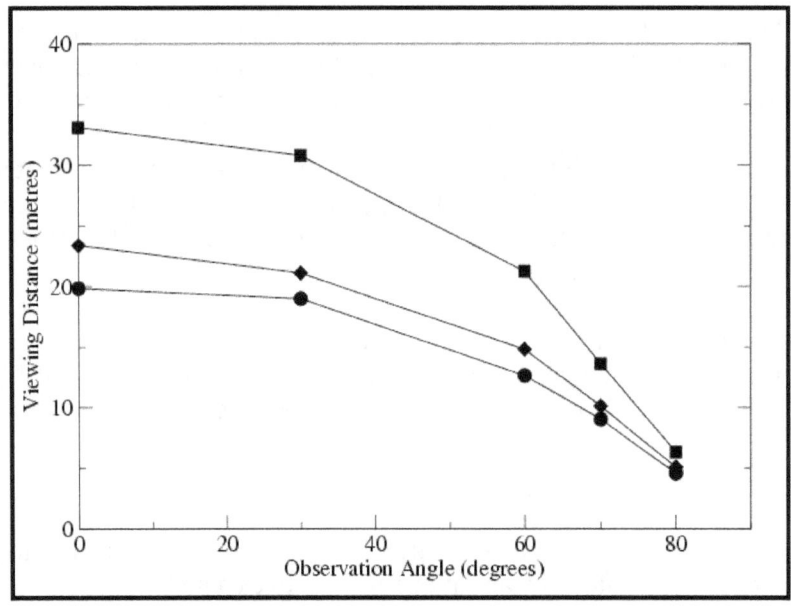

Figure 15: Empirical data of the mean viewing distances of Sign 1(♦), Sign 2 (■) and Sign 3 (●) at 5 observation angles; redrawn from original.

Key findings:
- [156_Pc_1] The legibility of signage is dependent upon the angle of observation.
- [156_Co_1] The number of exit signs required to ensure sufficient coverage may currently be underestimated in situations where the angle of approach of some evacuees may be at an oblique angle, which reduces access to the information.

157. Xiu H, Filippidis L, Galea ER, Blackshields D, Lawrence PJ. (2009, July). *Experimental study of the effectiveness of emergency signage. Proceedings of the 4th International Symposium on Human Behavior in Fire,* Cambridge, UK, Interscience Communications Ltd., London, UK, 289-300.

F		Pc	At	Co	Cr	Ps	Ac

The interaction between an occupant and emergency signage is influenced by cognitive and physical traits. The sign must be visible and the occupant must perceive it in order for the sign to be effective. The purpose of this experiment was to determine whether building occupants perceived an emergency exit sign during an evacuation and correctly interpreted its information.

The building used for the evacuation required the participants to walk down a hallway that ended in a 'T' junction. The participants then either went left or right to another hallway which led to an exit stairway. Exit signs were 0.1 meters × 0.3 meters in size. Sixty percent of the participants (41 of them) were completely unfamiliar with the building and 40 % (27 participants) were familiar with it. The evacuation sign, which was at a 0° angle (head on) to the line of sight, was detected by 39 % of the participants (16/41) that were unfamiliar with the building, with 100 % of those who saw it correctly interpreting the sign. Of participants familiar with the building, 33 % (9/27) detected the building sign.

At a second decision point, participants approached an exit sign at an angle. The detection probability was reduced in this case. Thirty seven percent (15/41) of occupants unfamiliar with the building but only 26 % (7/27) of those familiar with the building detected this exit sign. The probabilities were reduced for participants that were familiar with the building.

Key findings:
- [157_Pc_1] The probability of detecting an exit sign was less when an individual approached at an angle as compared to head on.
- [157_Pc_2] [157_At_1] People that were familiar with a building were less likely to notice exit signs than those who were unfamiliar with it.

158. Young SL, Wogalter MS. (1990). Comprehension and memory of instruction manual warnings: Conspicuous print and pictorial icons. *Human Factors: The Journal of the Human Factors and Ergonomics Society, 32*(6), 637-649.

Instruction manuals for assembling and operating equipment often contain safety information in the form of a warning. Because consumers may not always have the manual available, it is important that the warnings are designed to not only be understood at the first point of use, but also recalled whenever the object is used.

This study examines two factors that might influence comprehension and memory of warnings:
1. Increased conspicuity of printed test, and
2. The presence of pictorial icons.

Although written warnings can be made more conspicuous by highlighting them, no research had been carried out to investigate whether comprehension is improved for highlighted text. It was hypothesized that pictorial icons would facilitate memory of the warning.

In this experiment, participants performed a set of computer tasks using an instruction manual. Later they were given a manual for heavy machinery (a gas-powered electric generator) and were told to examine the instructions because they would have to operate the machinery later without the manual. The plain text of the manual was in 12-point Helvetica font. The manual contained eight warning messages, which were printed either in conspicuous print (18-point Times font covered with transparent orange florescent highlighting) or in plain text. The warnings were either text alone or accompanied by icons compatible with the warnings.

Instead of actually operating the machinery, participants were given tests that measured recall of the content of the warning messages and comprehension and recall of the icons. The results showed that conspicuous print with icons present in the warning was better understood and remembered than either text without the icons or plain text with icons.

Key findings:
- [158_Co_1] In an instruction manual, conspicuous print (18-point Times font covered with transparent orange florescent highlighting) accompanied by icons was better understood and remembered than either text without the icons or plain text with icons.

159. Young SL. (1991). Increasing the noticeability of warnings: Effects of pictorial, color, signal icon and border. *Proceedings of the Human Factors Society 35th Annual Meeting*, 580-584.

E		Pc	At	Co	Cr	Ps	Ac

The study looks at the noticeability (i.e., the ability to grab people's attention) of a variety of pictorials, color, signal icons and borders. Ninety six alcohol labels were constructed for fictional beer, wine and liquor products, including the government warning that all alcoholic beverage labels are required to carry. Half of the labels produced had warnings and half did not. The warnings were either red or black, and were designed with and without an icon or a border.

Seventy-two participants viewed the 96 alcohol labels in sequence and pressed a button if they recognized that a warning was present. The participant response time for each participant action was measured by the computer. Warnings with pictorials and icons resulted in significantly faster response times than those without. The presence of the signal word WARNING also improved response times as did the color red. Adding a border did not affect the response time.

Key findings: • [159_At_1] Pictorials, icons, color, and the signal word WARNING increase the likelihood of a warning being noticed.

 # 160. Zhang L, Sun X, Zhang K. (2006). Research of speech signal on fire information display interface. *China Safety Science Journal 16*(4), 13–18.

F		Pc	At	Co	Cr	Ps	Ac

The authors studied Mandarin speech for factors that aid in message comprehension. Four factors were studied: speech rate, voice type, the type of speech, and the ambient noise level. Actors with soprano, mezzo-soprano, tenor, and baritone voices recorded a set of messages of varying length and content, including words, numbers and sentences, in monotone speech. The participants were 36 young men, 20 to 27 years old, who were fluent in Mandarin. Participants listened through headphones to the recorded voices under three conditions of ambient noise level (0 dBA, 60 dBA, and 80 dBA) and were asked to repeat what they heard.

The ideal speech rate for recall was found to be five to nine characters per second. Since the experiment was carried out in Mandarin, it is uncertain how the results translate to English. The subjects were more accurate at repeating single words than entire sentences, and numbers were repeated less correctly than other parts of speech. Nouns were found to be the most easily comprehended, followed in order by numerals, general sentences, and numeral sentences. Participants responded most accurately in the absence of noise. The responses were more accurate with 60 dBA noise levels than with 80 dBA, but the results for these two noise levels were not statistically different. Accuracy decreased as the speech rate increased, with the effect found to be greater for higher levels of background noise. Female voices were generally easier to understand than male voices; this effect was also more pronounced in the presence of background noise.

Key findings:
- [160_Co_1] Comprehension of auditory messages deteriorates as speech rate increases.
- [160_Co_2] Single words in an auditory message were more accurately understood and repeated than entire sentences.
- [160_Co_3] Numbers in an auditory message were repeated less correctly than other parts of speech.
- [160_Co_4] Nouns were the most easily comprehended in an auditory message, followed in order by numerals, general sentences, and numeral sentences.
- [160_Co_5] Female voices are generally easier to understand than male voices.
- [160_Co_6] Participants repeated voice messages most accurately in the absence of noise.
- [160_Co_7] The effects of speech rate and voice type (male and female) on the comprehension of spoken messages were more pronounced in the presence of background noise.

Key recommendations:
- [160_Co_8] For good comprehension of a spoken message, background noise should be decreased as much as possible.

www.ingramcontent.com/pod-product-compliance
Lightning Source LLC
Chambersburg PA
CBHW080233180526
45167CB00006B/2264